ESO ASTROPHYSICS SYMPOSIA
European Southern Observatory

Series Editor: Philippe Crane

Springer
*Berlin
Heidelberg
New York
Barcelona
Budapest
Hong Kong
London
Milan
Paris
Santa Clara
Singapore
Tokyo*

David L. Clements
Ismael Pérez-Fournon (Eds.)

Quasar Hosts

Proceedings of the ESO-IAC Conference
Held on Tenerife, Spain,
24–27 September 1996

Springer

Volume Editors

David L. Clements
Institut d'Astrophysique Spatiale, Bâtiment 121
Université Paris XI
F-91405 Orsay Cedex, France
email: clements@ias.fr

Ismael Pérez-Fournon
Instituto de Astrofísica de Canarias
Vía Láctea, s/n
E-38200 La Laguna, Tenerife, Spain
email: ipf@ll.iac.es

Series Editor

Philippe Crane
European Southern Observatory
Karl-Schwarzschild-Strasse 2
D-85748 Garching, Germany

Cataloging-in-Publication data applied for

Die Deutsche Bibliothek - CIP-Einheitsaufnahme

Quasar hosts : proceedings of the ESO-IAC conference, held on Tenerife, Spain, 24 - 27 September 1996 / David L. Clements ; Ismael Pérez-Fournon (ed.). - Berlin ; Heidelberg ; New York ; Barcelona ; Budapest ; Hong Kong ; London ; Milan ; Paris ; Santa Clara ; Singapore ; Tokyo : Springer, 1997
(ESO astrophysics symposia)
ISBN 3-540-63793-1

ISBN 3-540-63793-1 Springer-Verlag Berlin Heidelberg New York

This work is subject to copyright. All rights are reserved, whether the whole or part of the material is concerned, specifically the rights of translation, reprinting, re-use of illustrations, recitation, broadcasting, reproduction on microfilms or in any other way, and storage in data banks. Duplication of this publication or parts thereof is permitted only under the provisions of the German Copyright Law of September 9, 1965, in its current version, and permission for use must always be obtained from Springer-Verlag. Violations are liable for prosecution under the German Copyright Law.

© Springer-Verlag Berlin Heidelberg 1997
Printed in Germany

The use of general descriptive names, registered names, trademarks, etc. in this publication does not imply, even in the absence of a specific statement, that such names are exempt from the relevant protective laws and regulations and therefore free for general use.

Typesetting: Camera ready by authors/editors
Cover design: *design & production* GmbH, Heidelberg
SPIN: 10552368   55/3142-543210 - Printed on acid-free paper

Preface

The ESO/IAC Workshop on Quasar Hosts was held in Puerto de la Cruz, Tenerife, from 24 to 27 September 1996 in the Conference Centre of the Cabildo Insular de Tenerife. The four days of the meeting were filled with fascinating new results and interesting discussions, and ranged from the centre of our own galaxy to some of the most distant objects known in the universe. Quasar Host studies are going through an exciting time, and are benefiting from new facilities, including the refurbished HST and the Keck, and from novel techniques, including adaptive optics and deconvolution methods. We also saw the first of hopefully many results from the ISO satellite. These results were presented during the many sessions and discussed in the gardens over coffee, and on the bus during our tour of the Canaries Observatories.

We would very much like to thank the secretaries of ESO and IAC, Christina Stoffer, Pamela Bristow, Monica Murphy, Judith de Araoz, and Beatriz Mederos, who we depended on for their expertise and efficiency.

Our colleagues on the scientific organising committee, Phil Crane, Bob Fosbury, Marie-Helene Ulrich, Peter Shaver and José Rodríguez-Espinosa, deserve considerable thanks for their contributions to the programme.

We must also thank the local organising committee, Fernando Cabrera-Guerra, Monica Murphy, Ismael Pérez-Fournon, Ana Pérez-García, Luis Ramírez-Castro, and Montserrat Villar-Martín, for all their efforts in making sure things ran smoothly on the day.

We were very lucky to secure financial support from Dirección General de Enseñanza Superior del Ministerio de Educación y Cultura in addition to the generous support of ESO, the European Southern Observatory, and IAC, the Instituto de Astrofísica de Canarias.

We must also express our gratitude to Cabildo Insular de Tenerife, Cabildo Insular de La Palma and Centros de Iniciativas Turísticas de La Palma y Tenerife for their assistance.

A special word of thanks is due to Pamela Bristow for her hard work in compiling the present volume. Without her, you wouldn't be reading this. Bob Fosbury also deserves a special thank-you for acting as unofficial conference photographer.

Finally we would like to thank all the people who attended the meeting for making it such an educational and stimulating experience.

Garching, August 1997

Dave Clements
Ismael Pérez-Fournon

Contents

PART 1.   UNIFICATION AND EVOLUTION

Remarks on the Extended Light in Quasars
  R.A.E. Fosbury 3

The Alignment Effect in Radio-Loud AGN:
Ten Years After
  S. di Serego Alighieri 5

The Unified Schemes and Radio Polarization Studies
  D.J. Saikia, S.T. Garrington, G.F. Holmes 13

Compact Steep-Spectrum Radio Sources
  D.J. Saikia, S. Jeyakumar 19

Host Galaxies of Low Luminosity Radio QSOs
  S.F. Sánchez, J.I. González-Serrano, R. Carballo,
  M. Vigotti, C. Benn 21

Quasar Host Luminosities
and the Evolutionary Unified Scheme
  F. Vagnetti 23

PART 2.   OBSERVING QUASAR HOST GALAXIES

Nearby Quasar Galaxies
  M.-P. Véron-Cetty, L. Woltjer 27

HST Images of Twenty Nearby Luminous Quasars
  J.N. Bahcall, S. Kirhakos, D.P. Schneider 37

Near-IR Properties of Quasar Host Galaxies
  K.K. McLeod 45

JHK Imaging of QSO Hosts and Companions
  J.B. Hutchings 51

Resolution of High Redshift QSO Hosts
  J.B. Hutchings 58

The Nature of Radio Emission in Radio-Quiet QSOs
P. Barthel, J. Gerritsen 63

Adaptive Optics Observations of Quasar Hosts
M.N. Bremer, G. Miley, G. Monnet, M. Wieringa 70

HST Planetary Camera Images of Quasar Host Galaxies
P.J. Boyce, M.J. Disney, J.C. Blades, A. Boksenberg,
P. Crane, J.M. Deharveng, F.D. Macchetto, C.D. Mackay,
W.B. Sparks 76

The Hosts of $z = 2$ QSOs
I. Aretxaga, B.J. Boyle, R.J. Terlevich 84

Observing the Galaxy Environment of QSOs
K. Jäger, K.J. Fricke, J. Heidt 90

Limitations of Differential CCD Photometry
Due to Weather Conditions
D. Sinachopoulos, A. Devillers, M. Geffert 92

The Host Galaxy of HE 1029–1401
L. Wisotzki 96

PART 3. STAR FORMATION AND THE ISM IN QUASAR HOSTS

Molecular Gas in Quasar Hosts
R. Barvainis 101

CO Emission Lines from a Quasar at $z = 4.7$
T. Yamada, K. Ohta, R. Kawabe, K. Kohno, K. Nakanishi,
M. Akiyama 110

Molecular Gas and Star Formation in I Zw 1
E. Schinnerer, A. Eckart, L.J. Tacconi, N. Thatte,
N. Nakai, S.K. Okumura 116

The Epoch of Major Star Formation in High-z Quasar Hosts
Y. Taniguchi, N. Arimoto, T. Murayama, A.S. Evans,
D.B. Sanders, K. Kawara 122

The Nuclear Stellar Cluster in NGC 1068
N. Thatte, R. Maiolino, R. Genzel, A. Krabbe,
H. Kroker 128

ISO Observations of Quasars and Quasar Hosts
B.J. Wilkes 136

ISO Observations of Seyfert Galaxies
    J.M. Rodriguez Espinosa, A.M. Pérez García 144

Origin of Spread in the $B - K$ Color of Quasars
    R. Srianand 150

Using HI to Probe AGN Hosts and Their Nuclei
    C.G. Mundell 156

The Large-Scale Environments
of Low Redshift Radio-Quiet Quasars
    P. Goldschmidt, L. Miller 162

High Velocity Resolution Observations of the ISM
in NGC 4151
    M.W. Asif, S.W. Unger, A. Pedlar, C.G. Mundell,
    A. Robinson, N.A. Walton 168

Mg_2 Index Map of the Centre of NGC 7331
    C. del Burgo, E. Mediavilla, S. Arribas,
    B. García-Lorenzo 171

Stellar Dynamics of Two AGNs
    F. Prada, C.M. Gutiérrez 173

PART 4. THE RADIO LOUD/QUIET DICHOTOMY

Infrared Imaging and Off-Nuclear Spectroscopy
of Quasar Hosts
    M.J. Kukula, J.S. Dunlop, D.H. Hughes, G. Taylor,
    T. Boroson 177

The Radio Properties of Radio-Quiet Quasars
    M.J. Kukula, J.S. Dunlop, D.H. Hughes, S. Rawlings 185

HST Imaging of Redshift $z > 0.5$ 7C and 3C Quasars
    S. Serjeant, S. Rawlings, M. Lacy 188

HST Imaging of BL Lac Objects
    R. Falomo, C.M. Urry, J. Pesce, R. Scarpa, A. Treves,
    M. Giavalisco 194

Near-IR Imaging of BL Lac Host Galaxies
    J. Heidt 200

HST Imaging of Quasar Host Galaxies Selected by Quasar
Radio and Optical Properties
    E.J. Hooper, C.D. Impey, C.B. Foltz 206

Optical Imaging and Spectroscopy of BL Lac Objects
R. Falomo, M.-H. Ulrich 212

Host Galaxies of Radio-Loud AGN
J.W. Fried, J. Heidt 215

Host Galaxies of Intermediate Redshift Radio-Loud
and Radio-Quiet Quasars
E. Örndahl, J. Rönnback, E. van Groningen 217

Some Examples of the Extremely Close Environments
of BL Lac Objects
A. Sillanpää, L.O. Takalo, T. Pursimo, P. Heinämäki,
K. Nilsson, J. Heidt 223

The Optical Jet in 3C 371
L.O. Takalo, K. Nilsson, T. Pursimo, A. Sillanpää,
J. Heidt 225

PART 5. LOW REDSHIFT POPULATIONS

Infrared QSOs
D.B. Sanders, J.A. Surace 229

HST Images of Warm Ultraluminous Infrared Galaxies:
QSO Host Progenitors
J.A. Surace, D.B. Sanders 236

ISO-SWS Results on Ultraluminous IRAS Galaxies
D. Rigopoulou 242

Quest for Type-2 Quasars with ASCA
T. Yamada 248

The Local Luminosity Function of Quasars –
Implications for Host Galaxy Studies
T. Köhler, L. Wisotzki 254

B–K Colours of Low-Luminosity Radio Quasars
J.I. González-Serrano, C. Benn, R. Carballo, S.F. Sánchez,
M. Vigotti 260

Cygnus A: Host Galaxy of a Nearby Quasar?
F. Cabrera-Guerra, I. Pérez-Fournon, J.A. Acosta-Pulido,
A.S. Wilson, Z.I. Tsvetanov 266

X-Ray Extended AGN?
P. Nass 272

A Search for Galaxies with a Variable Nucleus
    D. Trèvese, M.A. Bershady, R.G. Kron 278

Proper Motions in the Center of the Galaxy
    A. Eckart, R. Genzel 282

The Seyfert Nucleus in the S0 Galaxy NGC 5252
    J. Acosta-Pulido, B. Vila-Vilaró, I. Pérez-Fournon 288

NGC 3393 – Broad(er) Lines in a Nearby Seyfert 2 Galaxy
    A. Cooke, J. Baldwin, G. Ferland, H. Netzer, B. Wills,
    A. Wilson 291

AGNs with Composite Spectra
    A.C. Gonçalves, P. Véron, M.-P. Véron-Cetty 293

Nuclear Activity in Very Nearby Galaxies
    P. Lira, R. Johnson, A. Lawrence 295

The Spectral Variability of QSOs in the Optical Bands
    D. Trèvese, D. Nanni, R.G. Kron, A. Bunone 297

PART 6. RADIO GALAXIES AT HIGH REDSHIFT

Radio Galaxies at High Redshift:
Unification and Host Galaxies
    P. McCarthy 303

The Nature of the UV Excess
    C.N. Tadhunter, R. Dickson, R. Morganti,
    M. Villar-Martin 311

The Role of Shocks in the Extended Emission Line Regions
of Powerful Radio Galaxies
    N.E. Clark, C.N. Tadhunter, D.J. Axon, A. Robinson 320

The Hosts of High-z Powerful Radio Sources:
Keck Observations
    A. Cimatti 327

Author Index 335

Conference Participants

List of Participants

Name	Institution
Acosta-Pulido, José A.	ISOPHOT IDT, VILSPA jacosta@iso.vilspa.esa.es
Aretxaga, Itziar	Royal Greenwich Observatory itziar@ast.cam.ac.uk
Asif, Mirza Wasim	Isaac Newton Group, La Palma mwa@ing.iac.es
Bahcall, John	Institute for Advanced Study, Princeton jnb@sns.ias.edu
Balcells, Marc	Instituto de Astrofísica de Canarias balcells@ll.iac.es
Barthel, Peter	Kapteyn Astronomical Institute, Groningen pdb@astro.rug.nl
Barvainis, Richard	MIT Haystack Observatory reb@dopey.haystack.edu
Bellido, Iván	Instituto de Astrofísica de Canarias ibellido@iac.es
Benn, Chris	Isaac Newton Group, La Palma crb@ing.iac.es
Boyce, Peter	University of Wales, Cardiff pjb@astro.cf.ac.uk
Bremer, Malcolm	Leiden Observatory malcolm@strw.leidenuniv.nl
Burbidge, Margaret	Center for Astrophysics & Space Sciences, University of California, San Diego burbidge@cass05.span.nasa.gov
Cabrera-Guerra, Fernando	Instituto de Astrofísica de Canarias fcabrera@ll.iac.es
Cimatti, Andrea	Osservatorio di Arcetri cimatti@arcetri.astro.it

Clark, Neil	University of Sheffield n.e.clark@sheffield.ac.uk
Clements, Dave	European Southern Observatory dclement@eso.org
Cooke, Andrew	Institute for Astronomy, University of Edinburgh A.Cooke@roe.ac.uk
Crane, Philippe	European Southern Observatory crane@eso.org
Dallison, Jayne	Sheffield University php95jld@sheffield.ac.uk
del Burgo Díaz, Carlos	Instituto de Astrofísica de Canarias cburgo@ll.iac.es
di Serego Alighieri, Sperello	Osservatorio Astrofisico di Arcetri sperello@arcetri.astro.it
Disney, Mike	University of Wales, Cardiff mjd@astro.cf.ac.uk
Eckart, Andreas	Max-Planck-Institut für Extraterrestrische Physik eckart@mpe.mpe-garching.mpg.de
Falomo, Renato	Osservatorio Astronomico di Padova falomo@astrpd.pd.astro.it
Fosbury, Robert	Space Telescope – European Coordinating Facility rfosbury@eso.org
Fried, Josef	Max-Planck-Institut für Astronomie fried@mpia-hd.mpg.de
Goldschmidt, Pippa	Imperial College London p.goldschmidt@ic.ac.uk
Gonçalves, Anabela	Observatoire de Haute-Provence anabela@obs-hp.fr
González-Serrano, J. Ignacio	Instituto de Física de Cantabria gserrano@astro.unican.es
Heidt, Jochen	Landessternwarte, Heidelberg jheidt@hp4.lsw.uni-heidelberg.de
Hooper, Eric	Steward Observatory, University of Arizona ehooper@as.arizona.edu
Hutchings, John	Dominion Astrophysical Observatory Hutchings@dao.nrc.ca

Jäger, Klaus	Universitäts-Sternwarte Göttingen jaeger@uni-sw.gwdg.de
Kirhakos, Sofia	Institute for Advanced Study, Princeton sofia@sns.ias.edu
Kukula, Marek	Institute for Astronomy, University of Edinburgh m.kukula@roe.ac.uk
Lira, Paulina	Institute for Astronomy, University of Edinburgh plt@roe.ac.uk, P.Lira@roe.ac.uk
Marcha, Maria	University of Lisbon mmarcha@milkyway.cii.fc.ul.pt
McCarthy, Patrick	Carnegie Observatories pmc2@ociw.edu
McLeod, Kim	Harvard-Smithsonian Center for Astrophysics kmcleod@cfa.harvard.edu
Mundell, Carole	NRAL, Jodrell Bank cgm@jb.man.ac.uk
Nass, Petra	Max-Planck-Institut für Extraterrestrische Physik pnass@rosat.mpe-garching.mpg.de
Örndahl, Eva	Uppsala Astronomical Observatory eva@astro.uu.se
Pérez-Fournon, Ismael	Instituto de Astrofísica de Canarias ipf@iac.es
Pérez García, Ana M.	Instituto de Astrofísica de Canarias apg@iac.es
Prada, Francisco	Instituto de Astrofísica de Canarias fprada@iac.es
Ramírez-Castro, Luis	Instituto de Astrofísica de Canarias lcastro@iac.es
Regalado-Ojeda, Ana	Instituto de Astrofísica de Canarias regalado@iac.es
Rigopoulou, Dimitra	Max-Planck-Institut füer Extraterrestrische Physik dar@mpe.mpe-garching.mpg.de
Rodríguez Espinosa, José M.	Instituto de Astrofísica de Canarias jre@iac.es

Saikia, Dhruba Jyoti	NCRA, Tata Institute of Fundamental Research djs@gmrt.ernet.in
Sánchez, Sebastián F.	Instituto de Física de Cantabria sanchez@astro.unican.es
Sanders, David	Institute for Astronomy, University of Hawaii sanders@ifa.hawaii.edu
Schinnerer, Eva	Max-Planck-Institut füer Extraterrestrische Physik schinnerer@mpe-garching.mpg.de
Schober, Hans Josef	Institute for Astronomy, Univ. Graz schober@bkfug.kfunigraz.ac.at
Serjeant, Stephen	Imperial College London s.serjeant1@physics.oxford.ac.uk
Sillanpää, Aimo	Tuorla Observatory aimosill@sara.utu.fi
Sinachopoulos, Dimitrios	Royal Observatory of Belgium dimitris@oma.be
Srianand, Raghunathan	Inter University Center for Astronomy & Astrophysics anand@iucaa.ernet.in, visi3@iap.fr
Surace, Jason	Institute for Astronomy, University of Hawaii jason@galileo.ifa.hawaii.edu
Tadhunter, Clive	University of Sheffield c.tadhunter@sheffield.ac.uk
Takalo, Leo	Tuorla Observatory takalo@sara.utu.fi
Taniguchi, Yoshiaki	Astronomical Institute, Tohoku University tani@astroa.astr.tohoku.ac.jp
Thatte, Niranjan	Max-Planck-Institut für Extraterrestrische Physik thatte@mpe-garching.mpg.de
Trèvese, Dario	Istituto Astronomico Univ. di Roma "La Sapienza" trevese@astrm2.rm.astro.it
Vagnetti, Fausto	Università di Roma "Tor Vergata" vagnetti@roma2.infn.it
van Groningen, Ernst	Astronomiska Observatoriet, Uppsala University ernst@astro.uu.se

Véron-Cetty, Marie-Paule	Observatoire de Haute Provence mira@obs-hp.fr
Villar-Martín, Montserrat	Sheffield University m.villar-martin@sheffield.ac.uk
Wilkes, Belinda	Smithsonian Astrophysical Observatory belinda@cfa.harvard.edu
Wisotzki, Lutz	Hamburger Sternwarte lwisotzki@hs.uni-hamburg.de
Woltjer, Lodewijk	Observatoire de Haute Provence woltjer@obs-hp.fr
Yamada, Toru	Astronomical Institute, Tohoku University yamada@astr.tohoku.ac.jp

Part 1

UNIFICATION AND EVOLUTION

Remarks on the Extended Light in Quasars

Robert A. E. Fosbury[1,2]

[1] Space Telescope – European Coordinating Facility, D–85748 Garching bei München, Germany
[2] Affiliated to the Astrophysics Division, Space Science Department, European Space Agency

Abstract. In studies of these active objects, it is important to be able to distinguish between the stellar evolutionary processes and the AGN-induced extranuclear phenomena. Some comments are made about the importance of the rest-frame waveband and the likely appearance of 'pseudo-host' galaxies resulting from locally scattered quasar light and spatially-extended emission line regions.

1 Introduction

In the high redshift radio galaxies which exhibit the 'alignment effect', the observed properties above and below the 4000Å-break are different in character. The longer wavelength images are smoother, rounder and more 'elliptical-like' while the rest-frame images are clumpy, elongated along the radio axis and generally linearly polarized. Although dominated by the directly-viewed AGN, quasars must exhibit similar components in their extended structures which, especially at rest-frame ultraviolet wavelengths, could dominate any host galaxy structures.

2 Contributors to the Extended Light

From work on the radio galaxies, we know that the extended structures contain the following components:

- Starlight: seen clearly above the 4000Å-break but not unambiguously detected in the extended structures at shorter wavelengths.
- Nebular continuum — quantitatively related to the recombination line flux.
- Extended emission lines ionized by a hard continuum from the AGN and, possibly, from local shocks associated with radio jets.
- Dust and/or electron scattered AGN continuum.
- Scattered broad lines from the BLR, eg. Hα and Mg II.
- Scattered narrow lines, especially from high critical density transitions emitted close to the AGN.

The scattered components are measured and characterised using imaging- and spectro-polarimetry. There is generally a residual linear polarization with the E-vector perpendicular to the radio axis — even in spatially integrated light — due to the axial symmetry of these objects.

3 Implications for Quasar Host Observations

If the powerful radio galaxies and the radio quasars are from the same parent population, the quasars must exhibit 'pseudo-hosts' resulting from the spatial reprocessing of AGN light via scattering by — and photoionization of — an extended, dusty ISM. These hosts will be most apparent at shorter wavelengths where the AGN-induced extranuclear activity dominates over a cool stellar population. These pseudo-hosts will contain the same components that we see in the alighned radio galaxies but the scattered component may be relatively enhanced by the forward directed phase function of the scattering particles.

For the radio quasars, the magnitude of the effect can be estimated from the properties of the radio galaxies and quasars in the 3CR sample at a redshift around 1. We expect 10 – 15% of the quasar light to from a non-stellar pseudo-host which, by analogy with the radio galaxies, can have a spatial extent of some tens of kpc. This hypothesis may be difficult to test using polarimetry alone but high spatial resolution long-slit spectroscopy with HST may detect the extended broad line signature of scattered AGN light.

The Alignment Effect in Radio–Loud AGN: Ten Years After

Sperello di Serego Alighieri

Osservatorio Astrofisico di Arcetri, Largo E. Fermi 5, I–50125 Firenze, Italy

Abstract. The observational evidence for the alignment between the elongated optical structure and the radio axis of distant powerful radio galaxies, which accumulated in the 10 years since its discovery, is now strong and diversified. The alignment effect is likely due to a combination of anisotropic scattering and nebular continuum, with possible contributions from star formation in some objects. An important outcome of the study of the alignment is the realization of the multicomponent nature of the AGN environment.

1 Introduction

It is now 10 years since the remarkable alignment of the elongated optical structure of distant radio galaxies with their radio axis was discovered simoultaneously by two groups (McCarthy et al. 1987, Chambers et al. 1987). The first group examined about 80% of the 3CR galaxies with $z > 0.6$ using mostly V images; the other one looked at ultra steep spectrum ($\alpha < -1$, $S_\nu \propto \nu^\alpha$) radio galaxies from the 4C and 3C catalogues using R–band images. The average of the redshifts of the latter sample, which were only an estimate for the 4C objects, is around one. This simoultaneous discovery already indicated that the "alignment effect" is a general property of distant radio galaxies, and has in fact since been confirmed for the powerful ones (see McCarthy 1993 for an excellent review). This strong correlation between the optical and radio structures indicates that the presence of the radio source influences (or is influenced by) the light emitted by the host galaxy and recommends caution in using the colours of distant radio galaxies to infer the evolution of their stellar population.

The importance of the alignment effect in the context of this workshop rests on the fact that it is a fundamental observational phenomenon concerning the environment of radio galaxies and therefore, by virtue of the AGN unification, it is directly relevant for the quasar hosts. The attempts to understand the alignment effect have produced many interesting ideas and results on AGN and their environment which would have been impossible to obtain on quasar because of their overwhelming nuclear luminosity.

In the following I shall review the important observational evidence concerning the alignment effect, and discuss the various possible explanations, emphasizing the improvements which they have produced on our understanding of the multicomponent nature of the AGN environment.

2 Observational Evidence

Since the double discovery the alignment effect has been confirmed for the 3CR galaxies at high redshift (Dunlop & Peacock 1993, McCarthy et al. 1995) and for the 4C USS galaxies with $z > 2$ (Chambers et al. 1996), and has been found to exist also for a sample of southern USS radio galaxies with $1.9 < z < 3$ (Röttgering et al. 1997). Although nobody doubts its importance for distant and powerful radio galaxies, there are several aspects of the phenomenon which are very important for its understanding and are worth reviewing here.

2.1 Continuum and Emission Lines

The alignment holds both for the optical continuum (mostly emitted in the UV) and for the emission lines, as already stressed by McCarthy (1993). More recently excellent long-slit spectra obtained with the Keck telescope clearly demonstrate the presence of continuum and line emission along the radio axis for several distant radio galaxies (e.g. Cimatti et al. 1996). The morphology of the line emission is usually substantially different from that of the continuum, although they are both aligned (e.g. Rigler et al. 1992 and McCarthy et al. 1995).

2.2 Dependence on the Redshift

The alignment effect is not seen at low redshift (Baum et al. 1988). McCarthy & van Breugel (1989) by examining data from 3CR sources found a strong dependence of the alignment effect on redshift, mostly in the sense that it sets in for $z > 0.6$, particularly for the continuum, for which there is even evidence for counter alignment for $z < 0.1$. Emission lines are aligned also for objects at redshifts lower than 0.6. This analysis has been conducted at a fixed observed wavelength, mostly V: the sampled regions of the emitted spectrum depend strongly on redshift. Figure 1, derived from the extensive data set on 3CR galaxies compiled by McCarthy (1995), confirms a strong redshift dependence, but shows that one can hardly define a specific redshift threshold, and that there are non–aligned objects also at high redshift. Furthermore there are several objects, particularly at low redshift, with round optical structure. Many of these objects appear elongated in the HST snapshots of de Koff et al. (1996), who have imaged most of the 3CR galaxies with $0.1 < z < 0.5$ in a broad band filter close to the R-band. An histogram of the radio to optical position angle difference compiled from their data confirms that the alignment is not prominent at these redshifts, although many of their optical position angles are quite different from those of McCarthy et al. (1995). Some low redshift radio galaxies are aligned, particularly if looked at in the UV continuum below the 4000Å break (e.g. di Serego Alighieri et al. 1988, Dey & van Breugel 1994, Cimatti & di Serego Alighieri 1995). However Cimatti & di Serego Alighieri (1996) indicate that the redshift dependence is only partially a wavelength effect due to the K-correction. A systematic study of the UV alignment of low redshift radio galaxies is unfortunately still missing.

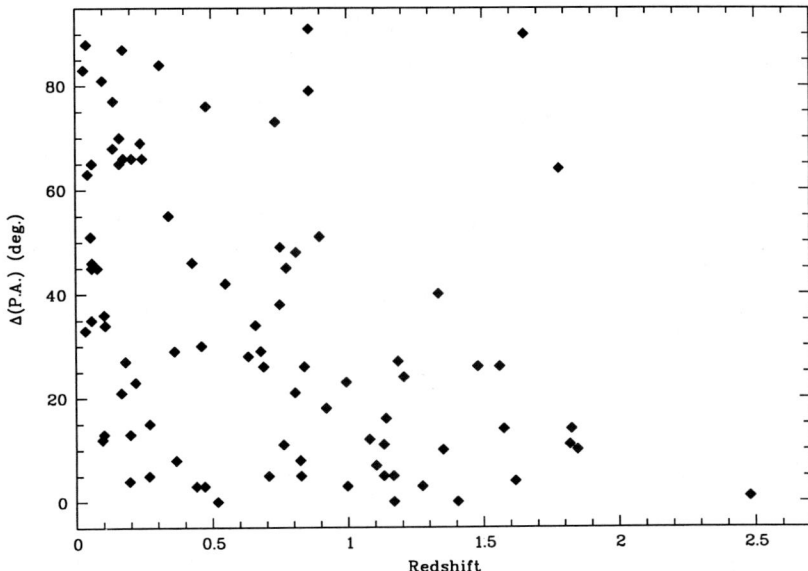

Fig. 1. Radio to optical continuum (R-band) position angle differences as a function of redshift for 3CR radio galaxies compiled by McCarthy et al. (1995).

2.3 Dependence on the Observed Wavelength

Rigler et al. (1992) found that the infrared structures of 13 3CR galaxies with $0.8 < z < 1.3$, as given by images taken in H or K to sample the rest frame around $1\mu m$, are rounder than the optical ones sampling the UV, and show a much weaker alignment. On the other hand Dunlop & Peacock (1993) found that the K-band structures of 19 3CR galaxies with $0.8 < z < 1.3$, despite being more nucleated than at optical wavelength, nevertheless display a clear infrared/radio alignment, which is even more precise than in the optical in terms of angular coincidence. The large differences in the infrared position angle in 5 out of the 10 objects in common between these two studies leave open the problem of the infrared alignment. What is reasonably sure is that the infrared structures are different than the optical ones (see also Eisenhardt & Chokshi 1990) and probably rounder, as emphasized by the suggestive collage of HST and K-band images presented by Best et al. (1995).

2.4 Dependence on the Radio Power

The importance of the alignment effect in the optical correlates with radio power. Dunlop & Peacock (1993) have studied this dependece on the radio power by imaging two matched samples of 3CR and Parkes radio galaxies with $0.8 < z <$

1.5, differing in mean radio power by a factor of 10. While the powerful 3CR galaxies show a strong alignment, particularly in the optical, the lower power Parkes galaxies show no evidence for alignment in R, J and K, with perhaps a weak evidence of alignment in B. Thompson et al. (1994) also find only a weak hint of alignment in the optical for a sample of 31 moderate power B3 galaxies with redshifts larger than 0.5 (mostly estimated redshifts), and remark that B3 galaxies tend to be rounder and more compact than 3CR galaxies. Now that many more redshifts are available for the B3 survey, Vigotti et al. (1996) find that B3 galaxies have on average a radio power an order of magnitude lower than that of 3CR galaxies at all redshifts. Studying the optical–radio alignment of a combined B3 and 3CR sample with $z > 0.8$, they find that the subset with radio power higher than the median one ($P = 4.17 \times 10^{34} erg/s/Hz$) shows the alignment effect, while the lower power subset does not. A subsequent analysis of the same data in a small redshift bin done to try to disentangle the redshift and power dependences is shown in Figure 2 and indicates a real correlation of the alignment effect on radio power (Vigotti, private communication).

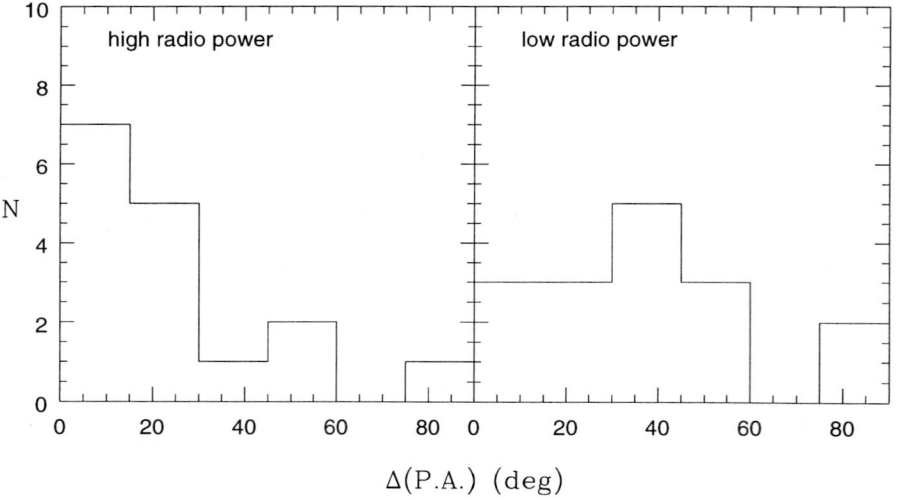

Fig. 2. Histograms of the radio to optical position angle differences for B3 and 3CR galaxies with $0.8 < z < 1.2$ and with radio power higher (left) and lower (right) than the median one ($P = 5 \times 10^{34} erg/s/Hz$) (Vigotti, private communication).

2.5 Perpendicular Polarization

Distant radio galaxies show strong, extended, perpendicular, optical linear polarization (di Serego Alighieri et al. 1989, Tadhunter et al. 1992, Cimatti et al.

1993). This polarization is directly related to the alignment effect since it concerns the same extended, rest–frame UV light, it is a similarly general property of powerful distant radio galaxies, and has the same redshift and wavelength dependence (Cimatti et al. 1993). The extended nature of the polarization has been confirmed by long-slit spectropolarimetry and imaging polarimetry with the Keck telescope (Cimatti et al. 1996, Cohen et al. 1996). Broad permitted lines are present in the polarized flux spectra of distant radio galaxies (di Serego Alighieri et al. 1994, Cimatti et al. 1996). Concerning the redshift/wavelength dependence of the polarization and alignment, two of the low redshift galaxies which are aligned in the UV are also polarized in the UV (di Serego Alighieri et al. 1988, Cimatti & di Serego Alighieri 1995). A large increase in the polarization below the 4000Å break is also seen in other low redshift powerful radio galaxies (e.g. Tadhunter et al. 1996). Broad permitted lines are present in the polarized flux spectra (di Serego Alighieri et al. 1994, Cimatti et al. 1996).

2.6 Dependence on the Radio Size

Best et al. (1996) have obtained HST broad–band images with two filters in the red (one containing [OII]3727 emission and one to the blue of it) and VLA high resolution maps for a sample of 8 3CR radio galaxies in a narrow redshift range ($1.0 < z < 1.3$) having all the same radio luminosity, but covering a wide range (a fator of 16) in projected radio size (according to unification, radio galaxies should have their radio axis closer to the plane of the sky than $45°$, so the projected size should correspond to the linear size within a factor $\sqrt{2}$). They find that their optical morphology depends strongly on the size of the radio source: the optical counterparts of small sources consist of a tightly aligned string of several bright knots, while larger sources have more compact counterparts with fewer bright components. Röttgering et al (1996) have examined the relationship between the alignment strength (the product of the optical ellipticity and the normalized complement to $90°$ of the radio to optical position angle difference) and the projected radio size for a sample of distant 3CR and USS galaxies. Although they fail to find a significant correlation, their figure 3 is suggestive: the five objects with the strongest alignment are all smaller than 200 kpc, and the 3 largest objects have poor alignment.

One unambigous result of the comparison of the radio and optical morphologies of powerful distant radio galaxies at high resolution is that they show very little overlap, although they are aligned (Best et al. 1995 and 1996). The alignment is a statistical effect, which is not necessarily observed on every single object, even for the samples which show it most clearly (Fig. 1).

3 Possible Explanations

The alignment effect clearly shows a connection between the active nucleus, which powers the radio structures, and the extended optical morphology. Therefore it bears strongly on the relationship between active nuclei and their hosts.

Several explanations have been proposed for the alignment in the past 10 years, and are reviewed here.

The observed polarization implies that at least part of the extended rest-frame UV light must be due to scattering of an anisotropic nuclear source. This model is attractive also because it allows to test directly the AGN unification scheme. It implies a close coupling between the radio jet axis — e.g. the rotation axis of the central black hole in the standard AGN model — and the rotation axis of the obscuring torus which collimates the optical/UV radiation. Dust and free electrons are both present in the environment of distant radio galaxies, and both contribute to scattering, with a relative importance which depends on the wavelength and on the distance from the nucleus. However anisotropic scattering is not the only cause for the alignment effect in all radio galaxies.

The recombination theory clearly predicts the presence of nebular continuum associated with ionized gas. Given that distant powerful radio galaxies have emission lines which have a very large equivalent width and are aligned with the radio axis, the nebular continuum must be a substantial component of the aligned UV continuum (Dickson et al. 1995). The high S/N ratio Keck spectra are producing evidence for the 3646Å break of the Balmer continuum (Fig. 3). The relevance of the nebular continuum for the alignment effect is however limited by the morphological differences between the aligned line and continuum emission.

Jet induced star formation was the explanation proposed by the discoverers of the alignment effect, which has since been examined theoretically (De Young 1989, Rees 1989). It is possible that young stars contribute with nebular continuum to dilute the polarization due to scattering in the UV (Tadhunter et al. 1996), but the direct evidence is still marginal (Chambers & McCarthy 1990).

Inverse Compton UV radiation can result from the scattering of microwave background photons by the radio–emitting relativistic electrons (Daly 1990). However its morphology would be identical to the radio one, which is not generally observed for the aligned light.

The optical/UV tail of the synchrotron radiation emitted by the relativistic electrons which produce the radio emission has been observed in jets and hot spots of nearby radio galaxies (e.g. Schlötelburg et al. 1988). However, again, it is bound to reproduce the radio morphology.

Selection effects favouring the presence in flux limited samples of radio sources expanding along the major axis of an asymmetric gas distribution have been called in by Eales (1992). However he predicts that the alignment resulting from these selection effects should not depend on radio power, contrary to observations.

4 Conclusions

The alignemnt effect is well established, but only in a statistical sense, for powerful radio galaxies at high redshift in the optical. More work needs to be done on complete samples in the IR and in the UV at low redshfit to disentangle

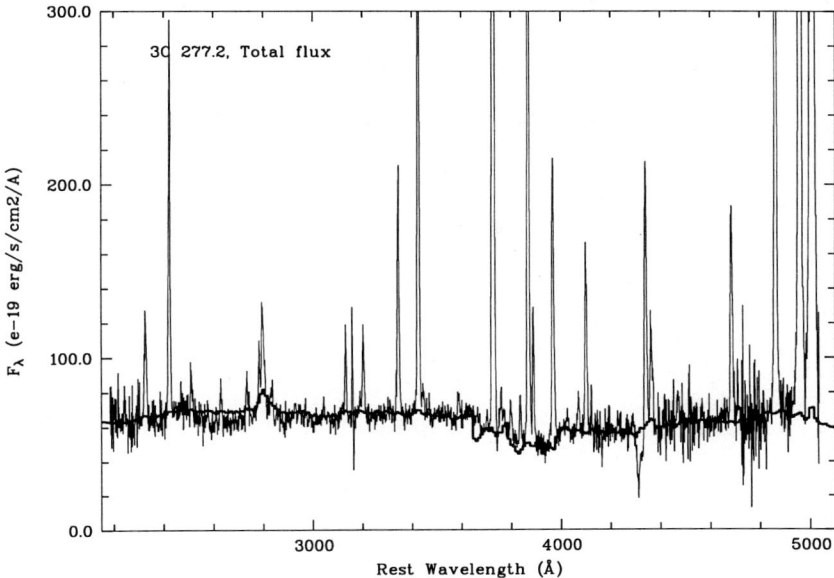

Fig. 3. The Keck spectrum of 3C 277.2 ($z = 0.763$) obtained by Cohen, Ogle and myself. The thick line is a model with scattered nuclear light, nebular continuum and evolved stars. The drop in the continuum around 3800Å, clearly seen under the strong emission lines, is likely due to a combination of the 3646Å break in the Balmer nebular continuum and of the 4000Å break in the spectrum of evolved stars.

the dependence of the alignment on redshift, wavelength and radio power. Spectroscopic and spectropolarimetric ground–based studies of objects with $z \sim 3$ would allow to reach the 1200-2500Å range, important for searching for absorptions from young stars and for dust signatures in the scattering function. Several factors contribute to the alignment effect, including scattering of an anisotropic nuclear source and nebular continuum from anisotropically illuminated ionized gas, with possible contributions from jet induced star formation. The study of the alignment effect has improved our understanding of the AGN unification and has demostrated that a radio galaxy contains many components: (1) evolved stars producing the red colours, the round IR morphology and the 4000Å break, (2) dust scattering the misdirected quasar in the nucleus, absorbing its radiation and reemitting it at millimetr wavelength, (3) ionized gas producing emission lines and nebular continuum, (4) a hot X-ray halo possibly contributing to scattering, (5) young stars possibly diluting the polarization in the UV. Disentangling these components is essential to understand the AGN environment and to use it for cosmological applications.

References

Baum, S.A., Heckman, T., Bridle, A., van Breugel, W, Miley, G., 1988, ApJSS **68**, 643
Best, P., Longair, M., Röttgering, H., 1995, Spectrum No. 8, p. 9
Best, P., Longair, M., Röttgering, H., 1996, MNRAS **280**, L9
Chambers, K.C., Miley, G.K., van Breugel, W., 1987, Nature **329**, 604
Chambers, K.C., McCarthy, P.J., 1990, ApJ **354**, L9
Chambers, K.C., Miley, G.K., van Breugel, W.J.M., Bremer, M.A.R., Huang, J.-S., Trentham, N.A., 1996, ApJSS **106**, 247
Cimatti, A., Dey, A., van Breugel, W., Antonucci, R., Spinrad, H., 1996, ApJ **465**, 145
Cimatti, A., di Serego Alighieri, S., 1995, MNRAS **273**, L7
Cimatti, A., di Serego Alighieri, S., 1996, in *Extragalactic Radio Sources*, R. Ekers et al. eds., Kluwer Academic Publ., p. 234
Cimatti, A., di Serego Alighieri, S., Fosbury, R.A.E., Salvati, M., Taylor, D., 1993, MNRAS **264**, 421
Cohen, M.H.,Tran, H.D., Ogle, P.M., Goodrich, R.W., 1996, in *Extragalactic Radio Sources*, R. Ekers et al. eds., Kluwer Academic Publ., p. 223
Daly, R. A., 1990, ApJ **355**, 416
de Koff, S., Baum, S.A., Sparks, W.B., Biretta, J., Golombek, D., Macchetto, F., McCarthy, P., Miley, G.K., 1996, ApJSS **107**, 621
Dey, A., van Breugel, W.J.M., 1994, AJ **107**, 1977
De Young, D.S., 1989, ApJ **342**, L59
Dickson, R., Tadhunter, C., Shaw, M., Clark, N., Morganti, R., 1995, MNRAS **273**, L29
di Serego Alighieri, S., Cimatti, A. & Fosbury, R.A.E., 1994, ApJ **431**, 123
di Serego Alighieri, S., Binette, L., Courvoisier, T.J.-L., Fosbury, R.A.E., Tadhunter, C.N., 1988, Nature **334**, 591
di Serego Alighieri, S., Fosbury, R.A.E., Quinn, P.J., Tadhunter, C.N., 1989, Nature **341**, 307
Dunlop, J.S., Peacock, J.A., 1993, MNRAS **263**, 936
Eales, S.A., 1990, ApJ **397**, 49
Eisenhardt, P., Chokshi, A., 1990, ApJ **351**, L9
McCarthy, P.J., 1993, ARAA **31**, 639
McCarthy, P.J., Spinrad, H., van Breugel, W., 1995, ApJSS **99**, 27
McCarthy, P.J., H., van Breugel, W., 1989, in *The epoch of Galaxy Formation*, C.S. Frenk et al. eds., (Kluwer Academic Publ.), p. 57
McCarthy, P.J., van Breugel, W., Spinrad, H., Djorgovski, S., 1987, ApJ **321**, L29
Rees, M.J., MNRAS **239**, 1P
Rigler, M.A., Lilly, S.J., Stockton, A., Hammer, F., Le Fevre, O., 1992, ApJ **385**, 61
Röttgering, H.J.A., van Oijk, R., Miley, G.K., Chambers, K.C., van Breugel, W.J.M., de Koff, S., 1997, A&A, in press
Röttgering, H.J.A., West, M.J., Miley, G.K., Chambers, K.C., 1996, A&A, **307**, 376
Schlötelburg, M., Meisenheimer, K., Röser, H.-J., 1988, A&A **202**, 23
Tadhunter, C.N., Dickson, R.D., Shaw, M., 1996, MNRAS **281**, 591
Tadhunter, C.N., Scarrott, S.M., Draper, P., Rolph, C., 1992, MNRAS **256**, 53P
Thompson, D., Djorgovski, S., Vigotti, M., Grueff, G., 1994, AJ **108**, 828
Vigotti, M., Djorgovski, S., Gregorini, L., Klein, U., Mack, K.H., Maxfield, L., Reuter, H.P., Thompson, D., 1996, in *Extragalactic Radio Sources*, R. Ekers et al. eds., Kluwer Academic Publ., p. 519

The Unified Schemes and Radio Polarization Studies

D.J. Saikia[1], S.T. Garrington[2], and G.F. Holmes[2]

[1] National Centre for Radio Astrophysics, Tata Insitute of Fundamental Research, Post Bag 3, Ganeshkhind, Pune 411 007, India
[2] Nuffield Radio Astronomy Laboratories, Jodrell Bank, Macclesfield, Cheshire SK11 9DL, England

1 Introduction

In the unified schemes for extragalactic radio sources, the apparent diversity of different types of sources is interpreted to be largely due to similar objects being seen with different angles of inclination to the observer. The orientation dependence arises due to bulk relativistic motion of the non-thermal jets squirting out from the active nucleus, and anisotropic distribution of circumnuclear material, which is possibly in the form of a torus. Since the extended radio emission from the lobes is believed to be largely isotropic, different unified schemes have been suggested for the high-luminosity Fanaroff-Riley or FR class II sources and the lower luminosity class I objects. In the unified scheme for the FRII radio sources, the radio galaxies are believed to be inclined close to the plane of the sky while the quasars are observed with small viewing angles. On the other hand the lower-luminosity FRI sources are believed to be the parent or misaligned population of the BL Lac objects (Barthel 1989; Antonucci 1993; Urry and Padovani 1995). In this paper we discuss some of the polarization studies which might be relevant for testing or constraing the unified schemes.

2 The Laing-Garrington Effect

In the powerful FRII radio sources the polarization of the two radio lobes is often very asymmetric and strongly correlated with jet sidedness. These sources usually have one-sided jets and the lobe on the jet side depolarizes more slowly than the one on the counter-jet side (Laing 1988; Garrington et al. 1988). This correlation can be interpreted in terms of depolarization of the receding lobe by a halo of hot gas surrounding the radio source and is in the expected sense if the jet asymmetry is due to Doppler boosting. This correlation is also relevant for testing the unified scheme in which radio galaxies are believed to be inclined close to the plane of the sky while the quasars or BL Lacs are observed with small viewing angles. For objects of similar size and redshift, the depolarization asymmetry would be small for galaxies which are seen closer to sky plane compared to quasars or BL Lacs which are observed at small viewing angles.

2.1 The FRI Sources

Low luminosity FRI sources have diffuse outer radio lobes or plumes. Their jets are reasonably symmetric on larger scales but often asymmetric closer to the nucleus. While the outer plumes may be transonic (Bicknell et al. 1990), several of the VLBI scale jets have moderately relativistic velocities (Giovannini 1996). Radio observations of a sample of about 30 largerly FRI sources observed to similar sensitivity limits with the VLA, show that the fractions of sources with detected radio jets are about 0.79±0.24 and 0.50±0.16 for sources with f_c, the fraction of emission from the core, ≥ 0.1 and <0.1 respectively (Saikia and Kapahi, in preparation). If the relative strength of the cores here is also believed to be due to relativistic beaming, it suggests that the asymmetry of the jets on arcsec scales is also due to bulk relativistic motion.

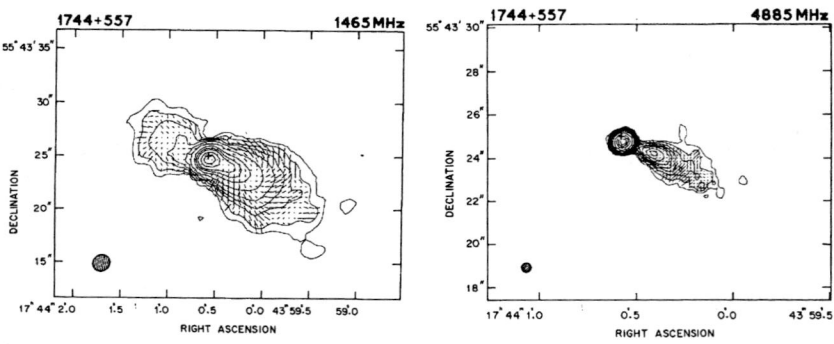

Fig. 1. The VLA A-array images of the low-luminsity radio galaxy 1744+557 at λ20 and 6cm show the jets which appear more asymmetric closer to the nuclues when observed with higher resolution.

We have observed a sample of low- and intermediate-luminosity sources showing a range of jet to counter-jet asymmetries and including several 'fat double' sources. Snapshot observations were made with the VLA at λ 6 and 18/20 cm in B, C and D configurations to give matched resolution of 5 or 15 arcsec. Larger sources were also observed with the WSRT at λ 49 cm.

Many sources (large and small) show little depolarization between λ 6 and 18/20 cm. Several sources show significant, symmetric depolarization: these sources are predominantly 'fat' or twisted FRI sources. Several sources show strong asymmetric depolarization, with less depolarization on the side of the source with the brighter jet. These sources include the fat doubles and FRI sources with the more asymmetric jets.

Figure 2 compares the measured depolarization ratio (low values of DP correspond to stronger depolarization) for the two sides of each source: where asymmetric jets have been seen in higher resolution maps, 'Side 1' has the brighter

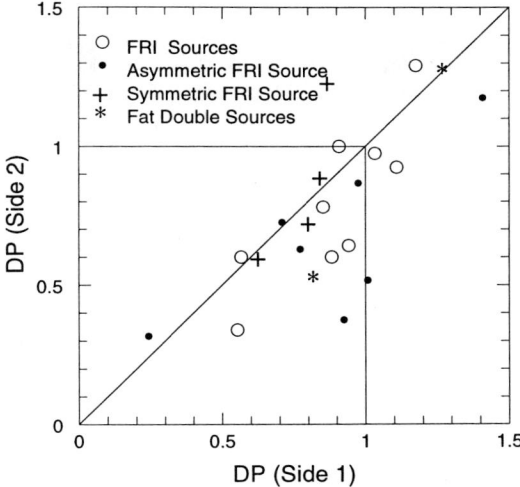

Fig. 2. Plot of DP ($= m_{18}/m_6$ or $= m_{49}/m_{18}$) for the two jets or lobes. Side 1 has the brighter jet at high resolution.

jet. The correlation between jet sidedness and depolarization asymmetry is seen in the cluster of sources below the diagonal line and is in the same sense as the more powerful sources.

The weak depolarization seen in these sources (consistent with their low luminosity) makes depolarization asymmetry more difficult to investigate. Our provisional results suggest that FRI sources are generally more symmetric in depolarization than FRII sources. The depolarization asymmetry seen in the sources with the most asymmetric jets is consistent with the effect seen in the FRII sources with one-sided jets. A similar conclusion has been reached by Parma et al. (1993). This would suggest that the observed asymmetry seen in the bases of the jets in the FRI sources might also be due to relativistic beaming.

2.2 The FRII Radio Galaxies and Quasars

In the unified model for radio galaxies and quasars, the radio galaxies should show depolarization asymmetries in the same sense as quasars but of a smaller magnitude owing to the larger angles of inclination to the line of sight. We observed a sample of radio galaxies, selected from the 3CR complete sample, with the WSRT at $\lambda 49$ and 21cm and with the VLA at $\lambda 20$, 18 and 6cm. Combining the results of our observations of FRII sources from the 3CR sample with information available in the literature, we find that 19 out of 27 radio galaxies with jets show stronger depolarization on the counter-jet side. Although the weaker trend seen for radio galaxies appears consistent with the unified scheme, it is important to investigate this further by considering samples of galaxies and quasars which are of similar luminosity and redshift. We have attempted to construct matched samples of galaxies and quasars by restricting ourselves to objects in the redshift range of about 0.3 to 1. This, however, leaves us with a sample of only 9 quasars with a median redshift of 0.56 and 9 galaxies with a median redshift of 0.64. Of the 9 quasars, 8 obey the effect while among the radio galaxies only 5 show stronger depolarization on the counter-jet side. The correlation of depolarization with jet sidedness is not significant in this sample of radio galaxies, except perhaps in the most asymmetric cases. It appears that in radio galaxies the correlation of depolarization with arm-length ratio is stronger than the relation with jet sidedness, while the reverse is true for quasars (c.f. Laing 1993). The correlation of depolarization with arm-length may be related to an intrinsic asymmetry in the distribution of gas surrounding the source (McCarthy et al 1991; Liu and Pooley 1991), but it could also be a consequence of a symmetric, centrally peaked distribution of depolarizing gas.

3 Depolarization Asymmetry and One-Sided Radio Sources

In low-frequency selected surveys, up to about a few per cent of the FRII sources could be one-sided with radio emission on only one side of the parent optical object, or highly asymmetric with the brightness ratio of the lobes exceeding a few 100:1. Although several statistical studies of these sources indicate that their apparent asymmetry is likely to be due to bulk relativistic motion of the lobes in sources inclined at small angles to the line of sight (Saikia et al. 1990 and references therein), depolarization observations of the lobes may be used to further test this possibility. We have made scaled-array observations with the VLA at $\lambda 20$ and 6cm of a sample of 13 such extremely asymmetric sources. However, only about half of them still appear to be one-sided. The extended components of these sources show little depolarization between $\lambda 20$ and 6cm, consistent with them being the approaching components of double-lobed sources seen through little of the depolarizing medium of the host galaxy. Some of the smaller sources, however, have low values of polarization at both $\lambda 20$ and 6cm.

4 Core Rotation Measures

In addition to depolarization observations of the lobes, polarization studies of the cores are also useful for investigating the unified scheme. From multifrequency studies of the cores of core-dominated quasars at $\lambda 18$, 20, 6 and 2cm, we find that the rotation measures (RM) determined from the long-wavelength data are generally less than about 30 rad m^{-2} while those between $\lambda 6$ and 2cm have significantly higher values, extending up to a few hundred rad m^{-2} (Figure 4). This is possibly due to contributions from more compact optically thick components buried deeper in the nuclear region.

For lobe-dominated quasars, the tendency for the core polarization E-vector to be perpendicular to the large-scale structure suggests that the RM for the core

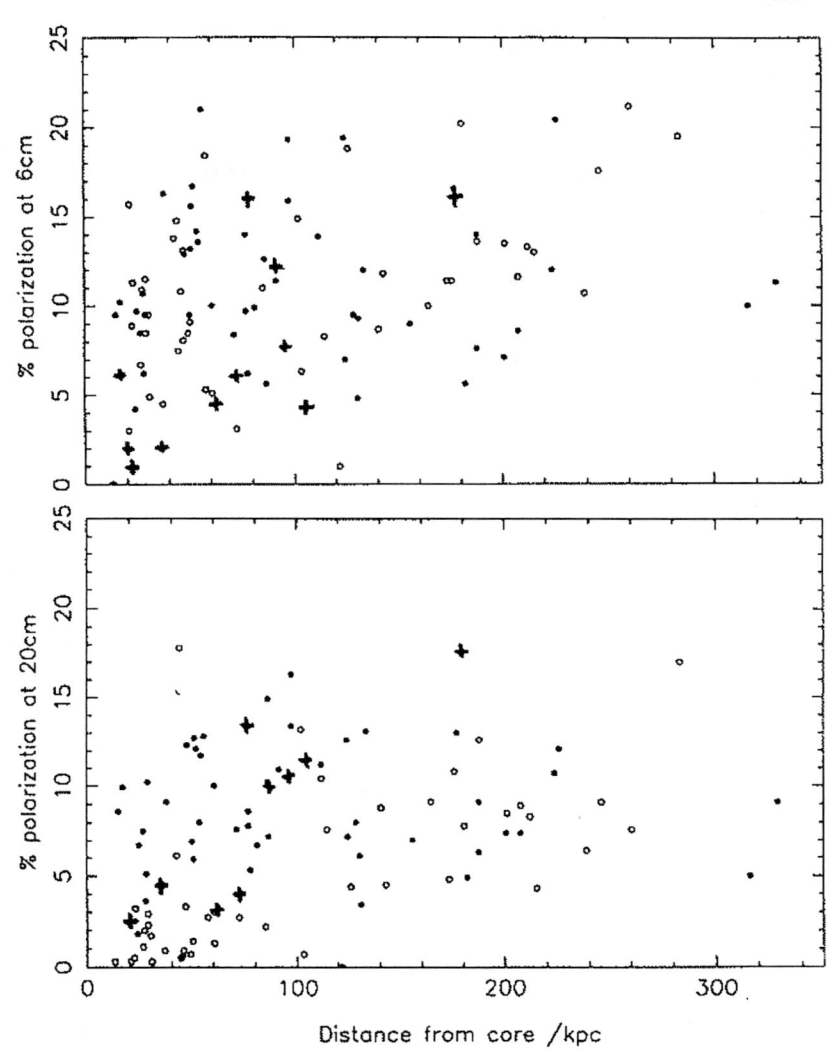

Fig. 3. Comparisons of the degree of polarization at λ6 and 20cm of one- and two-sided radio sources. The jet-side (•) and the counter-jet side components (o) of double radio sources are from Garrington et al. (1990). The one-sided sources (+) show little depolarization between 6 and 20cm, like the jet-side components.

component is also likely to be low. The core E-vector arises from a nuclear jet, and its overall orientation being along the jet axis would be consistent with the tendency for similar field structures in the jets in large-scale FRII sources. Large values of core RM would destroy this relationship. The large scatter in sources with strong cores is believed to be due to effects of aberration and projection.

In the unified model, the nuclear region of the quasar is surrounded by a torus-like structure which obstructs the view of the nucleus when the source is

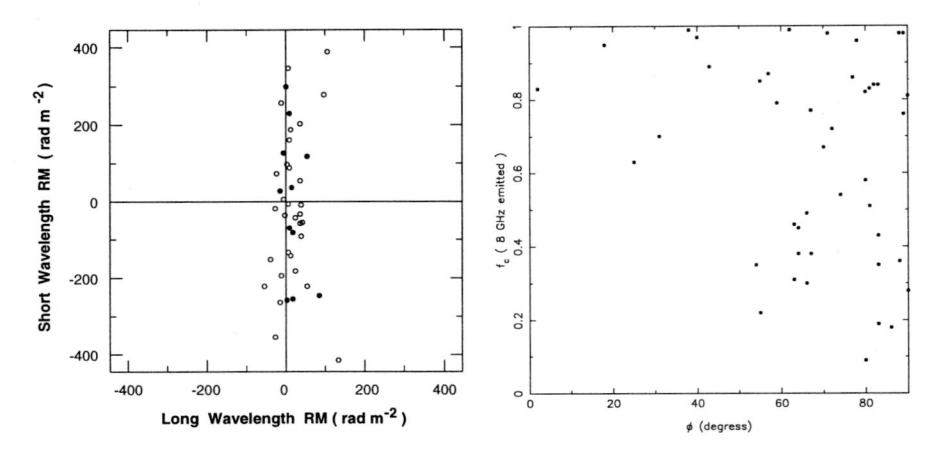

Fig. 4. Plot showing the rotation measures of cores in quasars from the long- and short-wavelength data (left); and the $\phi - f_c$ diagram, where ϕ is the relative orientation of the core polarization vector and the radio axis and f_c is the strength of the core.

inclined at a large angle to the line of sight; the source then appears as a radio galaxy. It would be interesting to determine whether the RM of the cores are different for radio galaxies and quasars. There has been some evidence that the cores of galaxies are much more weakly polarized than quasars (Saikia, Singal & Cornwell 1987), consistent with the possibility that the light from the nuclear region of the galaxy passes through a stronger depolarizing medium which could well be the obstructing torus.

References

Antonucci, R. (1993): ARAA, **31**, 473
Barthel, P.D. (1989): ApJ, **336**, 606
Bicknell, G.V., De Ruiter, H.R., Fanti, R., Morganti, R., Parma, P. (1990): ApJ, **354**, 98
Garrington, S.T., Leahy, J.P., Conway, R.G., Laing, R.A. (1988): Nature, **331**, 147
Giovannini, G. (1996): Extragalactic Radio Sources, IAU Symp. 175, eds Ekers, R.D. et al., Kluwer Academic Publishers, Holland
Laing, R.A. (1988): Nature, **331**, 149
Laing, R.A. (1993): Astrophysical Jets, STScI Symposium Series 6, eds Burgarella, D., Livio, M., O'Dea, C.P., Cambridge University Press, p. 95
Liu, R. and Pooley, G. (1991): MNRAS, **253**, 669.
McCarthy, P.J., van Breugel, W., Kapahi, V.K (1991): ApJ, **371**, 478.
Parma, P., Morganti, R. Capetti, A., Fanti, R., De Ruiter, H. (1993): A&A **267**, 31.
Saikia, D.J., Junor, W., Cornwell, T.J., Muxlow, T.W.B., Shastri, P. (1990): MNRAS, **245**, 408.
Saikia, D.J., Singal, A.K., Cornwell, T.J. (1987): MNRAS, **224**, 379.
Urry, C.M., Padovani, P., (1995): PASP, **107**, 803

Compact Steep-Spectrum Radio Sources

D.J. Saikia and S. Jeyakumar

National Centre for Radio Astrophysics, Tata Institute of Fundamental Research, Post Bag 3, Ganeshkhind, Pune 411 007, India

Abstract. A comparision of the radio properties of well-defined samples of CSSs with larger sources of similar luminsity, suggest that these are young sources advancing outwards through a dense and possibly asymmetric environment. We discuss evidence of interaction of radio jets with dense objects in the nuclear regions, and suggest that the radio properties of CSSs are consistent with the unified scheme.

Compact steep-spectrum radio sources (CSSs) have projected linear sizes of $\lesssim 20$ kpc, steep radio high-frequency spectra ($\alpha \geq 0.5$, where $S \propto \nu^{-\alpha}$), and make up about 25 per cent of steep-spectrum sources in flux-density limited samples. Morphologically they resemble the larger sources, and are believed to be similarly fed by jets from the nuclei of active galaxies. However, whether they are young objects seen at an early evolutionary stage or older objects heavily constrained by dense gas within the host galaxies is a matter of contention.

Over the last few years high-resolution observations of CSSs have significantly increased the number of sources with detected core components. This has made it possible to estimate the symmetry parameters of this class of objects reliably, and permits examining the effects of evolution in a dense asymmetric environment as well as the consistency of the symmetry parameters and relative core strengths of the CSSs with the expectations of the unification scheme. We have chosen the 3CR complete sample and the 3C and Peacock & Wall samples of CSSs with a radio luminosity at 178 MHz $\geq 10^{26}$ W Hz^{-1} sr^{-1}. We have also considered the complete sample of S4 radio sources with $\alpha_{408}^{5000} \geq 0.5$ and luminosity at 5000 MHz $\geq 5\times 10^{24}$ W Hz^{-1} sr^{-1}. All the sources in our sample were required to have a detected radio core.

The ratio of the separations of the two components from the nucleus, r_D, $= (1 + \beta\cos\phi)/(1 - \beta\cos\phi)$, where $v = \beta c$ is the velocity of advancement and ϕ is the angle of inclination of the source axis to the line of sight. Defining r_D to be ≥ 1, the median value of r_D for the quasars is about 1.4, which is marginally higher than ≈ 1.3 for the galaxies. However, there is a significant deficit of symmetric quasars (Fig. 1), which is possibly due to the smaller angle of inclination of the quasars to the line of sight. Considering only the CSS objects, the median values of r_D increases to about 1.8 and 2.0 for the galaxies and quasars respectively, with no galaxies having a value less than about 1.3. Thus although there is overall consistency with the unified scheme, the effects of an asymmetric environment are significant, particularly for the small sources. In addition, in about half the sources, the approaching component facing the jet is closer to the nucleus, suggesting again that the jets are propagating through an asymmetric environment.

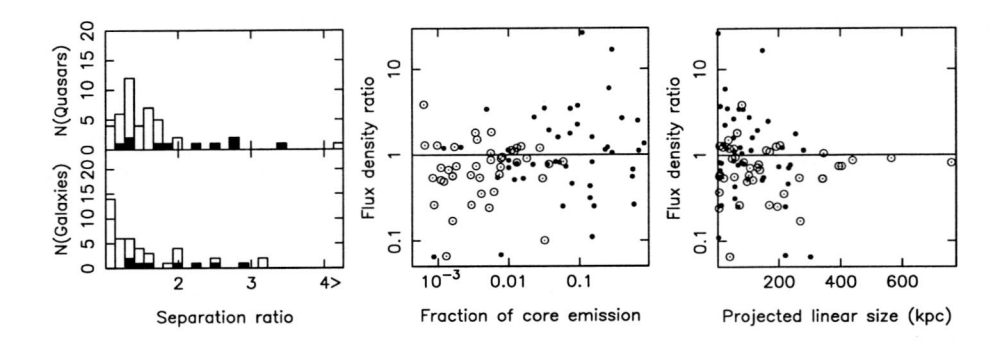

Fig. 1. The distributions of the separation ratio, r_D, for quasars and radio galaxies, where the CSS objects are shown shaded (left); the flux density ratio r_L vs the fraction of emission from the core for the quasars (●) and radio galaxies (⊙) (middle); and r_L vs the projected linear size for quasars and radio galaxies (right).

While examining the flux density ratio, $r_L = r_D^{3+\alpha}$ we find that the quasars are somewhat more asymmetric than the radio galaxies with median values of r_L of ~1.9 and 1.6 respectively, where r_L is defined to be ≥ 1. However, 31 of the 43 galaxies and 24 of the 47 quasars have the brighter component closer to the nucleus, which would be expected if there is greater dissipation of energy on the side with the higher density. The somewhat smaller fraction for quasars suggests that the effects of orientation and beaming become more significant for quasars than for radio galaxies, consistent with the ideas of the unified scheme (Fig. 1). The change in r_L with projected linear size shows that the scatter in r_L decreases as the sources evolve into a more homogeneous and less dense medium at larger distances from the parent nucleus. This is particularly noticeable for galaxies where the effects of beaming are not as important as for quasars.

The CSS quasars have stronger cores, and radio jets are detected more frequently in them than in the CSS radio galaxies, consistent with the unified scheme. However, the CSSs in both quasars and galaxies appear to be evolving in a dense asymmetric environment. A comparison with larger sources, and theoretical and numerical calculations of asymmetry parameters for jets propagating in an asymmetric environment suggest that most sources pass through such a phase. Evidence for dense gas in the central regions of CSSs also comes from radio polarimetric studies. A high proportion of objects with RM over about a 1000 rad m^{-2} are CSS sources. VLA observations of the well-known quasar CSS quasar 3C147 show that the southern component, which is on the jet side and is nearer the nucleus, is brighter and has an RM of about -3144 rad m^{-2} compared to about 630 rad m^{-2} for the northern lobe in the rest frame of the source (Junor et al. 1997, in preparation). It could be a good illustration of a source where the brighter nearer component is possibly interacting with a dense cloud responsible for the observed high RM.

Host Galaxies of Low Luminosity Radio QSOs

Sebastián F. Sánchez[1,2], J. Ignacio González-Serrano[1], Ruth Carballo[1,2], Mario Vigotti[3], Chris Benn[4]

[1] Instituto de Física de Cantabria-CSIC, Univ. de Cantabria, 39005 Santander, Spain
[2] Dpto. de Física Moderna, Universidad de Cantabria, 39005 Santander, Spain
[3] Instituto de Radioastronomía di Bologna-CNR, 40129 Bologna, Italy
[4] Royal Greenwich Observatory-La Palma, Spain

Abstract. Near-infrared imaging in K-Band has been obtained for 53 quasars from the B3-VLA survey. The host galaxy properties have been obtained for 16 quasars, that in two cases have $z \geq 2$. Low luminosity radio quasars inhabit luminous($< L > \sim 4.8L^*$) and large ($r_{1/2} > 10Kpc$) galaxies. The analysis of the restored host galaxies shows that the $K-z$, $\mu_{1/2}-r_{1/2}$ and $\alpha_{408}^{1460} - M_{K,gal}$ distributions are similar to the same distributions for radio galaxies, in a large range of redshifts (almost to $z \sim 2.5$). These results support the unification of these two families of radio-sources.

1 The Observed Sample

Our sample was extracted from the B3-VLA quasar sample (Vigotti et al. 1989). We present the study of 53 objects (all with radio flux >500mJy at 408MHz), observed on February the 5th and 6th, 1996 using the infrared camera WHIR-CAM at the Nasmyth focus of the 4.2m WHT. The filter used was very similar to standard K.

Some focusing problems were present on the second night images, which reduced the useful sample to 32 objects. We have applied our method to study the host galaxies associated with the quasars.

2 Description of the PSF Subtraction Method

Our method consists basically in deconvolve the target images by the PSF (Point Spread Function) and obtain a surface brightnes profile for each deconvolved object. The profile was then fitted with a model profile.

Due to the small field of view of our images, they hardly contain stars. Therefore the PSF was built averaging the four brightest field stars along the night. This mean PSF has a FWHM=1.9", that was the mean seeing along the night.

With a perfect PSF determination, the deconvolution process would have produce a narrow($\sim$ 1 pixel width) point source surrounded by an extended source or galaxy. In our case, however, we expect to find a broader point source, with its FWHM reflecting the seeing variations.

We have fitted the targets profiles using an interactive least-square method. We fitted four differents models:(i) A gaussian function,(ii) gaussian+King profile,(iii) gaussian+$r^{1/4}$ law and (iv) gaussian+exponential disk profile.

3 Results

After making different simulations and checks, like applying all the procedure to standard calibration and field stars, we have set limits to our substraction method.

We have restored the host galaxy for 16 sources. These hosts are luminous ($\langle L \rangle > 4L^*$) and large ($r_{1/2} > 10 Kpc$) galaxies, with poorly known morphologies.

We have indications that the host galaxies are similar to radio-galaxies: (i) the $K-z$ distribution is consistent with the known law for radio galaxies (Lilly& Longair 1984), almost until $z \sim 2.5$ (Fig.1), (ii) there is a $\mu_{1/2} - r_{1/2}$ correlation with a slope of ~ 3.2, similar that obtained for radio galaxies (Ledlow et al. 1995) and (iii) flatter radio-sources inhabit more luminous host galaxies, tendency that has been observed for radio galaxies (Hutchings 1987).

There are weak correlations between $M_{K,gal}$ and $M_{K,nucl}$ and between $M_{K,gal}$ and core radio power. These correlations could indicate a feeding process of the nuclear engine by matter of the host galaxy.

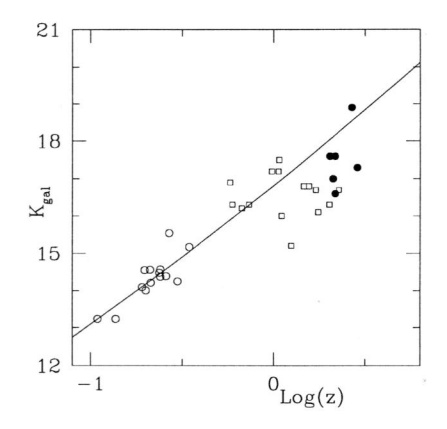

Fig. 1. $K_{gal} - z$ relation of our sample(open squares), Dunlop et al. 1993(open circles) and Lehnert et al. 1992 (solid circles). The line is the $K-z$ relation taken from Lilly& Longair 1984

References

Dunlop J.S., Taylor G.L., Hughes D.H. and Robson E.I., 1993, MNRAS, 264, 455
Hutchings J.B, 1987, ApJ, 320, 122
Ledlow M.J. & Frazer N.O., 1995, AJ, 110, 1959
Lehnert M.D., Heckman T.M., Chambers K.C. & Miley G.K., 1992, ApJ, 393, 68
Lilly S.J. & Longair M.S., 1984, MNRAS, 211, 833
Vigotti M., Grueff G., Perley R., Clark B.G. & Bridle, A.H., 1989, AJ, 98, 419

Quasar Host Luminosities and the Evolutionary Unified Scheme

Fausto Vagnetti

Dipartimento di Fisica, Università di Roma Tor Vergata, via della Ricerca Scientifica, I-00133 Roma, Italy

1 Introduction

The *evolutionary unified scheme* (Vagnetti et al. 1991, Vagnetti and Spera 1994) describes the radio source populations on the basis of the changing balance between three optical luminosities: (i) a nuclear isotropic component, L_{is}, (ii) a relativistic beam, $L_{beam} = L_j \delta^{2+\alpha_{opt}}$, and (iii) the luminosity of the host galaxy, L_G. Here $\delta = [\Gamma(1 - \beta \cos\theta)]^{-1}$ is the Doppler factor, and L_j, the intrinsic jet luminosity, is assumed to have the same cosmic evolution as L_{is}, while a weak increase of Γ, the bulk Lorentz factor of the beam, is able to account for the slower evolution of flat-spectrum quasars (Vagnetti et al. 1991). In turn, the comparison of the total nuclear luminosity $L_N = L_{is} + L_{beam}$ with the non evolutionary galactic luminosity L_G predicts the appearance of a source as a radio galaxy if $L_G > L_N$ (Vagnetti and Spera 1994), in addition to cases determined by obscuration for viewing angles $\theta > \theta_{obsc} \simeq 45°$ (see e.g. Barthel 1989). To explore the behavior of the *evolutionary unified scheme* separately for quasars and radio galaxies, I present here the distributions of two orientation indicators, namely the jet asymmetry (in the radio band) $J = (\delta_{jet}/\delta_{counterjet})^{2+\alpha_r} = [(1 + \beta \cos\theta)/(1 - \beta \cos\theta)]^{2+\alpha_r}$ and the apparent transverse velocity $\beta_{app} = \beta \sin\theta/(1 - \beta \cos\theta)$. Results are presented for the 2nd and 3rd of the three models by Vagnetti and Spera (1994). Both models (hereinafter referred to as VS2 and VS3) include distributions of Γ (with Γ_{max} weakly increasing with cosmic time); model VS3 includes also convolutions through the luminosity functions of host galaxies and quasars. Technically, the luminosity function of steep-spectrum quasars at $z = 0$ is required, as dependence on z and θ can then be appropriately described (see Vagnetti and Spera 1994). Following the analysis by Yee (1992), a gaussian luminosity function is adopted for the host galaxies, with $\langle M_B \rangle = -21.8$, $\sigma = 0.45$. For steep-spectrum quasars, the optical luminosity function is computed using the conditional distribution function and the radio luminosity function, resulting again in a gaussian distribution, with $\langle M_B \rangle = -20$, $\sigma = 1.2$. For model VS2, the central luminosities of such distributions are adopted (details in Vagnetti and Spera 1994).

2 Results

Model VS2. See Vagnetti (1996), Vagnetti (1997). The separation between radio galaxies and quasars is determined by obscuration at redshift high enough ($z_{crit} \simeq 0.4$ for our choice of the relevant parameters). If the obscuration angle is constant, cut-offs are produced in the J and β_{app} distributions of radio galaxies at high redshift. The cut-off values are $J^* = [(1 + \cos\theta_{obsc})/(1 - \cos\theta_{obsc})]^{2+\alpha_r}$ and $\beta^*_{app} = \sin\theta_{obsc}/(1 - \cos\theta_{obsc})$ and are attained asymptotically when $\beta \to 1$. Taking for example $\theta_{obsc} = 45°$ and $\alpha_r = -0.17$ (which is the average spectral index of flat-spectrum quasars in the samples analyzed in Vagnetti et al. 1991), $J^* \simeq 25$ and $\beta^*_{app} = 1 + \sqrt{2}$ are obtained. For the same critical values, breaks are produced in the quasar distributions. At low redshift, radio galaxies reach high values of J and β_{app}, although less extreme than quasars.

Model VS3. See Fig. 1. The cut-offs J^* and β^*_{app} in the radio galaxy distributions become effective at $z > z_{crit}$. In turn, z_{crit} is an increasing function of the ratio $L_G/L_{is}(z=0)$, so that the cut-offs J^* and β^*_{app} are smoothed out by the luminosity functions. A weak sign is however left at $z \gtrsim 1$, while the corresponding breaks for quasars are present at all redshifts.

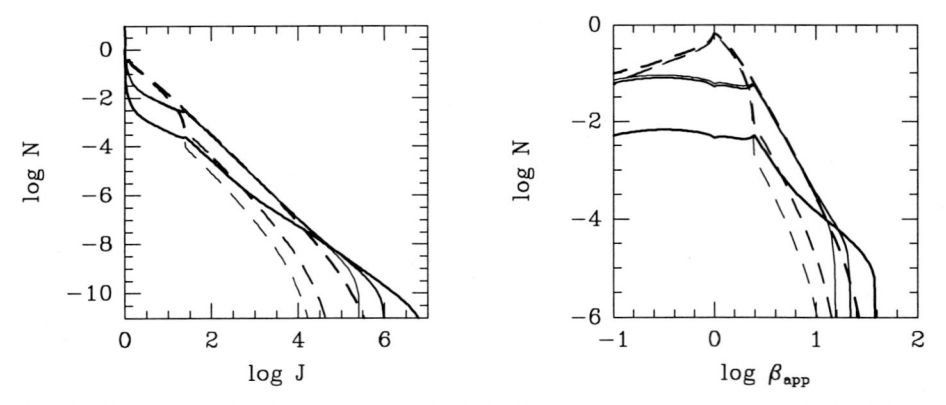

Fig. 1. J and β_{app} distributions for model VS3. Continuous lines: quasars; dashed lines: radio galaxies. $z = 2, 1, 0$ represented with increasing thickness.

References

Barthel, P.D. (1989): Ap. J. **336**, 606
Vagnetti, F. (1996): IAU Symp. **175**, 391
Vagnetti, F. (1997): in *Blazars, Black Holes and Jets* (Kluwer), in press
Vagnetti, F., Giallongo, E., Cavaliere, A. (1991): Ap. J. **368**, 366
Vagnetti, F., Spera, R. (1994): Ap. J. **436**, 611
Yee, H.K.C. (1992): in *Relationships Between Active Galactic Nuclei and Starburst Galaxies*, ASP Conf. Ser. **31**, 417

Part 2

OBSERVING QUASAR HOST GALAXIES

Nearby Quasar Galaxies

M.-P. Véron-Cetty, L. Woltjer

Observatoire de Haute Provence, F-04870 Saint-Michel l'Observatoire, France

1 Abstract and Introduction

In this paper four questions will be addressed:
(a) Are the best ground based and HST data on quasar hosts in agreement ?
(b) Do radio quiet quasars occur in elliptical galaxies as well as in spirals ?
(c) Are the colours of quasar galaxies on average bluer than typical gE galaxies, indicating that recent star formation has taken place ?
(d) Is the optical luminosity function of radio quasars and strong radio galaxies more or less the same ?
The answer to all questions will turn out to be affirmative. The implications for unified models will be discussed and a scenario for quasar evolution proposed.

2 HST and Ground Based Data

Much has been made in the recent literature and also in the press about the discovery of "naked" quasars and more in general about the large differences between quasar hosts observed with HST and from the ground. On closer inspection, however, a very satisfactory agreement is found, although of course the HST data show fine structure not seen in the ground based images. We shall here make a detailed comparison between the data we obtained previously (Véron-Cetty & Woltjer, 1990, VW) and the data obtained by Bahcall et al. (1995a,b) and by Disney et al. (1995) with HST for the radio quasars PKS 1302-102 (z=0.286) and PKS 2349-014 (z=0.173).

2.1 PKS 1302-102

We recall that VW determined the amount of light in an annulus with radii of 12.5 and 25 kpc and then adopting the best fitting de Vaucouleurs law calculated the difference between the total magnitude of the galaxy (0-∞) and the magnitude of the annulus which turned out to be not too sensitive to the effective radius of the model. For PKS 1302-102 the radii of the annulus are 1."5 and 3" and VW found i = $19.^m2$. All the i magnitudes in VW have to be corrected by $-0.^m16$ because earlier measurements of the calibration star (LTT 3218) were erroneous (see appendix). Adopting a normal colour for a gE we have V_0-i_0 = $0.^m49$, k_i = $0.^m25$ from Schneider et al. (1983) and k_V = $0.^m72$ from Pence (1976) and V = 20.0 for the annulus. Since the redshifted V band is not very far from the i band, errors in the adopted colours would not have much effect.

The HST data of Bahcall et al. (1995b) show that PKS 1302-102 has two close companions at 1."1 (m = 20.6) and at 2."2 (m = 21.9) where the magnitudes refer to the F606W filter. According to these authors, for a gE type spectrum at z = 0.286, the V values should be $0.^m38$ higher than these m values; we then find for the total contribution of the two to the light in our annulus (taking into account the 1" seeing) $V = 21.^m8$ and upon correction for this, our annulus would have V = 20.2, which, with the effective radius $r_e = 1$", corresponds to V = 18.5 for the host galaxy and $M_V = -23.^m6$.

Bahcall et al. (1995b) have made simulations to see what galaxies they should be able to detect. Simulation 5e in table 3 corresponds to a galaxy with a de Vaucouleurs $r^{1/4}$ law and $r_e = 1.$"2 and shows that a galaxy fainter than m = 17.9, corresponding to V = 18.3, would not have been noticed. There is, therefore, no contradiction with our result. The apparent problem has only arisen because the limits given in table 1 of Bahcall et al. are based on the *average* of eight simulations. However, what counts in establishing a definite limit is not this average, but the simulation which yields the smallest m.

The same quasar has been observed with HST by Disney et al. (1995) who obtain with the F702 Filter $m = 21.^m3$ and $22.^m0$ for the two companions and M_{host} = -23.5, where M_V "could be a few tenths of a magnitude brighter". Therefore, within the uncertainties due to the different photometric systems there is agreement with our result and no contradiction with that of Bahcall et al. There is, however, a substantial difference with the ground based result of Hutchings & Neff (1990) who obtain $M_R = -25.^m2$ which, with V-R = 0.7 for gE galaxies, would yield $M_V = -24.^m5$, about a magnitude brighter than the other results.

2.2 PKS 2349-014

While in the preceding object Bahcall et al. only found an upper limit, for PKS 2349-014 a positive result was obtained (Bahcall et al. 1995a). It is interesting to compare the HST images for this source with the ground based image of VW. The asymmetry of the galaxy with respect to the quasar and the two (tidal ?) arms to the north are evident in both cases, though, of course, the HST image shows more detail in the inner parts. The HST image shows a companion at 1."8 which should not affect the photometry in the 12.5-25 kpc (2."5-5") annulus of VW. They found in the annulus i = 17.6, to be corrected with $-0.^m16$ because of the calibration change, and with $k_i = 0.^m16$, $k_V = 0.^m33$ and V_0-$i_0 = 0.^m49$, V = 18.1 in the annulus and V = 16.3 for the galaxy. Bahcall et al. found in an annulus between 0."5 and 9", m = 16.8 which at z = 0.173 should correspond to $V = 16.^m6$. With $r_e = 2.$"6 as found in VW, $V = 16.^m4$ is found for the whole galaxy, in close agreement with the VW result. Since the galaxy is asymmetric and perturbed (like many quasars in the VW sample) some deviations from an $r^{1/4}$ law could be expected.

We conclude therefore, that the HST data are in very good agreement with the quantitative results obtained by VW for the luminosities of the quasar galaxies.

In passing we also note the claim of Abraham et al. (1992) that VW had underestimated the effects of the subtraction of the PSF. Since VW only used PSF's determined on the CCD frame containing the quasar this cannot be the case. The conclusion of Abraham et al. resulted from the application of a PSF determined at La Palma to the La Silla data. In fact, it appears that for the same half power seeing disk diameter the wings were frequently higher in their data.

3 "Radio Quiet" Quasars in Ellipticals

It has frequently been stated that since spirals are generally not very strong radio sources, radio quasars may be taken to be in ellipticals, while since Seyfert nuclei usually are found in spirals, the same might be assumed to be the case for the radio quiet quasars (QQ). However VW showed that the difference in M_V between radio quasars and QQ could also be understood if the latter were in ellipticals and if their relatively weak radio emission followed the standard relation between radio power and M_V for moderately weak radio galaxies. Such an argument shows that nothing stands in the way of QQ being ellipticals, but, of course, does not prove this either.

Meanwhile some more direct evidence has become available. Véron-Cetty et al. (1991) studied the object F9 (z = 0.045) which at discovery had a luminosity in the quasar range and at other times has been weaker. The radio emission was found to be weak ($P_{5GHz} < 5\ 10^{22}$ W Hz^{-1}), but the luminosity profile follows that of a gE with M_V = -23.0. In passing we note that the image of F9 shows a remarkable similarity with that of PKS 2349-014 discussed before. Also our new data for the nearby QQ Ton S180 (z = 0.062) show a rather accurately followed $r^{1/4}$ luminosity profile with $r_e \approx$ 17kpc for a modest galaxy (M_V = -22.0).

With HST, Disney et al. showed the QQ 2215-037 host to have the general appearance of an elliptical and to follow an $r^{1/4}$ law quite well, while Bahcall et al. (1996) found the QQ PHL 909 to be associated with an elliptical.

Not all QQ are associated with ellipticals. Véron-Cetty et al. (1991) found 09149-6206 (z = 0.057) to have a host with a profile like an Sa (M_V = -23.0), while Bahcall et al. (1996) found with HST that PG 0052+251 was situated in a rather normal looking Sb with M_V = -22.8. Apparently both luminous ellipticals and spirals can produce quasars; in what proportion as a function of the luminosity of the quasars remains to be seen.

4 Colours of Host Galaxies

Much evidence has accumulated that quasar hosts and radio galaxies are frequently bluer than gE. Sandage (1972) has already noticed this in photoelectric observations of radio galaxies, but the role of a possible blue nucleus made the situation unclear. Smith & Heckmann (1989) subsequently showed that B-V colours too blue by $0.^m2$ are not at all exceptional. If we average their data for the colours of 38 strong radio galaxies ($\log P_{5GHz} > 24.7$) outside a 10 kpc radius, $<B_0\text{-}V_0> = 0.^m83$ is found, $0.^m14$ bluer than brightest cluster galaxies.

Quasar galaxies appear to be relatively blue. Malkan (1984) found the quasar galaxy of 3C 48 to be very blue with v, g, r colours corresponding to an Sc or Sd galaxy. VW measured R-i colours for 5 radio quasar galaxies. The average is $0.^m17\pm0.^m09$ bluer than a gE, corresponding with typical spectral distributions to $(B_0\text{-}V_0) \approx 0.^m6$. Boroson and Oke (1984) and Boroson et al. (1985) obtained spectral scans for 9 radio quasar galaxies and 3 radio quiet quasar galaxies which yielded rough colours averaging $(B_0\text{-}V_0) = 0.^m5$. Taking these data at face value, the quasar galaxies would on average be bluer than the radio galaxies.

Important evidence that quasar galaxies have spectral distributions which are very different from those of giant ellipticals comes from observations of absorption lines by Miller (1981), who observed the Mg Ib (5180Å) line in BL Lacs, N-galaxies and quasars. Assuming the galaxies in question to have normal gE spectra, he determined the amount of galaxy light present or an upper limit to it in case the feature was so diluted as to be invisible. For 5 BL Lac objects (Mark 180, AP Lib, Mark 501, 3C 371, BL Lac) with recent imaging data we obtain from his measurements $<M_V> = -23.^m0$, while from the imaging data by Abraham et al. (1991), Baxter et al. (1987) and Stickel et al. (1993) for the same objects we have $<M_V> = -22.^m9$. Apparently these galaxies have the standard Mg Ib absorption for their luminosity. If indeed the BL lacs are FR I radio galaxies as proposed in the unified models, this should be no surprise since the latter are mainly associated with fairly normal ellipticals. In 6 N systems with detected Mg Ib Miller's data give $<M_V> = -22.^m0$, while the imaging data of Smith & Heckman (1989) for the same objects yield $<M_V> = -23.^m1$. The simplest way to interpret this result is to say that the Mg Ib line is a factor 2.8 too shallow for a gE galaxy. For quasars, Miller only finds upper limits. For four quasars with $M_V < -23.0$ and with imaging data, his results yield for the hosts $<M_V> > -22.^m0$, while the imaging data give $<M_V> = -23.^m4$. Apparently the Mg Ib absorption is a factor of > 3.6 too shallow. In addition, in the quasar 3C 459 Miller found the Balmer lines in absorption, corresponding to a young stellar population. Further confirmation comes from the scans of Boroson and Oke (1984) who measured the Mg_2 index (Mg I + Mg H) in some radio quasar galaxies. In 3C 37.43 and 3C 48 values of $0.^m098$ respectively $0.^m058$ were measured, while the values for gE are typically between $0.^m25$ and $0.^m35$; however, in the quasar 4C 31.63 a value of $0.^m344$ was found. Also in 3C 48 Balmer lines were seen in absorption. The straightforward interpretation of these data is that the spectra of the quasars (and N-galaxies) are different from those of gE, with a much larger contribution from young stars - consistent with the evidence for bluer colours.

Recently Dunlop et al. (1993) and McLeod & Rieke (1994a,b) have observed quasar hosts in the K respectively H bands with array detectors. As pointed out by these authors, the IR is a particulary suitable place to image quasar hosts because the contrast galaxy/quasar nucleus is favorable.

For six of the radio quiet quasars observed by McLeod & Rieke (1994 a,b) reliable optical data (assembled in table 7 of VW) are available which yield a very blue mean V_0-H_0 = $2.^m1$, while for the RQ PKS 1302-102 V_0-H_0 = $2.^m7$ is found from VW's observations. Normal early type galaxies have about V_0-H_0 = $2.^m9$. From comparisons between different samples McLeod & Rieke (1994b) concluded that QQ averaged up to about half a magnitude too blue, but they concluded later (McLeod & Rieke 1995) from a comparison with HST data that, within the errors, a small sample of RQ and QQ was not noticeably different from "normal" early type galaxies. Probably it has some advantage to compare ground based IR samples with ground based optical samples in which very nearby companions are included in the same way and in which precise photometry in a well calibrated system can be more easily obtained.

Very recently, Taylor et al. (1996) published an improved version of the Dunlop et al. (1993) data. For 8 RQ treated as ellipticals, also studied by VW or Smith et al. (1986),<V_0-K_0> = $3.^m1$, while for 3 QQ, treated as spirals, <V_0-K_0> = $2.^m3$. The average of all 11 yields <V_0-K_0> = $2.^m85 \pm 0.^m21$. For 5 radio galaxies in common with Smith & Heckman (1989), <V_0-K_0> = $3.^m42 \pm 0.^m35$; of particular interest in these samples is the QQ 2215-037 which Disney et al. showed (section 3) to have the properties of an elliptical, but which has V_0-K_0 = $2.^m5$, nearly a magnitude less than "old" gE's.

For radio galaxies, Eisenhardt & Lebofsky (1987) obtained approximate V and K magnitudes simultaneously through identical apertures. They found radio galaxies with z < 0.2 to have V-K colours $0.^m12$ bluer than brightest cluster galaxies. The result could still be influenced by contributions from undetected nuclei.

In fact in many cases only small deviations from "normal" V-K colours should be expected. For example recent models by Tantalo et al. (1996) show that a burst of star formation occuring 1-5 Gyr ago should have a Δ(V-K) equal to about 0.8Δ(B-V), where Δ denotes the deviation from the colour of an old gE. The Δ(B-V) = $0.^m14$ found by Smith and Heckman for radio galaxies therefore is consistent with the Δ(V-K) result of Eisenhardt & Lebofsky. The Δ(B-V) of $0.^m3$-$0.^m5$ noted earlier for quasar galaxies would then correspond to Δ(V-K) of $0.^m2$-$0.^m4$ entirely compatible with the results discussed before.

Some quasar galaxies are very blue. The quasar Mark 1014 is generally regarded as a QQ though with P_{5GHz} = 10^{24} W Hz^{-1}, its radio emission is well above that for normal spirals. Smith et al. (1986) found the associated galaxy to be extraordinarily luminous, with M_V = $-25.^m1$ while McLeod & Rieke (1994b) found it to have a very blue colour V_0-H_0 = $1.^m4$. A mass of molecular gas of $10^{11.1}$ $M_\odot$ has been found in this object (Sanders et al. 1988a). Apparently extensive star formation is taking place. From the models of Tantalo et al. (1996) it follows that a burst of star formation has very blue V-K colours only until about 10^8 years, after which a drastic reddening takes place very rapidly. If the light of some of the quasar galaxies is, in fact, dominated by such young stars, the optical light to mass ratio would also be large, and the masses need not be larger than for more typical gE.

The overall impression from the colour and spectral data is that quasar galaxies are rather blue, on average more so than radio galaxies. If so, quasars may well represent statistically more recent merger products than strong radio galaxies, with the FR I radio galaxies and BL Lac's being still more ancient. While this is not in agreement with a strict unification model, it does not exclude that many radio galaxies are, in fact, obscured quasars. A mixture of evolutionary and orientation effects certainly can account better for the observations, including those on luminous infrared galaxies, some of which might represent an even earlier evolutionary stage (Sanders et al. 1988b).

Another important aspect in evaluating the effects of evolution is the gas content. The high surface densities of the gas in objects like Mark 1014 make it clear that star formation must be taking place. However, also atomic hydrogen may be an important constituent, though more difficult to observe in strong radio sources. An example is the gE radio galaxy PKS 1718-649 in which Véron-Cetty et al. (1995), following earlier observations of Fosbury et al. (1977), detected an asymmetric disk of gas with a diameter of 180 kpc containing $3\ 10^{10}\ M_\odot$ of atomic hydrogen. However, recently Morganti et al. (1996) discovered $2.4\ 10^{10}\ M_\odot$ of HI in the luminous elliptical or S0 galaxy NGC 5266 spread out over 280 kpc; this object is not a radio galaxy. Until larger samples will be available it is difficult to be sure about the implications of such curiosities.

5 Radio Quasars and Strong Radio Galaxies

In table 1, we summarise the data on quasar galaxies for a sample of luminous quasars ($-24.6 > M_V > -26.6$), the high luminosities reducing selection effects also for the QQ. The zero point error of $0.^m16$ has been corrected and we have adopted B_0-$V_0 = 0.^m6$. If a spectral distribution appropriate for a standard gE had been adopted, the mean M_V would have been only $0.^m06$ fainter. The results for radio quiet quasars are based on exponential disk models; ellipticals would be $0.^m4$ more luminous. Also given in table 1 are the results for a sample of strong radio galaxies based on CCD imaging by Smith & Heckman (1989). Results for some other samples of AGN may be found in VW.

Table 1. Numbers, mean galaxian M_V and rms dispersion in samples of radio quasars, radio quiet quasars and radio galaxies.

Sample		N	$<M_V>$*	Dispersion
VW	RQ	20	-23.48 ±0.12	±0.50
VW	QQ	16	-22.55 ±0.15	±0.8
Smith & Heckman	RG	72	-23.34 ±0.08	±0.65

*) for galaxy from 0-∞ and with $H_0 = 50$ km s^{-1} Mpc^{-1} and $q_0 = 0$; the quoted errors do not include systematic photometric and modelling errors.

From the results in table 1 we conclude that the average luminosity and its dispersion for strong radio galaxies and radio quasars galaxies are not very different. This would appear to be consistent with unified models since the absorption effects invoked in such models affect only a very small region ($\approx$ 1pc) near the centre. In passing we note that the comparison between radio galaxies and radio quasars from ground based data includes for both the possible effects of unresolved companions. However, in the unified models these effects should be expected to be statistically the same for quasars and radio galaxies.

Also in the comparison between quasars and strong radio galaxies, the BL Lac objects should be excluded. If the unified models are correct, most BL Lac's are intrinsically relatively weak radio galaxies which are known to have lower luminosities than the strong radio galaxies (see Ulrich, 1989, Woltjer 1989).

6 Conclusion

They are three ways in which radio quasars and radio galaxies may be related to each other.

a) *Orientation.* Depending on the angle of the line of sight with respect to the axis of an absorbing torus around the centre, the nucleus would be visible (quasars) or obscured (radio galaxies). If a relativistic jet is present and if the angle of the jet axis with respect to the line of sight is small, a BL Lac or a OVV might result. In these cases, the properties of the host galaxies should be the same, unless an absorbing layer pervades large parts of the galaxy. The negligible difference in M_V between radio quasars and radio galaxies supports such a model (table 1), but the colour differences are difficult to accomodate.

b) *Variability* is generally neglected. However most quasars are optically variable ($\approx 0.^m1$-$0.^m3$ or more) on timescales of a decade or less. Variations might be larger on longer time scales and one could easily envisage the change of a radio quasar into a radio galaxy by the temporary extinction of the nucleus. The host galaxies would again remain unchanged, as in 6.a.

c) *Evolution* must occur in quasars and radio galaxies on time scales no longer than that for the evolution of the population of such objects. The fact that many quasar galaxies have spectral distributions that are rather blue suggests that they evolve on timescales of one or a few times 10^8 years, the less blue radio galaxies being, on average, somewhat older than quasars. Since the space density of strong radio galaxies is only three times that of radio quasars of the same radio power, their lifetimes could not be much longer than those of the quasars. In agreement with this, in some cases, the synchrotron lifetimes of the relativistic electrons are seen to be of the order of 10^8 years. The somewhat smaller sizes (Barthel 1989) of the quasar radio sources, if not an accidental sample effect, would fit an evolutionary model as well as the orientation dependent models.

Undoubtedly all three effects play a role. Only infrared observations of the effects of an absorbing torus can demonstrate the importance of 6.a, while spectral observations with HST should further establish various aspects of 6.c.

The relation of the QQ to other AGN is more uncertain. The quasars in spiral galaxies may well represent a high luminosity extension of the Seyferts. Seyfert galaxies seem too abundant to represent recent mergers and, in fact, most Seyferts without the nuclei would not look very different from non-Seyfert spirals. It therefore seems plausible that Seyferts and quasars associated with spirals represent an accretion process fuelled by gas from within the galaxy, perhaps slightly perturbed by neighbouring galaxies.

The quasars in ellipticals appear to be very different. The galaxies frequently look perturbed and are too blue. In many cases, even without the nucleus, one would be able to see that the galaxy is not a normal gE. Here it seems that a short lived phenomenon has shaken up the galaxy, presumably a major merger. Since violents events tend to produce objects with an $r^{1/4}$ profile (Barnes 1988), the antecedents of the merger are not obvious, but at least one gas rich galaxy must be involved. There then would be two different kinds of objects: the internally fuelled Seyfert sequence involving at most minor mergers and the catastrophic mergers producing young looking elliptical like objects.

Blandford (1990) has suggested that strong radio sources are associated with rapidly rotating black holes and that only accretion at rates close to the Eddington limit could spin up the black holes to such rotation. In Seyferts the gas will presumably be supplied rather slowly, accretion rates would be low, radio emission weak and the nuclear luminosity due to dissipation in the accretion disk. In major mergers involving a gas rich galaxy, the situation would be very different. Star formation would result in an elliptical with a blue colour for a few 10^8 years (e.g. QQ 2215-037, sect. 3&4), while much gas could be transferred to the nucleus (Barnes & Hernquist 1991), the hole spun up on an Eddington time scale (few 10^8 years) and a strong radio source produced. *Strong radio emission would not so much be a property of ellipticals, as a consequence of the process that forms (some of ?)them.* It would only occur if in the later phases enough accreting matter remains available to complete spin up. Radio quiet quasars then would be on average younger with more perturbed hosts, and potentially more numerous than RQ. Also the evolution of the quasar population fits naturally into such a scenario. After the termination of the intense accretion phase the quasar luminosity would diminish, while radio emission could continue for a longer time in aging, normal looking ellipticals.

7 Appendix

The calibration of the data in VW was based on observations of the star LTT 3218 for which Eggen (1979) obtained in the Kron system $R_K = 12.08$ and $I_K = 12.04$. With the formulae of Eggen (1971) and Cousins (1980) this becomes $R_{KC} = 11.98$ and $I_{KC} = 11.85$ in the Kron-Cousins system now generally used. Subsequent data show the Eggen result to be erroneous. Landolt (1992) obtained

$R_{KC} = 11.753$ and $I_{KC} = 11.643$, while Hamuy et al. (1992) found $R_{KC} = 11.762$ and $I_{KC} = 11.651$. We then obtain i = 12.31 instead of i = 12.47 adopted by Moorwood et al. (1986). Consequently, all i-magnitudes in VW should be diminished by $0.^m16$. Since the R-magnitude of Eggen also should be corrected, the effect on the R-i colours given in VW is small.

References

Abraham R.G.,McHardy I.M.,Crawford C.S. 1991,MNRAS **252**,482
Abraham R.G.,Crawford C.S.,McHardy I.M. 1992,ApJ **401**,474
Bahcall J.N.,Kirhakos S.,Schneider D.P. 1995a,ApJ **447**,L1
Bahcall J.N.,Kirhakos S.,Schneider D.P. 1995b,ApJ **486**,237
Bahcall J.N.,Kirhakos S.,Schneider D.P. 1996,ApJ **457**,557
Barnes J.E. 1988,ApJ **331**,699
Barnes J.E.,Hernquist L.E. 1991,ApJ **370**,L65
Barthel P.D. 1989,ApJ **336**,606
Baxter D.A. Disney M.J.,Phillips S. 1987,MNRAS **228**,313
Blandford R.D. 1990,in Active Galactic Nuclei (Saas-Fee Course 20),Springer, p.261
Boroson T.A.,Oke J.B. 1984,ApJ **281**,535
Boroson T.A.,Persson S.E.,Oke J.B. 1985,ApJ **293**,120
Cousins A.W.J. 1980,South African Astron. Obs. Circ. **1**,234
Disney M.J.,Boyce P.J.,Blades J.C. et al. 1995,Nature **376**,150
Dunlop J.S.,Taylor G.L.,Hughes D.H.,Hughes E.I. 1993,MNRAS **264** ,455
Eggen O.J. 1971,ApJS **22**,389
Eggen O.J. 1979,ApJS **39**,89
Eisenhardt P.R.,Lebofsky M.J. 1987,ApJ **316**,70
Fosbury R.A.E.,Mebold U.,Goss W.M.,van Woerden H. 1977,MNRAS **179**,89
Hamuy M.,Walker A.R.,Suntzeff N.B. et al. 1992,PASP **104**,533
Hutchings J.B.,Neff G. 1990,AJ **99**,1715
Landolt A.U. 1992,AJ **104**,372
Malkan M.A. 1984,ApJ **287**,555
McLeod K.K.,Rieke G.H. 1994a,ApJ **420**,58
McLeod K.K.,Rieke G.H. 1994b,ApJ **431**,137
McLeod K.K.,Rieke G.H. 1995,ApJ **454**,L77
Miller J.S. 1981,PASP **93**,681
Moorwood A.F.M.,Veron-Cetty M.P.,Glass I.S. 1986,A&A **160**,39
Morganti R.,Sadler E.M.,Oosterloo T.,Bertola F.,Pizelli A. 1996, Australia Tel. Nat. Facility Newsletter **30**,7
Pence W. 1976,ApJ **203**,39
Sandage A. 1972,ApJ **178**,25
Sanders D.B.,Scoville N.Z.,Soifer B.T. 1988a,ApJ **335**,L1
Sanders D.B.,Soifer B.T.,Elias J.H. et al. 1988b,ApJ **325**,74
Schneider D.P.,Gunn J.E.,Hoessel J.G. 1983,ApJ **264**,337
Smith E.P.,Heckman T.M. 1989,ApJ **341**,658
Smith E.P.,Heckman T.M.,Bothun G.D. et al. 1986,ApJ **306**,64
Stickel M.,Fried J.W.,Kuhr H. 1993,A&AS **98**,393
Tantalo R.,Chiosi C.,Bressan A.,Fagotto F. 1996,A&A **311**,361
Taylor G.L.,Dunlop J.S.,Hughes D.H.,Robson E.I. 1996,MNRAS **283**,930

Ulrich M.-H. 1989,Lect. Notes in Phys. **334**,45
Veron-Cetty M.P.,Woltjer L. 1990,A&A **236**,69
Veron-Cetty M.P.,Véron P. 1996,ESO Scientific Report N17
Veron-Cetty M.P.,Woltjer L.,Roy A.L. 1991,A&A **246**,L73
Veron-Cetty M.P.,Woltjer L.,Ekers R.D.,Staveley-Smith L. 1995, A&A **297**,L79
Woltjer L. 1989,Lect. Notes in Phys. **334**,460

HST Images
of Twenty Nearby Luminous Quasars

John N. Bahcall[1], Sofia Kirhakos[1], and Donald P. Schneider[2]

[1] Institute for Advanced Study, School of Natural Sciences, Princeton, NJ 08540, USA
[2] Department of Astronomy and Astrophysics, The Pennsylvania State University, University Park, PA 16802, USA

Abstract. We present observations with the Hubble Space Telescope for a representative sample of 20 intrinsically luminous quasars with redshifts smaller than 0.30. These observations show that luminous quasars occur in diverse environments, including ultra-luminous ellipticals, normal ellipticals, spirals with H II regions, complex systems of gravitationally interacting components, and faint nebulosity. The quasar host galaxies are centered on the quasar to the accuracy of our measurements, 400 pc. Contrary to expectations, there are more radio quiet quasars in ellipticals than in spirals. Most of the radio loud quasars are, as expected, in ellipticals. These 20 luminous quasars occur preferentially in galaxies that are more luminous than the typical field galaxies. Eight companion galaxies lie within projected distances of 10 kpc from the quasar nuclei.

1 Observations and the Unprocessed Data

To study the nature of the environment of quasars, we have observed a sample of 20 of the most luminous ($M_V < -22.9$, for $H_0 = 100$ km s^{-1} Mpc^{-1}, $\Omega_0 = 1.0$) nearby ($z < 0.30$) radio-quiet and radio-loud quasars, using the Wide Field Camera (WFPC2) of the Hubble Space Telescope (HST). The quasar images were placed at the center of WF3 and the visual-band filter F606W was selected for its high throughput. The size of WF3 is 800×800 pixels and its image scale is $0\rlap.{''}0996$ pixel^{-1}.

Figure 1 shows images of eleven of the 20 quasars, plus the image of a blue star (upper left-hand panel) for comparison. The only image processing performed on these images is cosmic ray removal and pipeline STScI flatfielding. The exposure times for the quasar images are 1100 s or 1400 s. The exposure time of the star image shown is 20 s, which yields a total number of counts similar to that obtained in the quasar images. The star is MMJ 6490 in the M67 cluster. Its apparent V magnitude is 10.99 and its $B - V = 0.11$ (Montgomery, Marschall & Janes 1993). As can be seen in Figure 1, some of the quasars are noticeably non-stellar, and features of host galaxies are visible in the unprocessed *HST* images. In some cases, like NAB 0205+02, PG 0953+414, and PG 1307+085, it is difficult to distinguish the quasar image from the star.

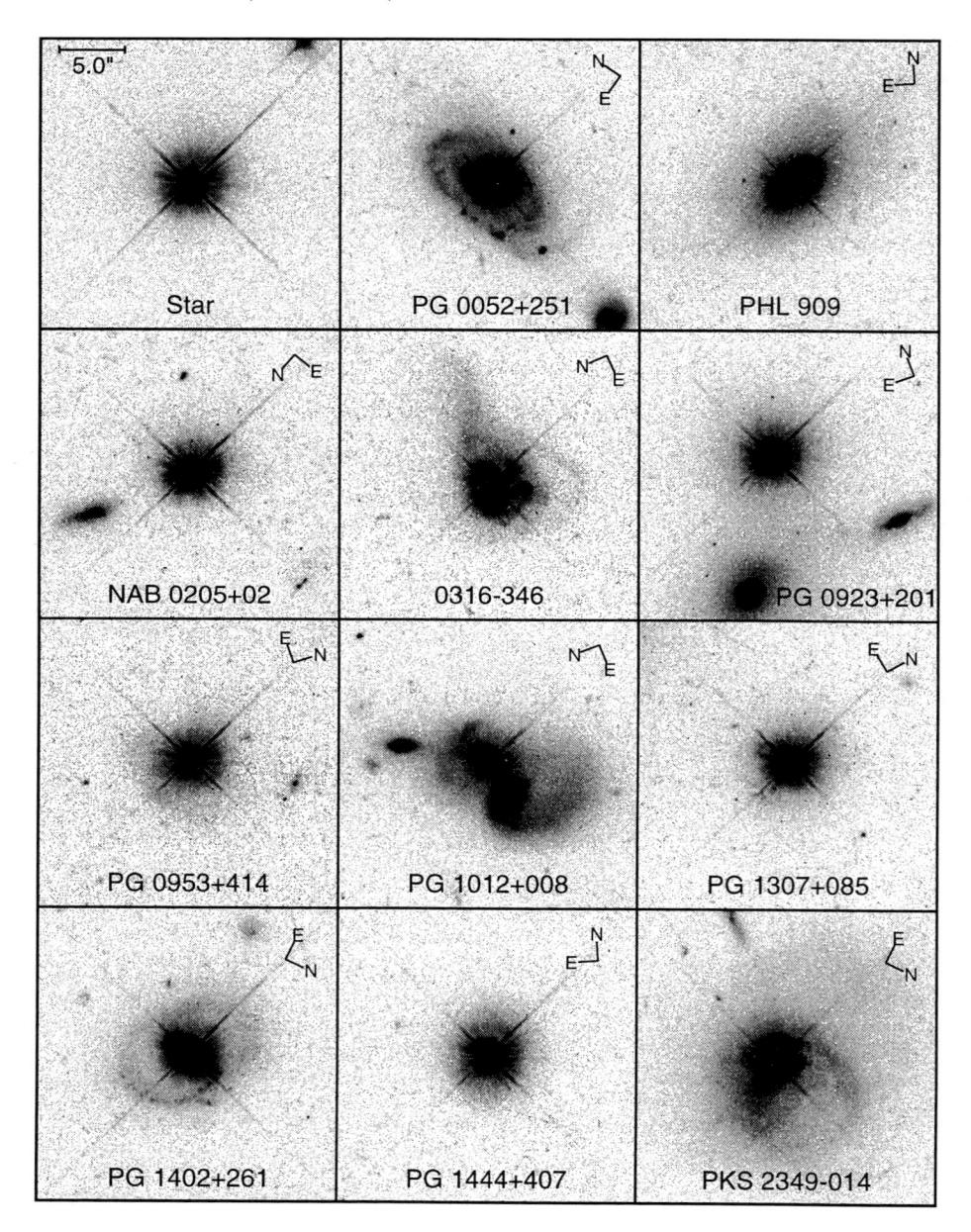

Fig. 1. Unprocessed *HST* images of 11 luminous nearby quasars and a blue star for comparison (upper left-hand panel). Each image is $23'' \times 23''$.

2 Subtraction of a Stellar Point-Spread Function

We have observed four blue stars in the M67 cluster to use as point-spread function (PSF); the calibration stars (and their $B-V$ colors) are: MMJ 6481(-0.073),

MMJ 6490(0.11), MMJ 6504(0.22), and MMJ 6511(0.34). Their apparent magnitudes range from $V = 10.03$ to $V = 10.99$. For each star, a series of four images were used by Krist and Burrows (1996) to produce a PSF that samples the full dynamic range of the star image and covers the saturation range found in the quasar images.

We subtracted all four PSFs determined by Krist and Burrows from each of the quasars, and then selected the result that gave the cleanest subtraction.

We fit a stellar point-spread function to each quasar image and subtracted a multiple of the normalized PSF to search for underlying diffuse light from hosts. The quasar images with the PSF subtracted presented were obtained by minimizing the χ^2 in unsaturated annular regions (typical inner and outer radii of $1''$ and $3''$), centered on the quasar. Adjustments of the amount of PSF subtraction were often made after visual inspection of the "best-fit" PSF-subtracted images.

Figure 2 shows the images for all the quasars in our sample after a best-fit stellar point-spread function (PSF) was subtracted from the original images, many of which are shown in Figure 1.

3 Methods of Analysis

We have used three different methods of analysis to determine the properties of the 20 quasar hosts.

3.1 Aperture Photometry

We performed aperture photometry in circular annuli centered on the quasars, after a best-fit PSF was subtracted. An inner radius of $r = 1.0''$ was used for all quasars, except for HE 1029−140 and 3C 273. In most observations, the region $r < 1''$ is contaminated by artifacts left by the PSF subtraction. The saturated areas in the images of HE 1029−140 and 3C 273 are larger; for those two cases, the inner radii used were $1.5''$ and $2.0''$, respectively. The outer radii chosen in general represent how far we could see the host galaxy.

3.2 One-Dimensional Radial Profiles

The one-dimensional azimuthally-averaged surface brightness profiles of the host galaxies were constructed from the *HST* data after subtraction of a best-fit stellar PSF. Regions affected by saturation, diffraction spikes, or residual artifacts from the PSF subtraction were not included in the azimuthally averaged profiles. For each galaxy, we obtained a best-fitting exponential disk (henceforth Disk) and a de Vaucouleurs (1948, henceforth GdV) profile that fits the observed data in the region $r \geq 1''$.

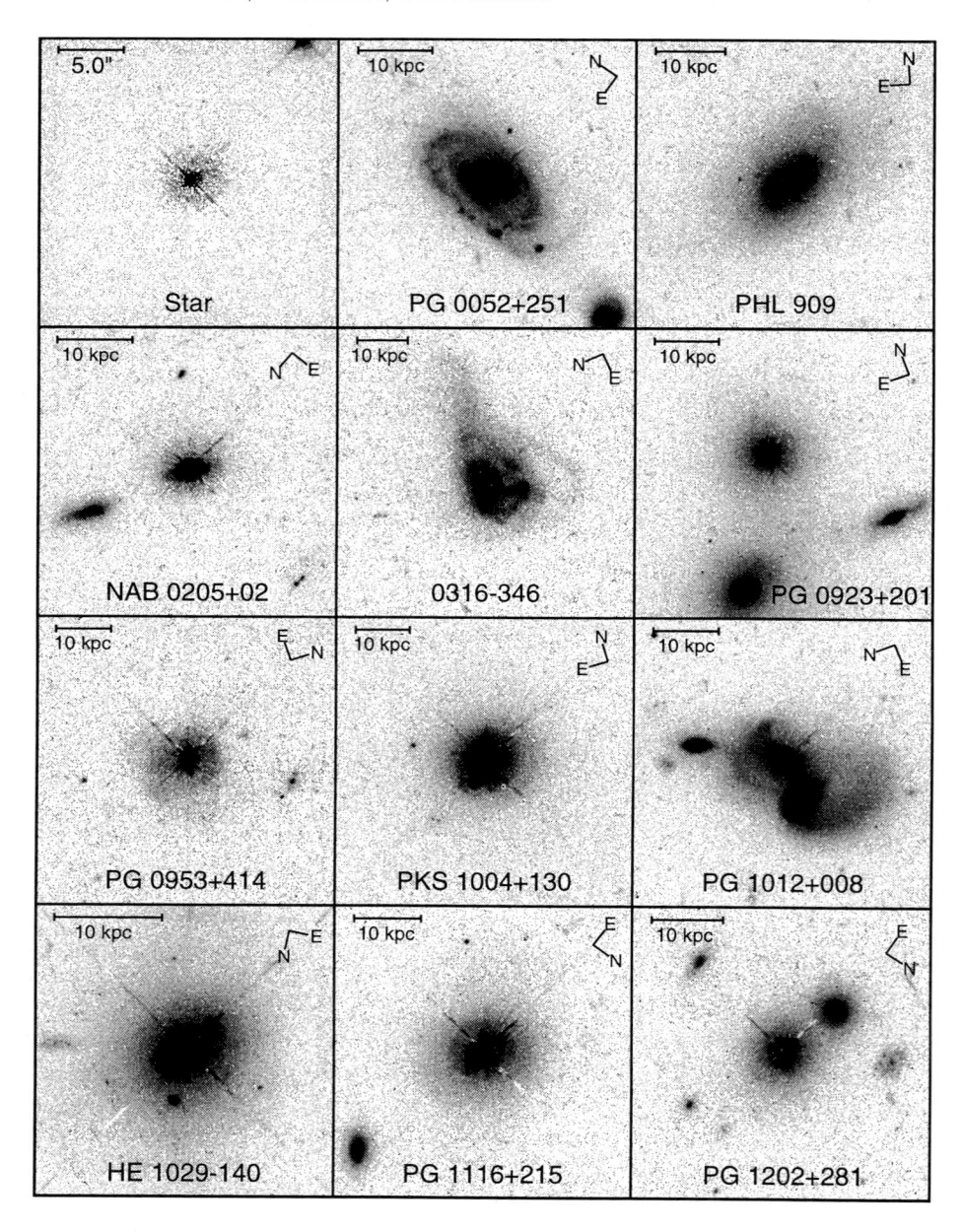

Fig. 2. *HST* images of 20 luminous nearby quasars and a blue star for comparison (upper left-hand panel), with a best-fit PSF for a standard blue star subtracted. Each image is $23'' \times 23''$.

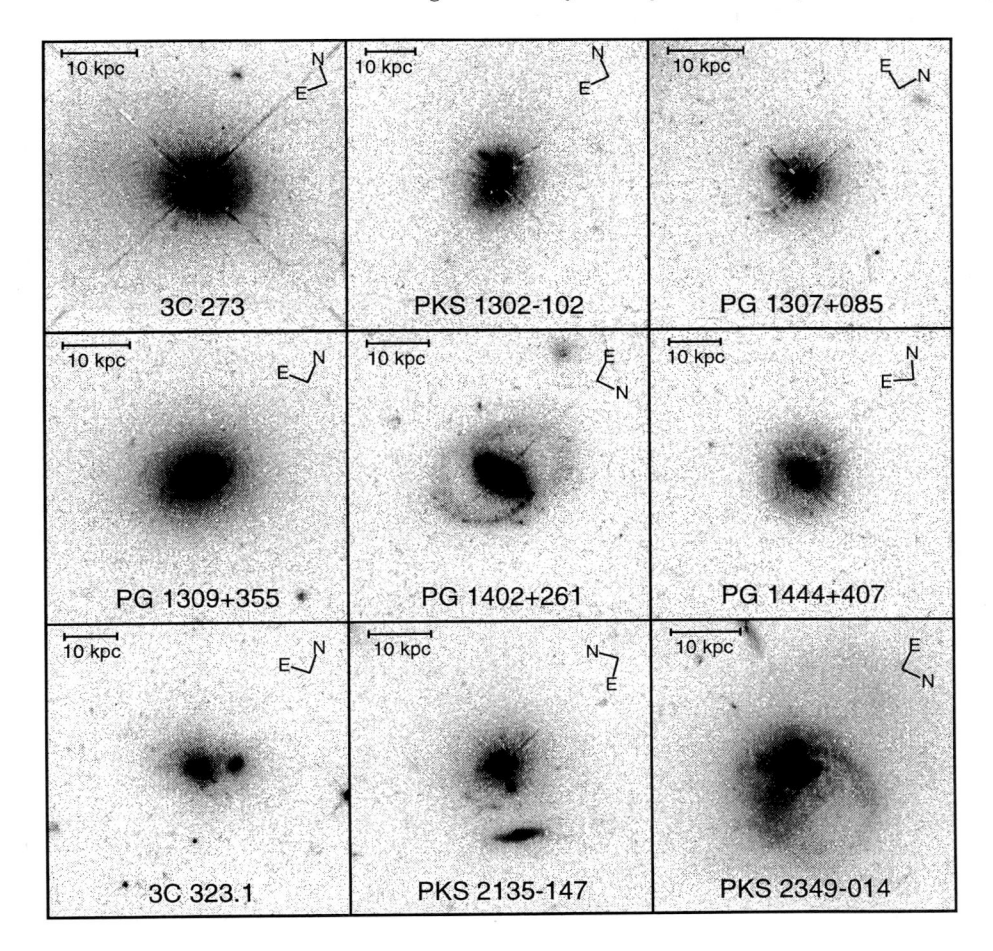

Fig. 2. Continued.

3.3 Two-Dimensional Fit

The *HST* imaging provides greater detail than has been available previously in ground-based images of luminous quasars. Traditionally, the properties of host galaxies have been determined in ground-based studies by making model fits to azimuthally-averaged radial profiles.

To take advantage of the *HST* resolution, we have developed software to fit a two-dimensional model to the PSF-subtracted quasar images. For each quasar, we fit an analytic galaxy model (exponential disk or de Vaucouleurs profile) to the data. The area used for the fit was approximately an annular region, centered on the quasar, that excluded the central area ($r < 1.0''$), and the remnants of the diffraction spikes or other artifacts clearly due to improper PSF subtraction. We fit four parameters: the (x,y) pixel position of the center, the total number of counts, and the radius (effective radius or scale length) in the galaxy model.

4 Magnitude of the Host Galaxies

Table 1 lists our best estimates for the magnitudes of the host galaxies in our sample; these magnitudes were obtained by fitting a two-dimensional analytic galaxy model to the data. We also list the effective radius or exponential scale length, and give the morphology of the host based on visual inspection of the images.

A difference between aperture and model magnitudes is expected because the aperture magnitudes do not include the area within 1" of the quasar; all of the models we considered have surface brightnesses that increase monotonically toward the center.

The two-dimensional models are somewhat fainter than the corresponding one dimensional fits. Specially, we find:

$$\langle m_{606W,2-D} - m_{606W,1-D} \rangle = \begin{cases} 0.4 \pm 0.2, & \text{GdV} \\ 0.3 \pm 0.1, & \text{Disk} \end{cases} . \quad (1)$$

Similarly, the magnitudes of the hostphotome galaxies estimated by fitting a GdV model to the azimuthal averaged radial profile of the residual light are on average 1.0 mag brighter than the magnitudes obtained from aperture photometry; fitting an exponential disk model gives magnitudes that are on average 0.5 brighter than the results from aperture photometry:

$$\langle m_{606W,1-D} - m_{606W,\text{aperture}} \rangle = \begin{cases} -1.1 \pm 0.2, & \text{GdV} \\ -0.4 \pm 0.2, & \text{Disk} \end{cases} . \quad (2)$$

For the two-dimensional model fits, we have on average

$$\langle m_{606W,2-D} - m_{606W,\text{aperture}} \rangle = \begin{cases} -0.7 \pm 0.2, & \text{GdV} \\ -0.2 \pm 0.1, & \text{Disk} \end{cases} . \quad (3)$$

The 2-D (1-D) GdV model gives magnitudes for the host that are on average 0.6 mag (0.5 mag) brighter than the exponential disk estimates.

The results from the two-dimensional fits indicate that the host galaxies are typically centered within 400 pc of the location of the quasar point source.

5 Summary

The most luminous nearby quasars exist in a variety of environments. There are three hosts that apparently are normal spirals with H II regions (PG 0052+251, PG 1309+355, and PG 1402+261), seven ellipticals (PHL 909, PG 0923+201, PKS 1004+130, HE 1029−140, PG 1116+215, 3C 273, and PKS 2135−147), as well as three obvious cases of current gravitational interaction (0316−346, PG 1012+008, and PKS 2349−014). There are also five other hosts that appear to

Table 1. Magnitudes and Morphology for Quasar Host Galaxies

Object	z	m_{606}	Two-Dimensional M_V^b	$r('')^a$	Morphology
PG 0052+251	0.155	17.2	−20.9	1.3	Sb
PHL 909	0.171	17.2	−21.0	2.3	E4
NAB 0205+02	0.155	19.0	−19.1	0.7	S0?
0316−346	0.265	18.3	−20.8	1.2	Inter.
PG 0923+201	0.190	17.5	−21.0	2.9	E1
PG 0953+414	0.239	18.8	−20.2	1.1	?
PKS 1004+130	0.240	16.9	−22.0	1.2	E2
PG 1012+008	0.185	17.7	−20.7	1.6	Inter.
HE 1029−140	0.086	16.2	−20.5	3.2	E1
PG 1116+215	0.177	16.9	−21.4	1.4	E2
PG 1202+281	0.165	17.7	−20.5	1.4	E1
3C 273	0.158	16.0	−22.1	3.7	E4
PKS 1302−102	0.286	18.2	−21.1	1.1	E4?
PG 1307+085	0.155	17.8	−20.2	1.3	E1?
PG 1309+355	0.184	17.3	−21.1	1.2	Sab
PG 1402+261	0.164	18.3	−19.9	1.6	SBb
PG 1444+407	0.267	18.4	−20.5	1.0	E1?
3C 323.1	0.266	18.1	−21.0	1.6	E3?
PKS 2135−147	0.200	17.4	−21.1	2.6	E1
PKS 2349−014	0.173	16.2	−22.1	4.8	Inter.

^{a} Effective radius or exponential scale length.
^{b} (F606−V) from Fukugita et al. (1995).

be elliptical galaxies and are listed as En(?) in Table 1. The hosts for two of the quasars (NAB 0205+02 and PG 0953+414) are faint and difficult to classify.

There are more radio quiet quasars in galaxies that appear to be ellipticals than in spiral hosts, contrary to expectations. However, the HST images confirmed the expectations based upon ground-based observations that nearly all radio loud luminous quasars occur in elliptical galaxies.

The *HST* images frequently reveal companion galaxies that are projected very close to the quasar, in some cases as close as $1''$ or $2''$. We found 20 galaxy companions that are projected closer than 25 kpc to the center-of-light of a quasar and brighter than $M(\text{F606W}) = -16.4$. The amplitude for the quasar-galaxy correlation function determined from our data is 3.8 ± 0.8 times larger than the galaxy-galaxy correlation function (Fisher et al. 1996).

Our results are inconsistent with the hosts having a Schechter luminosity function. The average absolute magnitude for a field galaxy is about 1.8 mag fainter than $M_V(L^*) = -20.5$ (for an assumed minimum luminosity of $M_V(L^*) = -17$; see, e. g., Efstathiou, Ellis,& Peterson 1988 for a discussion of the field galaxy luminosity function). In our sample (see Table 1) the average host is $M_V(L^*) = -20.9$. We conclude that on average, the host galaxies of the luminous quasars in our sample are about 2.2 magnitudes more luminous than typical field galaxies.

The presence of very close companions, the images of current gravitational interactions, and the higher density of galaxies around the quasars suggests that gravitational interaction plays an important role in triggering the quasar phenomenon. Details of the observations, data analysis, and results of this program can be found in Bahcall, Kirhakos and Schneider (1995, 1996) and Bahcall *et al.* (1997).

For electronic versions of the images and papers discussing individual objects, see: http://www.sns.ias.edu/~jnb.

References

Bahcall, J. N., Kirhakos, S., & Schneider, D. P. (1995): *ApJ*, **447**, L1
Bahcall, J. N., Kirhakos, S., & Schneider, D. P. (1996): *ApJ*, **457**, 557
Bahcall, J. N., Kirhakos, S., Schneider, D. P., & Saxe, D. H. (1997): to appear in *ApJ*, **479**, April 20
de Vaucouleurs, G. (1948): *Ann. d'Ap.*, **11**, 247
Efstathiou, G., Ellis, R. S., & Peterson, B. A. (1988): *MNRAS*, **232**, 431
Fisher, K. B., Bahcall, J. N., Kirhakos, S., & Schneider, D. P. (1996): *ApJ*, **468**, 469
Montgomery, K. A., Marschall, L. A., & Janes, K. A. (1993): *AJ*, **106**, 181

Near-IR Properties of Quasar Host Galaxies

Kim K. McLeod

Smithsonian Astrophysical Observatory, 60 Garden St., Cambridge, MA 02140, USA

Abstract. We have obtained deep, near-IR images of nearly 100 host galaxies of nearby quasars and Seyferts. We find the near-IR light to be a good tracer of luminous mass in these galaxies. For the most luminous quasars there is a correlation between the maximum allowed B-band nuclear luminosity and the host galaxy mass, a "luminosity/host-mass limit". Comparing our images with images from HST, we find that the hosts of these very luminous quasars are likely early type galaxies, even for radio-quiet objects whose lower-luminosity counterparts traditionally live in spirals. We speculate that the luminosity/host-mass limit represents a physical limit on the size of black hole that can exist in a given galaxy spheroid mass. We discuss the promises of NICMOS for detecting the hosts of luminous quasars.

1 Introduction

Although most of the attention given to AGN is concentrated on the "N," there are compelling reasons to understand the "G." The central engine and the host galaxy must influence each other, and the exact connections hold crucial clues for understanding the quasar phenomenon. Moreover, it is plausible that nuclear activity has played a role in the evolution of a significant fraction of all galaxies; Seyferts account for $\gtrsim$ 10% of galaxies today (Maiolino & Rieke 1995; Ho 1996), and AGN were even more important in the past. Therefore, to understand the evolution of galaxies, we must understand the host galaxies of AGN.

By now it is well established that the redshift range $2 \lesssim z \lesssim 3$ represents a critical period in the evolution of both "normal" galaxies and quasars. It is likely that galaxies at that epoch were starting to turn their gas into stellar disks. The mass in neutral hydrogen gas in damped Lyman$-\alpha$ absorbers at that redshift is comparable to the mass in disk stars today and shows strong evolution since that time (Lanzetta et al. 1995; Storrie-Lombardi et al. 1996). Furthermore, a photometric-redshift analysis of the Hubble Deep Field shows that the luminosity density from star-forming galaxies peaks near that same redshift (Sawicki et al. 1996). The AGN luminosity function shows a similarly strong evolution; if described in terms of luminosity evolution (but see Wisotzki et al. in these proceedings), the "characteristic" luminosity of quasars increases as $\sim (1+z)^{3.4}$ to $z \sim 2$ and flattens between $2 < z < 3$ (Boyle 1993). Thus *quasars were an energetically important component of galaxies at a critical period in their evolution and could have influenced their growth.* Understanding the relationships between AGN and their hosts thus holds clues to the processes of galaxy formation and evolution.

HST and near-IR imaging together have become especially powerful tools for studying quasar hosts. The spatial resolution of HST has allowed us, for the first time, to determine morphological types for hosts of luminous quasars (McLeod and Rieke 1995b; Bahcall et al. and Disney et al. in these proceedings). Near-IR imaging has done a good job of showing the host galaxy starlight with less contamination from the nucleus than at visible wavelengths (McLeod & Rieke 1994ab,1995ab; Dunlop et al. 1993; Kotilainen & Ward 1994). As shown in Fig. 1, the host galaxy starlight peak coincides with the near-IR minimum in the nuclear energy distribution. Furthermore, the near-IR light highlights the old, red, mass-tracing stars in the population while suffering less from extinction and emission-line contamination than the visible images.

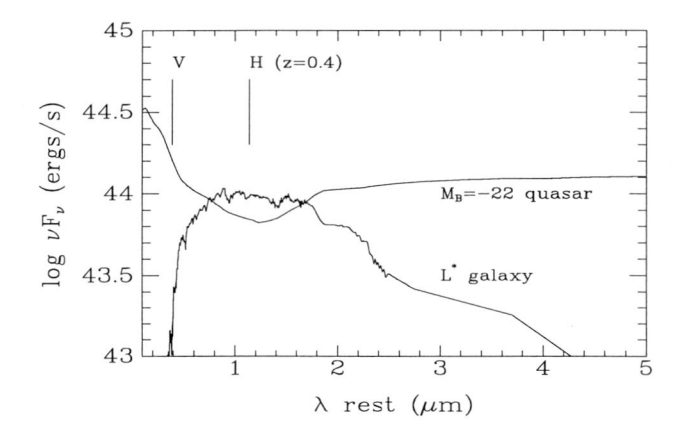

Fig. 1. Energy distributions of a typical quasar (from J. McDowell) and galaxy (from M. Rieke) for $H_0 = 80$. The H-band gives good contrast of starlight to nuclear light and is a good tracer of luminous mass in the host galaxy.

2 The IR Images

We have exploited this wavelength range by obtaining deep near-IR images for nearly 100 AGN using a 256x256 NICMOS array camera on the Steward Observatory 2.3m telescope. We chose two samples of AGN that allowed us to investigate host galaxy properties over 10 B magnitudes in nuclear luminosity. For low-luminosity AGN we used the CfA Seyfert sample, which is selected on the basis of the nuclear spectrum and which has roughly equal numbers of Sy1's and Sy2's. For high-luminosity AGN, we chose the lowest redshift ($z < 0.3$) quasars from the PG sample, to ensure a sample selected on the basis of nuclear properties and close enough so that the host galaxies would be resolved. The quasars were imaged in the H-band ($1.65\mu m$) and we were able to measure the isophotes down to a 1σ level of $H \approx 23$mag arcsec^{-2}, which corresponds approximately to $B \approx 26.7$mag arcsec^{-2} for typical galaxy colors.

The results have been presented in McLeod and Rieke (1994ab,1995ab). We describe in these papers the relationships between host galaxy luminosities (and masses) and nuclear properties; the existence of substantial obscuration coplanar with the disks of host spirals of Seyferts; and the search for signs of disturbances that could aid the flow of fuel towards the centers of the galaxies. The reader is encouraged to consult these papers for details; here we will concentrate on one of the most intriguing results.

3 The Luminosity/Host-Mass Limit: Observations

We find the near-IR light to be a good tracer of luminous mass in these galaxies. In Fig. 2 we plot host near-IR luminosity against nuclear luminosity, for the objects in our samples and from other near-IR studies in the literature. As shown in this figure, Seyferts are found in galaxies with a range of H-band luminosity, and hence luminous mass, from ~ 0.1 to $\sim 5L^*$. Their morphological types range from S0 to Sc. The lowest-luminosity quasars live in similar kinds of galaxies spanning the same range of H-band luminosity centered around L^*.

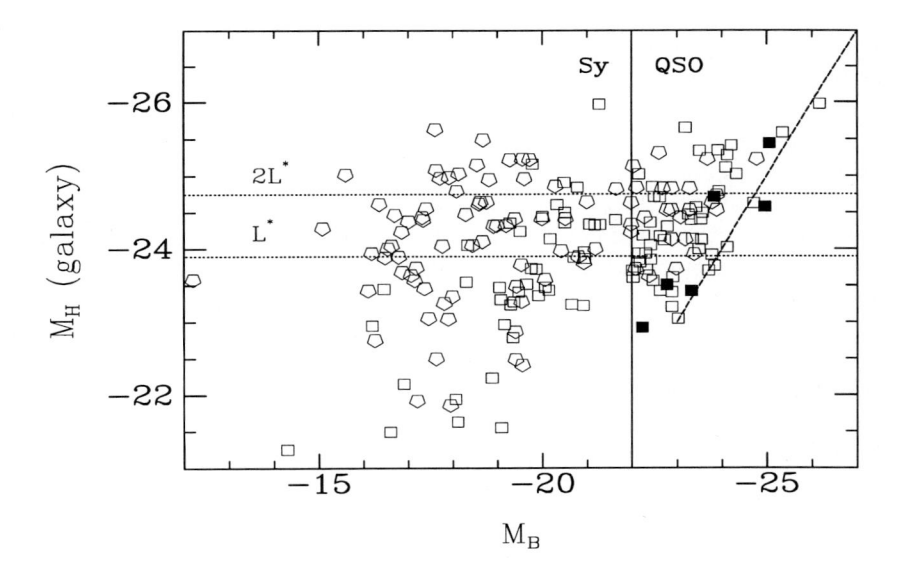

Fig. 2. Host v. nuclear luminosity for Seyferts and quasars with $z < 0.3$ for $H_0 = 80$ (McLeod & Rieke 1995a and references therein). For quasars brighter than $M_B \lesssim -23$, there is a minimum host galaxy luminosity that increases with nuclear power. The dotted lines show the positions of L^* and $2L^*$ galaxies. Filled squares are objects for which the host galaxy luminosities are upper limits or weak detections–we will obtain NICMOS images for these and other high-luminosity objects. The diagonal line shows the luminosity/host-mass limit.

We have found, however, that for the highest-luminosity quasars, there is a *minimum host H-band luminosity* that increases with nuclear power, shown as a diagonal line in Fig. 2. This is reminiscent of a similar trend previously noted in the visible (Yee 1992). The relationship has the functional form M_H(galaxy) $\approx$ M_B(nucleus) with an onset approximately one magnitude brighter than the traditional (arbitrary) quasar/Seyfert boundary. Because H-band light probes the galaxy *mass* by highlighting the old, red, mass-tracing stars of the stellar population, this "luminosity/host-mass limit" reflects a relation between fundamental physical parameters that govern the process of hosting a quasar: galaxies with more mass can sustain more activity.

WFPC2 images have been published for $\sim$ 30 luminous, nearby quasars including many of our highest-luminosity objects (e.g. Hutchings et al. 1994; Disney et al. 1995; Bahcall et al. 1996ab). Of these, only a few ($\sim$ 3) appear to be spiral hosts like those of Seyferts. The rest are plausibly smooth, early-type galaxies or ellipticals in the making (mergers), despite being radio quiet objects that are traditionally assumed to be spirals (McLeod & Rieke 1995b; see also Taylor et al. 1996).

4 The Luminosity/Host-Mass Limit: Wild Speculation

Intriguingly, the transition from spiral hosts to spheroid-dominated hosts appears to occur approximately at the nuclear luminosity where the luminosity/host-mass limit becomes apparent. A simple explanation is that the objects along the diagonal line in Fig. 2 are early-type galaxies that have a maximum allowed black hole mass for their galaxy mass and that the black hole is accreting at the Eddington rate. In this case and for typical galaxy mass-to-light ratios, the diagonal line then represents a line of *constant fraction of black hole mass to stellar spheroid mass*, with value $f_{BH} \equiv M_{BH}/M_{stars} \approx 0.0015$.

This suggestion is especially exciting because central compact objects (presumably dead quasars) discovered in nearby spheroids show a similar relation with $f_{BH} \approx 0.002$ (Kormendy & Richstone 1995), and an HST study of bulges and ellipticals indicates that compact objects following approximately this same relation are likely required to produce the cores seen in the starlight (Faber et al. 1996). Thus, the luminosity/host-mass limit possibly results from general physical processes that govern the formation and evolution of the spheroid components of galaxies.

5 The Luminosity/Host-Mass Limit: Wilder Speculation

If the luminosities of most powerful quasars really do trace the potentials of the most massive spheroids, then we might be able to use the evolution of the quasar luminosity function to trace the history of spheroid formation in the universe. The observed evolution of the quasar luminosity function can be nicely reconciled with the model of galaxy evolution recently summarized by Fukugita et al. (1996), in which spheroid components of very massive galaxies formed at $z > 3$, followed by less massive objects at later epochs (see also Sawicki et al. 1996). We speculate that the $z \sim 5$ quasars formed as parts of the most massive spheroids, the peak in the quasar population at $z \sim 2$ occurred when massive disks were being added, and less powerful AGN formed later in less massive galaxies. Thus, *central black holes themselves form as the spheroids are being assembled, and the changing AGN population reflects the changing population of galaxies.* The luminosity functions would then require nearly every large galaxy to go through an AGN phase that lasts a modest portion of the galaxy's lifetime (e.g. Weedman 1986).

6 Future Work: All Hail NICMOS!

Before we can exploit the luminosity/host-mass limit we must answer several questions. We are already at work on the theoretical basis for such a limit, and we are beginning a groundbased imaging and spectroscopy program to determine whether the limit applies to AGN in smaller spheroids (Seyferts in bulges).

Importantly, we need to know whether the limit is strictly correct especially at high nuclear luminosities where our statistics are poor. We have been granted Cycle 7 HST time to address this question with NICMOS. NICMOS will combine the advantages of near-IR imaging with the superior spatial resolution of HST. We will look to find exceptions to the luminosity/host-mass limit by imaging objects whose hosts have been thus far elusive from the ground. We will also extend our sample to slightly higher redshifts to allow a look at higher luminosity objects. As shown in Fig. 3, NICMOS will be very good at detecting the extended emission from smooth hosts that are difficult to see with WFPC2.

Eventually, we need to probe nucleus–host relationships at high redshifts where galaxies are being assembled. The NICMOS GTO program (Weymann et al.) will provide near-IR images of radio-quiet quasar hosts at $z \approx 1$. We will probe further by obtaining near-IR images using adaptive optics on the soon-to-be-upgraded MMT. Preliminary adaptive optics results (John Hutchings, this conference) suggest that this method holds great promise for host studies at high redshift. By measuring morphologies and near-IR luminosities of the distant quasar hosts from the images and then applying models of stellar and dynamical evolution, we will predict what those galaxies look like today.

Acknowledgements: Thanks to the conference organizers for conference organizing, thanks to Avi Loeb for very fruitful discussions, thanks to Dr. Yamada for pointing out the Kormendy ratio, and thanks to B. Wilkes for financial support.

References

Bahcall, J. N., Kirhakos, S., & Schneider, D. P. 1996a, ApJ, 457, 557
Bahcall, J. N., et al. 1996b, preprint
Boyle, B. J. 1993, in "The Environment and Evolution of Galaxies" p. 433 (Kluwer; eds. J. M. Shull & H. A. Thronson, Jr.)

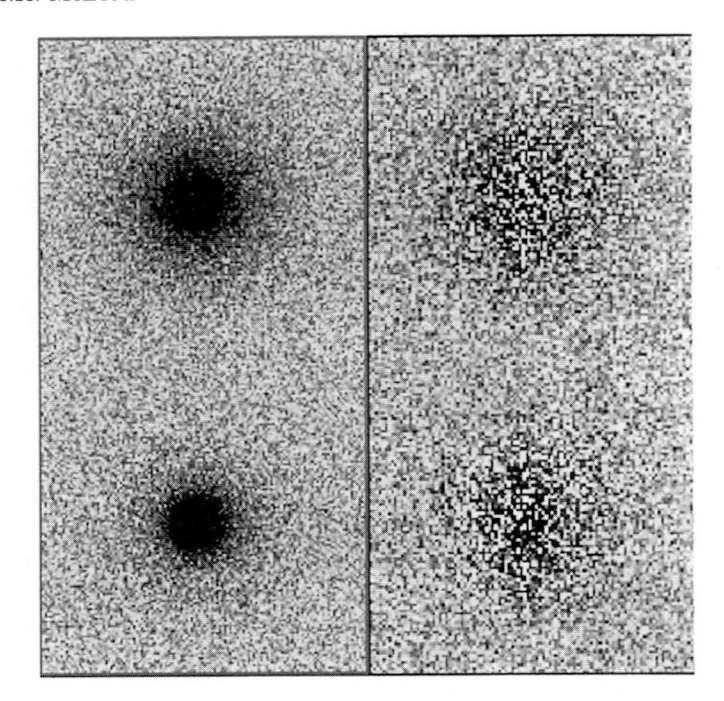

Fig. 3. Simulated 2800s images of $z = 0.4$ quasars with NIC2 (left) and WFC2 (right). The hosts at top are L^* spirals, the ones at bottom are L^* ellipticals. In both cases, the nucleus has been removed and the linear greyscale stretch runs from -1 to 10 times the 1σ noise. The frame is 19.2" on a side (the size of one NIC2 frame).

Disney, M. J. et al. 1995, Nature, 376, 150
Dunlop, J. S., Taylor, G. L., Hughes, D. H., & Robson, E. I. 1993, MNRAS, 264, 455
Faber, S. M. et al. 1996, AJ, preprint
Fukugita, M., Hogan, C. J., & Peebles, P. J. E. 1996, Nature, 381, 489
Ho, L. 1996 preprint
Hutchings, J. B. et al. 1994, ApJL, 429, L1
Kormendy, J. & Richstone, D. O. 1995, ARAA, 33, 581
Kotilainen, J. K. & Ward, M. J. 1994, MNRAS, 266, 953
Lanzetta, K. M., Wolfe, A. M., & Turnshek D. A. 1995, ApJ, 440, 435
Maiolino, R., & Rieke, G. H. 1995, ApJ, 454, 9
McLeod, K. K., & Rieke, G. H. 1994a, ApJ, 420, 58
McLeod, K. K., & Rieke, G. H. 1994b, ApJ, 431, 137
McLeod, K. K., & Rieke, G. H. 1995a, ApJ, 441, 96
McLeod, K. K., & Rieke, G. H. 1995b, ApJL, 454, L77
Sawicki, M. J., Lin, H., & Yee H. C. K. 1996, AJ, preprint
Storrie-Lombardi, L. J., McMahon, R. G., & Irwin M. J. 1996, MNRAS preprint
Taylor, G. L., Dunlop, J. S., Hughes, D. H., & Robson, E. I. 1996, MNRAS preprint
Weedman, D. W. 1986, "Quasar Astronomy" (Cambridge University Press)
Yee, H. C. K. 1992, in "Relationships Between Active Galactic Nuclei and Starburst Galaxies" ASP Conference Series, Vol. 31, A. V. Filippenko (ed.), 417–422

JHK Imaging of QSO Hosts and Companions

J.B. Hutchings

Dominion Astrophysical Observatory, National Research Council of Canada,
5071 W. Saanich Rd., Victoria B.C. V8X 4M6, Canada

Abstract. Deep J,H,K imaging is presented of 90 arcmin fields around QSOs at redshifts in the range 0.06 to 2.4, including images, luminosity profiles, and NIR 2-colour diagrams. The low z hosts are all resolved, and NIR colours are measured. These suggest that they all live in groups of companion galaxies. The two radio-loud objects live in richer cluster environments than the others. Gissel population models indicate association with the QSOs, and some systematic offsets at $z \lesssim 0.3$. The QSO luminosity profiles are complex and reveal some of their tidal disturbance and star-formation history. At high redshift, the models match well, and indicate compact groups of galaxies with young populations.

1 Introduction and Data

The host galaxies of QSOs and AGN in the z range 0.1 to 0.5 have been resolved and studied quite extensively at optical wavelengths (e.g. Hutchings and Neff 1991, Bahcall et al 1996). These images are dominated by the nuclear light and also by young stellar populations. In many cases the host galaxies are interacting systems which appear irregular because of tidal distortions, as well as associated star- and dust-formation. Host galaxies at higher redshifts have been marginally resolved (e.g. Hutchings 1995a) and appear to have very high luminosity. Compared with optical wavelengths, the NIR offers the advantage of a lower brightness contrast between the active nucleus and the host galaxy stellar population, and as redshift increases it offers a lower k-correction of the host galaxy. Dunlop et al (1993) and McLeod and Rieke (1994) have published NIR imaging investigations of QSOs.

This paper reports extension of NIR imaging to 3 colours: J, H, and K, enabling colour-colour plots which are diagnostic of stellar populations. The sample includes previously resolved objects at low z, and some high z objects not previously observed. The field size of 90-120 arcsec also includes nearby (possible companion) galaxies. The observations were obtained with the Redeye wide-field camera at the prime focus of the Canada France Hawaii telescope with 0.5" pixels on 0.8" images. Total exposure times with each filter were approximately 32 minutes. Details of the data and reductions are given in Hutchings and Neff (1996). Overall, we achieve 1σ errors in magnitudes of 0.1 at 18.8; 19.6; 20.3 in K, H, and J. At 0.2 mag error, the magnitudes are 19.9, 20.7, 21.5 respectively. Figure 1 shows subimages centred on some of the low z objects, showing that they are all clearly resolved.

J, H, and K magnitudes were obtained for all objects in the fields with measuring error 0.15 in all bands, shown as H-K/J-H diagrams, with error bars.

Stars are omitted, identified in a plot of peak signal against total. All or parts of the QSO host galaxy were measured, where uncontaminated by nuclear light. Finally, the azimuthally averaged or ellipse-fitted profile of the QSO was derived.

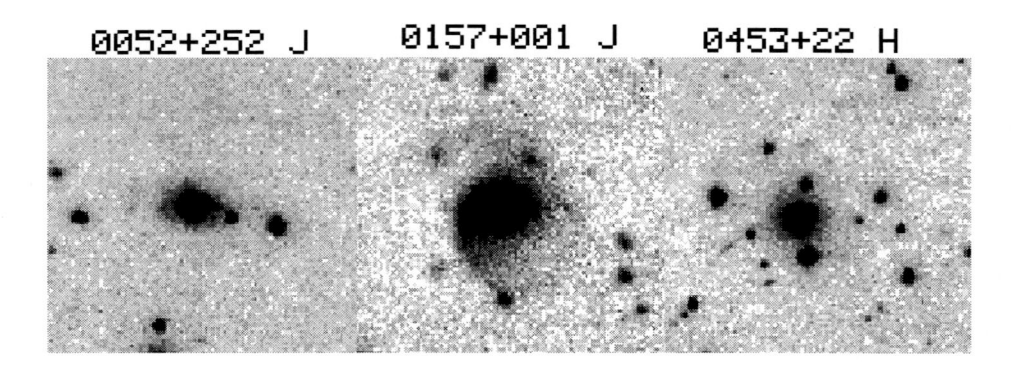

Fig. 1. 50 arcsec subimages of 3 program objects

2   Stellar Population Models

Model values for the J, H, K fluxes were calculated using the Gissel stellar population models (Bruzual and Charlot 1993), and compared with the measures for each field and appropriate redshift. The models are all for a passively evolving population after a 1 Gyr starburst. Continued star-formation, renewed star-formation, or a truncated IMF were also modelled.

The tracks evolve very quickly from their lower left to the upper portion after cessation of star-formation. At lower redshifts (0.06 to 0.30) the tracks distinguish redshifts with a resolution of about 0.05 in z Since any field will contain background and foreground objects, we can use the diagram to determine which are at the QSO redshift. Reddening moves galaxies along almost the same locus as redshift increases above $z = 0.15$. Thus, we need to consider evidence for dust in interpreting the data.

While there are indications of systematics in the models for $z \leq 0.3$, at $z=0.4$ and higher, they match the observations very well. This is because at higher redshift, the rest wavelengths are in the more reliably modelled range.

3   Low z QSOs

Two of the sample are mentioned here to illustrate some typical results. Figure 2 shows their 2-colour plots.

0157+001 = Mkn1014 This is a radio-quiet QSO at redshift 0.16. Stockton and MacKenty (1987) published [O III] images, and Hutchings and Crampton

(1990) discussed off-nuclear spectra. The host galaxy has strong spiral features. There are many faint small objects near the QSO: in particular objects 5 and 17 appear at the end of the two major spiral features, and 3 lies in a dense region of one arm. These condensations are more prominent in the NIR than at optical wavelengths.

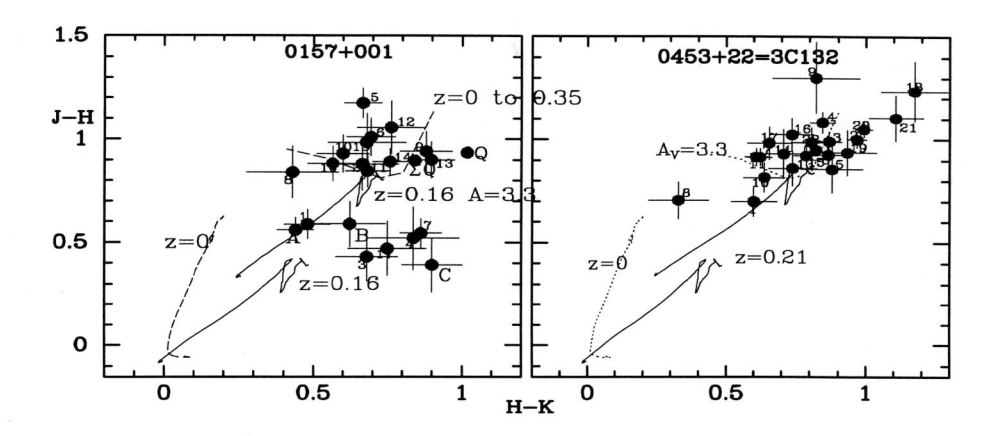

Fig. 2. NIR 2-colour diagrams of two fields, with stellar population models for their redshift. Each shows groups of galaxies with very similar colours

Figure 2 shows the 2 colour diagram of the objects in the field, together with some relevant model tracks. The best fits indicate the model shown, which has $A_v=3$ at the rest wavelength, and a local (z=0) $A_V=0.3$. These are likely to be model systematics but the relative positions of the data are the important result to note. Both the QSO nucleus and whole (nucleus plus host galaxy), are marked: also the colours of the two major arms (A, the lower one - excluding object 5, and C, the upper one, and the brightest region within A, as B - see Fig 1). The objects fall in two main groups in the diagram. The arms and associated object 17 appear to belong to a younger population, along with objects 1, 4, and 7, all of which are close to the QSO. With the exception of object 8, all others appear to have an older population. The QSO nucleus lies to the right of the group in the 2-colour plot, while the total flux (nucleus plus host) has old population colours. Within the host galaxy, the object 5 has high J-H colours while the others, 3, 4, 17 are low enough to be young population candidates.

The luminosity profiles (Fig 3) show the mixed properties that are common in QSOs (see Hutchings and Neff 1991). The diagram includes the PSF from the bright nearby star, as this is our best illustration of a typical PSF. The QSO profile is well resolved outside of ∼1.8 arcsec. There is a good fit to an $r^{1/4}$ profile in the inner 5 arcsec, which is the bright inner galaxy. From radius 5 to 17 arcsec there is a reasonable exponential profile, as suggested by the spiral structure seen in this range. The exponential slope is steeper with increasing

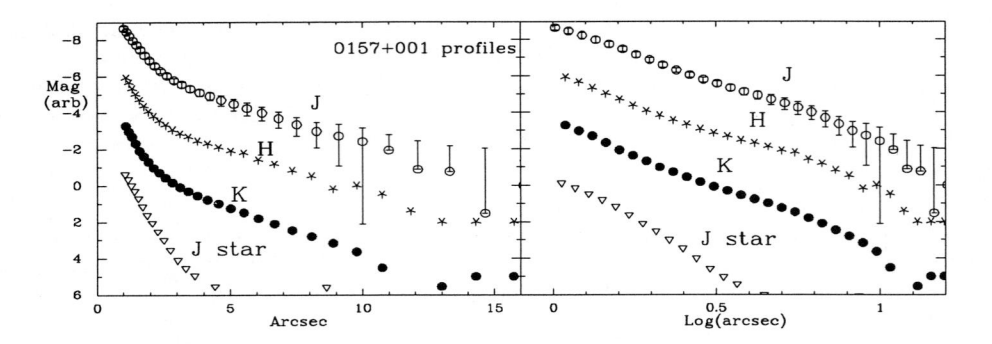

Fig. 3. Luminosity profiles showing power law inner galaxy and exponential outer parts

wavelength, while the inner bulge slope is the same in all our bands. The bright features at the end of the arms suggest that these are tidal features rather than features of a normal disk galaxy, and the high concentration of faint galaxies of old population that appear to be real and close QSO companions suggest that there has been a merger or close encounter.

0453+22 = 3C132 This object is a strong double-lobed radio source identified with a galaxy at z = 0.21. Optical imaging was reported by Hutchings, Jansen, and Neff (1987). The galaxy is in a crowded field and its faint outer parts extend over a number of superposed faint objects. The object is at fairly low galactic latitude, so the field contains more stars than the others in the program.

The field is more crowded than those of the radio-quiet QSOs, and almost all of the measured galaxies form a tight group in the 2 colour diagram around the radio galaxy (Figure 2). Thus, we have evidence for a richer group or cluster around the radio source. Unfortunately the two brightest nearby galaxies, could not be measured in our K-band image, because of a small telescope pointing offset with respect to the other observations. They may be the more luminous members of this cluster. The radio galaxy has no distinct nucleus at our resolution, but was measured for colour in two spearate regions away from the centre and the colours are within measuring errors the same as the whole galaxy.

4 High z QSOs

z=2 QSO in Abell 851 field. While studying the z=0.4 rich cluster Abell 851 (0939+4713) with HST images, Dressler, Oemler, Gunn, and Butcher (1993) detected an apparent cluster of compact faint galaxies around a QSO with z=2.06. They suggest that these galaxies may be in the very early stages of formation, and thus contain important clues for galaxy evolution. A recent WFPC2 exposure, taken through the F702W filter, totalled 21000 seconds; we have obtained the data from the HST archive, and photometric measurements from this image (referred to as 'R') were coupled with those from the J and K bands of this program. Since the field lies in the rich z=0.4 Abell 851 cluster, this provides a model calibration as well as contamination.

Galaxy counts were made in 20 arcsec square boxes at various locations in the HST images. The box centred on the QSO contains 2 bright galaxies and 24 faint galaxies. Boxes at other locations had counts of 2.7 bright, and 12 ± 2.7 faint (R>19) galaxies. The bright galaxies are large and are either members of the z=0.4 main cluster or the foreground. These measurements show that the QSO subfield has a faint galaxy excess at the 3-4σ level. The QSO companions are also different from other faint galaxies in the field by being brighter, more compact, and showing a wide range of size. If they are at z=2, they are a more luminous population than that of the z=0.4 cluster.

Two subsets of the photometric measurements are considered: 1) the faint galaxies near the QSO ; 2) all other galaxies in the field with measuring errors less than 0.15 mag. Figure 4 shows the distribution of the galaxies superposed on the model tracks. The main cluster population forms a very well-defined group. The main population of z=0.4 cluster galaxies is seen at a position suggesting ages in the range 3-10 Gyr, with no reddening.

The z=2 track lies unfortunately close to that for z=0.4, so it is not easy to distinguish galaxies should they exist at both redshifts. However, at redshift 2, galaxies evolve rapidly to higher value of R-J. The QSO companion galaxies lie close to those of the main cluster in this diagram so, if they are at redshift 2, they must be very young (during or just after an initial starburst). Note that the QSO, whose redshift is known, lies in the centre of the group.

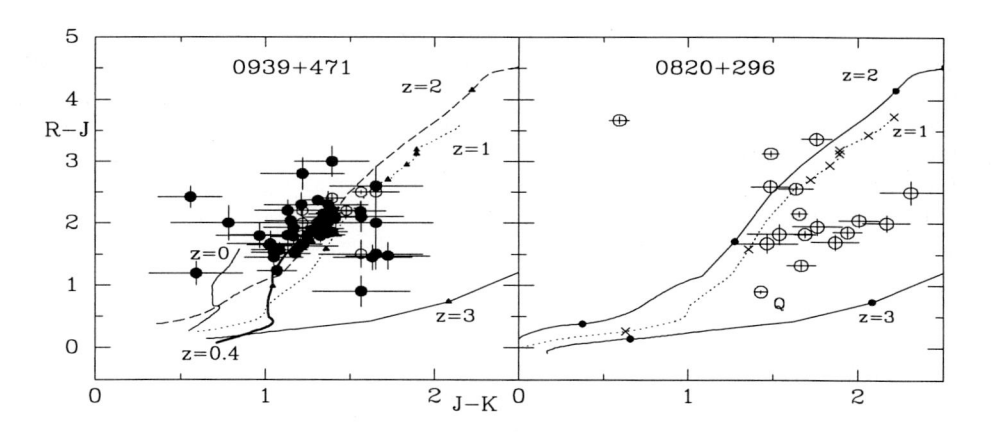

Fig. 4. 2-colour diagrams of high z fields. 0939+471 shows z=0.4 cluster and z=2.0 QSO group, and 0820+296 shows z=2.4 QSO and group

z=2.39 QSO 0820+296. R-band imaging of this field was obtained at the CFHT HR Cam, as reported by Hutchings (1995a,b) with ~0.1 arcsec pixels and

~0.8 arcsec FWHM images. These indicated an excess of faint galaxy counts near the QSO, and that the QSO itself is marginally resolved. The R-band data are used as a fourth colour for this field.

Figure 4 shows some 2-colour plots of the measured galaxies and is remarkable in the tight clumping of the objects. The QSO itself is an outlier, corresponding to a very young population (or power-law continuum) at its redshift. The region populated by the data corresponds to a 1-2 Gyr old population at z=2.4. The size and brightness of the galaxies suggests that they are at large redshift, so that the plot is consistent with, and suggestive of a group of galaxies with young stellar populations, associated with the QSO.

As projected on the sky, the QSO lies centrally in the group defined by Figure 4. The distance modulus for the QSO is ~46, so the uncorrected absolute magnitudes are -23 and brighter in R, to ~-28 in K. If they are young populations this corresponds to $M_R=21$. The ten 'best' companions are found within an area projected as 50 arcsec in the sky, which is 250Kpc at z=2.4. This is a very high density of bright galaxies if they are all associated. Hutchings (1995) reported increased galaxy counts near all of a small sample of z=2.4 QSOs. Pascarelle et al (1996) have recently confirmed a high density of faint companion galaxies in the field of a z=2.4 radio galaxy. Thus, there is mounting evidence that high z AGN are found in compact groups of very blue young-population galaxies.

5 Summary

There are several principal results. 1) The low z QSO hosts are all resolved. 2) The host galaxy colours are consistent with stellar populations at the QSO redshift. 3) The colours of nearby field galaxies suggest that all live in groups of galaxies. 4) The population models may indicate some reddening, but match the data well only at $z \geq 0.3$. 5) The QSO luminosity profiles are non-standard and reveal some of their tidal disturbance and star-formation history. 6) The radio-loud objects live in richer cluster environments. These results are described in full detail by Hutchings and Neff (1996), Hutchings and Davidge (1996) and Hutchings (1996).

6 References

Bahcall J.N., Kirhakos S., Schneider D.P., 1996, ApJ, 457, 557
Bruzual G., Charlot S., 1993, ApJ, 405, 538
Dressler A., Oemler A., Gunn J.E., Butcher H., 1993, ApJ, 404, L45
Dunlop J.S., Taylor G.L., Hughes D.H., Robson E.I., 1993, MNRAS, 264, 455
Hutchings J.B., 1995a, AJ, 110, 994
Hutchings J.B. 1995b, AJ, 109, 928
Hutchings J.B. 1996, AJ, submitted
Hutchings J.B., and Crampton D., 1990, AJ, 99, 37
Hutchings J.B., Janson T., and Neff S.G. 1989, ApJ, 342, 660

Hutchings J.B., and Neff S.G., 1991, AJ, 101, 434
Hutchings J. B. and Davidge T. 1996, PASP submitted
Hutchings J. B., and Neff S. G., 1996, AJ, submitted
McLeod K.K. and Rieke G.H. 1994, ApJ, 431, 137
Pascarelle S.M., Windhorst R.A., Keel W.C., Odewahn S.C., 1996, Nat, 383, 45
Stockton A., and MacKenty J.W., 1987, ApJ, 316, 584

Resolution of High Redshift QSO Hosts

J.B. Hutchings

Dominion Astrophysical Observatory, National Research Council of Canada, 5071 W. Saanich Rd., Victoria B.C. V8X 4M6, Canada

Abstract. At z>2 a surprising number of QSOs have been resolved in visible light. These results imply that these hosts have extremely high luminosity and blue populations. Improved and quantitative data require a combination of good signal and high resolution. Existing data are discussed and the initial results of adaptive optics imaging with CFHT are presented.

1 Introduction

The study of QSO host galaxies becomes more difficult with redshift up to z~1, due to decreasing angular size and increasing k-correction. QSO host galaxies to redshifts of ~0.4 are generally luminous and often in tidal interaction or merging with a smaller galaxy: tidal events appear to play a major role in their fuelling. Also, the difference between radio-loud and radio-quiet QSOs seems to depend on whether the host galaxy is elliptical-like or disk-like (see Hutchings and Neff 1992 and references therein).

At redshifts 0.5 to 0.9, host galaxies of radio-loud QSOs are resolvable with good signal and image quality (<0.7 arcsec: see Hutchings 1992). At high redshifts (≥2), only those of very high radio luminosity have been well resolved (Lehnert et al 1992). Lowenthal et al (1995) report marginal K-band detection of radio-quiet host galaxies at z~1 but no detection at z=2.5. In addition to the host galaxies themselves, radio-loud QSOs often have extended emission-line gas, sometimes of large size and luminosity, and often associated with the extended radio structure (Heckman et al 1991). In recent years, radio-quiet QSOs at higher redshifts have also been optically resolved (Hutchings 1995a, Aretxaga et al 1995). The radio-quiet QSOs appear to have less luminous and more symmetrical hosts.

These results were unexpected and imply unprecedented luminosity, size, and blue colour of the host galaxies. It is thus of importance to know whether they pose a challenge to ideas of heirarchical galaxy formation. If they are young mergers of luminous subcomponents, they might represent early stages of galaxy formation. However, the observations to date suffer from poor seeing, signal, and lack of colour information, so that the possibilities are very poorly constrained.

Ground-based data show that at redshifts above 2, with mean FWHM of 1.3 arcsec, ~45% of radio-loud QSOs have been resolved. This rises to > 65% with FWHM 0.7 arcsec. 55% of radio-quiet QSOs have been resolved with FWHM 0.8 arcsec. These results suggest strongly that with subarcsecond seeing and deep exposures, most high redshift QSOs can be resolved. Few of these observations have been made under good seeing conditions, and deep colour imaging has not been attempted. The results to date are thus not homogeneous and some are dubious. Thus it is essential that further work be done to try to obtain a good database on these important objects.

2 HST and Ground-Based Results

Since angular scale does not shrink with increasing redshift beyond $z\sim1$, the main problem with better observations is signal level: this requires long integrations, and/or large telescopes. Naturally, high angular resolution is desirable, to discover morphological details, but there is always a tradeoff between pixel size and signal level. For this reason HST observations at high redshift have not produced much yet.

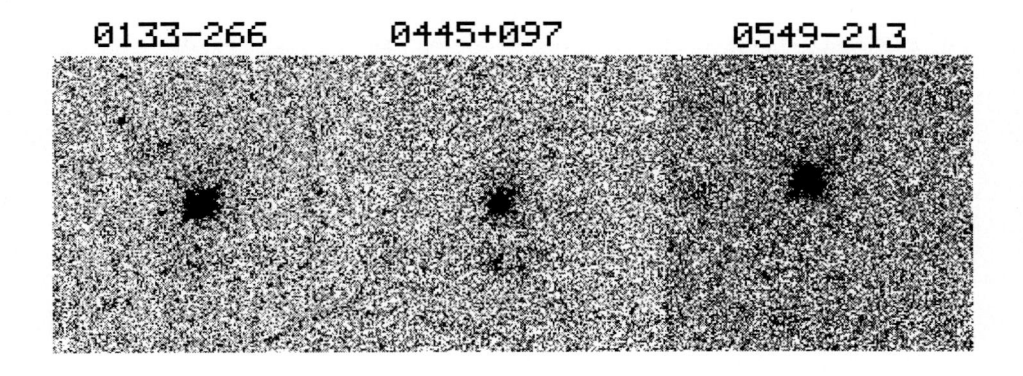

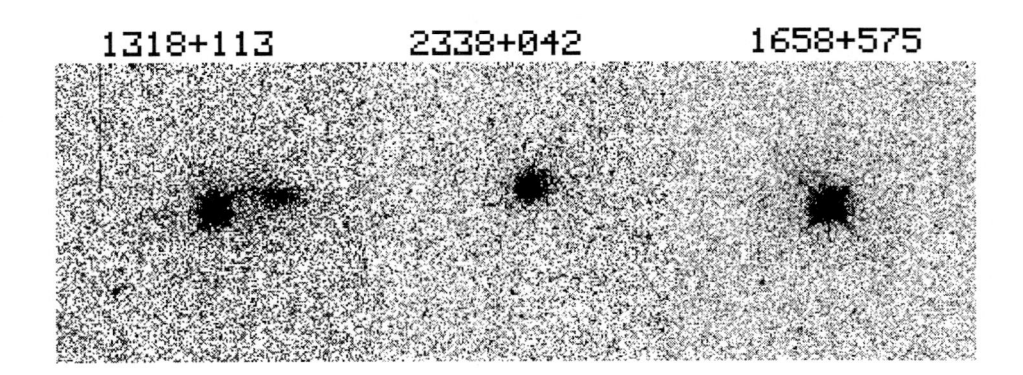

Fig. 1. 10arcsec fields with V-band images of Z>2 QSOs with HST

A program of HST imaging at redshifts near 2 was undertaken in cycle 4 by Lehnert, and the data are now in the archive. I have a cycle 6 program with much longer exposures which still awaits execution. Figure 1 shows several of the existing images: with total exposure of 1-2 hours most of these show little more than the nuclear PSF, but some clearly have close companions. Ground-based exposures with a 4m telescope (e.g. Lehnert et al 1992) and similar exposure and 10-15 times worse resolution do reveal faint extended luminosity in all these objects. Figure 2 shows a comparison. The HST data do show up bright nearby objects or features within the host galaxy in 3 of the cases illustrated, and these show up as asymmetries in the ground-based data. Thus, with very small number statistics, it appears that an appreciable fraction of these (radio-loud) objects have very close companions. Clearly, we need to obtain a larger sample with much better signal, to investigate the actual host galaxies.

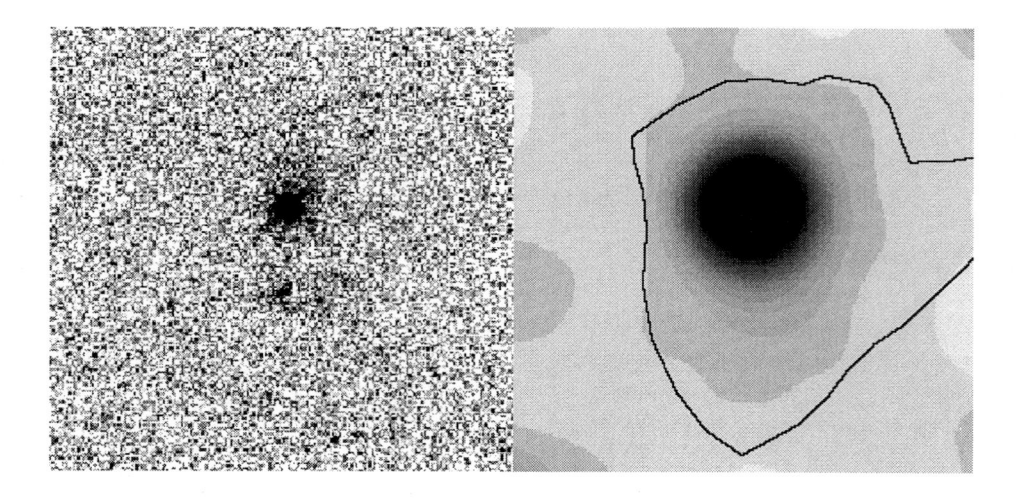

Fig. 2. QSO 0445+097:left HST image; right same smoothed to 1.2", with independent ground-based outer contour sketched in.

Our best results may be obtained from using adaptive optics on large ground-based telescopes for the details, and sacrificing resolution for signal where the signal is faintest. The recently commissioned Adaptive Optics Bonnette (AOB) at the CFHT is being used in such programs.

3   AOB Results

The AOB system uses a natural guide star. Thus, the degree of correction depends on the brightness and angular offset of the guide star. It also depends on the wavelength of observation. With guide stars of 15 mag or brighter, maximum correction is obtained. The corrected field is good over a diameter of 20-30 arcsec in H and K band, and the images are essentially diffraction limited with strehl 0.7 or more. At shorter wavelengths the correction is less, but images with FWHM 0.06" have been obtained in I-band, and better than 0.2" as short as B-band.

During commissioning time with this camera, four z∼1.7 QSOs were imaged with short exposure. Figure 3 shows the luminosity profile of one, compared with the PSF. While it appears that the QSO is resolved, further modelling is required of the PSF as the QSO lies some 20 arcsec from its guide star. Figure 4 shows the measured image widths of the four QSOs compared with typical stellar image widths as a function of offset from the guide star. The full program of QSO host galaxy observations will require longer exposures and coordinated PSF observations. They promise to complement or surpass HST imaging of QSO hosts at all redshifts, and through both CCD and NIR detector wavelengths.

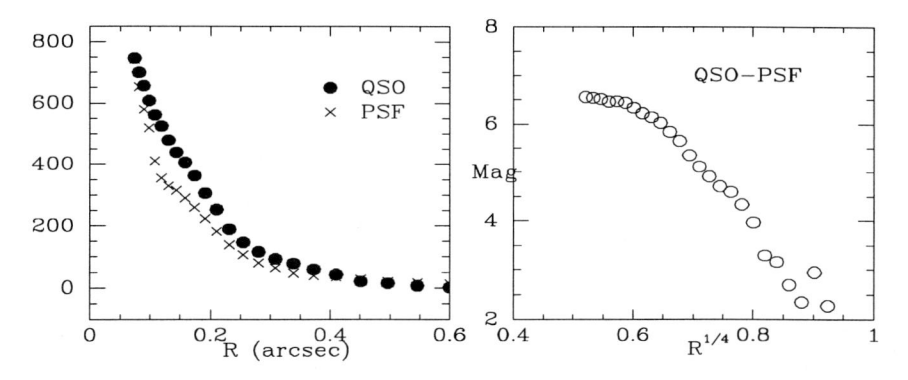

Fig. 3. Azimuthally averaged luminosity profiles of z=1.6 QSO 1337-013, PSF star, and difference.

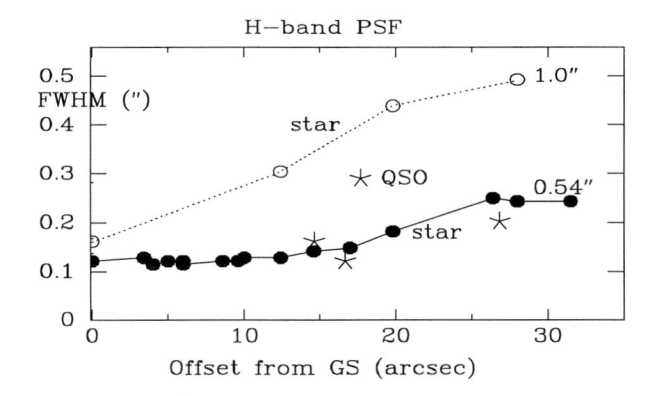

Fig. 4. FWHM of PSF with offset from guide star, and four z∼1.6 QSOs. Data are shown for two values of uncorrected seeing.

The building and commissioning of the CFHT AOB was the combined work of teams in Canada and France. More information may be obtained on the AOB at http://cdsweb.u-strsbg.fr/cfht/aob/wht.html.

4 References

Aretxaga I., Boyle B.J., Terlevich R.J. 1995, MNRAS, 275, L27
Heckman T.M., Lehnert M.D., van breugel W., Miley G.K. 1991, ApJ, 370, 78
Hutchings J.B. 1992, AJ, 104, 1311
Hutchings J.B., 1995a, AJ, 110, 994
Hutchings J.B., and Neff S.G., 1992, AJ, 104, 1
Lehnert M.D., Heckman T.M., Chambers K.C., Miley G.K. 1992, ApJ, 393, 68
Lowenthal J.D., Heckman T.M., Lehnert M.D., Elias J.H. 1995, ApJ, 439, 588

The Nature of Radio Emission in Radio-Quiet QSOs

Peter Barthel and Jeroen Gerritsen

Kapteyn Astronomical Institute, University of Groningen,
P.O. Box 800, 9700 AV Groningen, Netherlands

Abstract. Investigations of the origin of the weak radio emission in radio-quiet QSOs are reviewed. A picture emerges where – in varying degrees – optically thick radio emission from active nuclei combines with optically thin emission from galactic disks with large scale and/or circumnuclear star formation and optically thin emission from weak AGN-fed components. High resolution radio observations in combination with far-infrared (FIR) data are suggestive of an evolutionary link with ultraluminous far-infrared galaxies.

1 Introduction

Though QSOs were originally discovered in samples of luminous radio sources, it became soon clear that strong radio emission is exception rather than the rule (e.g., Schmidt 1969). Typical radio-loud QSOs (or QSRs – some use 'quasars' to denote this class) combine optical magnitudes ~15–18 with radio flux densities of order 0.1Jy at cm wavelengths, while radio-quiet QSOs have flux densities which are typically of order 1 mJy or less. Radio maps at arcsec scale resolution generally show double lobed structures (with jets) for steep, and core-halo morphologies for flat radio spectrum QSRs. In their important study of the radio-loudness distribution of the sample of ultraviolet excess Palomar-Green QSOs (the Bright Quasar Survey, or BQS) Kellermann et al. (1989) define the R-parameter quantifying radio-loudness as the ratio of 6cm radio and optical B-band (4400Å) flux density:

$$R = S_{6cm}/S_{4400\text{Å}} \qquad (1)$$

The R-distribution in the BQS suggests $R \sim 10^2$ to 10^3 for radio-loud QSOs while $R < 10$ for their radio-quiet counterparts. The fraction of $R > 10$ QSOs in the BQS (having M_B brighter than -23) appeared to be about 20%. We stress that *radio-quiet* does not mean *radio-silent*: very few BQS QSOs remain undetected and the inferred radio luminosities go down to 10^{21} W/Hz (at 5 GHz). Recall that normal galaxies have radio luminosities $10^{17} \lesssim P_{5GHz} \lesssim 10^{21}$ W/Hz. Hence, radio-loudness may alternatively be expressed in absolute terms, where the division between radio-quiet and radio-loud objects is commonly taken at $P_{5GHz} = 10^{25}$ W/Hz, using a Hubble constant of 50 and deceleration parameter of 0.5[1]. The influence of its environment on a QSR radio-luminosity however

---

[1] $H_0 = 50$ and $q_0 = 0.5$ used throughout

should not be underestimated in this respect (Barthel and Arnaud 1996). Prime questions of course are the physical origin of the radio-loudness, the connection – if any – between the radio-quiet and -loud subpopulations, and possible interrelationships with other active galaxies such as Seyfert and starburst galaxies. As for the latter, Heckman (1991, 1994) discussed the issue of a starburst-AGN connection.

2 Radio-Loudness and its Cosmic Evolution

As mentioned above, Kellermann et al. (1989) investigated the distribution of the radio-loudness R-parameter in the Bright Quasar Survey, consisting of 92 $M_B < -23$ QSOs from the Palomar-Green sample of stellar uv-excess objects. Because the BQS cuts at $m_B = 16.2$ the objects are mostly at small redshifts, with median value $z \sim 0.2$. The completeness of the BQS at the bright end is discussed by Wisotzki, elsewhere in this volume. Kellermann et al. (1989) made VLA observations of the BQS at low resolution (18″) in order to be sensitive to low surface brightness radio emission, and to determine reliable integrated radio flux densities. K-corrections were computed using known redshift values and adopted spectral slopes. The resulting distributions of both R-parameter and radio luminosity showed bimodality with dividing values $R \sim 10$ and $P_{5GHz} \sim 10^{25}$ W/Hz. The radio luminosity function $G(> R)$ for the BQS appeared similar to earlier determinations (Weedman 1986). Radio observations of fainter QSOs by Miller, Peacock, and Mead (1990) confirmed the bimodal radio luminosity function and suggested it to evolve cosmologically.

The Large Bright Quasar Survey (e.g., Foltz et al. 1987) allowed Visnovsky et al. (1992) and Hooper et al. (1995) to investigate the radio-loudness issue in more detail, including cosmic evolution. It was found that the fraction of radio-loud objects decreases with increasing z and fainter absolute magnitude. Some doubt was raised however regarding the bimodal nature of the R-distribution. Larger samples will be needed to determine whether QSOs display an (evolving) continuous R-distribution, or whether the bimodality is real. Radio observations of optically selected QSOs at very high redshift support the picture of differential evolution for the two subpopulations (Schmidt et al. 1995).

3 The Issue of the Radio-Loudness

Are radio-quiet QSOs:

1. Misdirected radio-loud QSRs, with relativistic radio jets pointing close to the sky plane, and effective flux de-boosting?
2. Scaled-down versions of radio-loud QSRs, with weak nonthermal active nuclei?
3. Scaled-up versions of Seyfert galaxies or starburst galaxies, without dominant AGN? Such objects are seen to roughly follow the radio-FIR correlation, as established earlier for star forming objects (e.g., Soifer, Houck and Neugebauer 1987).

4. A combination of 2. and 3. such as is for instance observed in the Seyfert galaxies NGC 1068 (AGN-starburst symbiosis, with radio core and lobes) or 3C 120 (Seyfert having double lobed radio source with superluminal motion in the bright radio core/jet).

In order to address these questions one needs to examine the radio emission of radio-quiet QSOs in detail. Good examples of previous work in this respect are: Condon, Gower, and Hutchings (1987), Kellermann et al. (1994), and Sopp and Alexander (1991). The first paper examined the radio continuum morphology and luminosity of half a dozen $z < 0.08$ PG (=BQS) QSOs, and found the data to be consistent with spiral galaxy hosts, with the radio/FIR ratios indicating star formation as the major contributor to the radio emission. Kellermann et al. (1994) presented high resolution (0.5″) observations of the BQS and concluded that at least some of the radio-quiet QSOs may contain compact central engines as observed in the radio-loud QSRs. Sopp and Alexander (1991) constructed a radio/FIR plot for late type galaxies, Seyfert galaxies, QSOs, QSRs, and radio galaxies and concluded on the basis of this plot that the FIR and radio emission from radio-quiet QSOs is from star-forming host galaxies and not from active nuclei. The fact that the QSOs were seen to roughly follow the radio/FIR correlation for late type galaxies, Seyferts and ultraluminous FIR galaxies indeed argues for late-type, gas-rich hosts with ongoing star formation, but the Sopp and Alexander (1991) analysis is too simplistic as we will see below.

4 Radio and FIR Radiation in Active Galaxies

Realistic treatment of the radio and FIR emission in normal and active galaxies requires taking into account the multi-component nature of the FIR (and radio) emission. The following components need to be considered (e.g., Soifer, Houck, and Neugebauer (1987), Telesco (1988), Hes, Barthel, and Hoekstra (1995)):

1. Cold ISM dust, heated by the interstellar radiation field. The dust grains are in quiescent clouds and in filamentary structures ("cirrus"), and the emission is characterized with dust temperature $T_d \sim 10 - 20K$.
2. Cool ISM dust having $T_d \sim 30 - 50K$, heated by recently formed massive stars in the galactic disks or circumnuclear regions.
3. Warm ($T_d \sim 75 - 150K$) circumnuclear dust, heated by the hard continua of the nuclei in active galaxies.
4. Hot ($T_d \sim 200 - 300K$) dust from mass loss around evolved stars, and possibly connected to circumnuclear star bursts.
5. Beamed non-thermal radiation from small-scale jets emanating from active nuclei.

Component #2 is found to scale with the radio continuum emission through the already mentioned radio/FIR correlation, component #3 is inferred to have an anisotropic distribution in order to explain AGN unifications schemes (Antonucci 1993), and the strength of component #5 is governed by the level of relativistic beaming (for instance this component swamps all others in the radio-loud blazar class). Component #4 is generally weak but may be strong in certain objects. As to the relative strength of the star formation (#2) and the AGN (#3) component, Wilson (1988) has demonstrated elegantly that by using detailed radio information, the AGN radio component(s) can be subtracted from the integrated radio emission leaving the diffuse disk emission, which then in turn is found to scale with the FIR emission for several Seyfert galaxies.

5 The 1996 Picture

Detailed examinations of the radio and the FIR emission in QSOs and other active galaxies, taking the different mechanisms into account were recently presented by Crawford et al. (1996), Rush, Malkan, and Edelson (1996), Barvainis, Lonsdale, and Antonucci (1996), and Falcke, Sherwood, and Patnaik (1996). Crawford et al. (1996) made sensitive VLA observations of a substantial sample of ultraluminous FIR galaxies and examined the radio/FIR ratios for these objects. It was found that the ratios were similar to those of less luminous FIR galaxies and normal spirals. AGN powered radio emission may contribute to some extent, but is not dominant. This is entirely consistent with the detection of high brightness temperature radio cores in ultraluminous FIR galaxies by Lonsdale, Smith, and Lonsdale (1993) when taking the fairly low fractional contribution of the high T_B cores to the total radio emission into account. Rush, Malkan, and Edelson (1996) showed that the average radio/FIR ratio for Seyfert galaxies is about twice the value for spirals, indicating the presence of a moderate AGN component. Giuricin, Fadda, and Mezzetti (1996) showed more evidence favouring an AGN origin for the radio core emission in Seyfert galaxies, rather than supernovae. Barvainis, Lonsdale, and Antonucci (1996) measured integrated radio spectra for a sample of radio-quiet QSOs, and found similar spectral shapes as those observed for the radio-loud QSRs. Barvainis, Lonsdale, and Antonucci (1996) attribute the observed spectra to three components in varying degree: optically thin emission from regions of star formation and weak jets or lobes, and optically thick emission from a compact core. Falcke, Sherwood, and Patnaik (1996) draw attention to radio-intermediate QSOs (i.e., having $R \gtrsim 10$), and suggested that their moderate radio luminosity is due to a relativistically beamed AGN component. The presence of such high T_B core emission in these radio-intermediate QSOs has indeed been demonstrated, using VLBI, by Blundell et al. (1996) and Falcke, Patnaik, and Sherwood (1996): relativistic jets must be present in some radio-weak QSOs. It will not be a surprise that such objects do *not* obey the radio-FIR correlation.

6 A New Radio-FIR-NIR-Optical Study

Radio-quiet QSOs are possibly spawned by the merging of gas-rich galaxies, such that the QSO phase is preceded by a luminous FIR/starburst phase where the newborn QSO is enshrouded by dust. This idea, advanced by Sanders et al. (1988), is supported by the similarities between ultraluminous far-infrared galaxies, discovered by IRAS, and the QSOs. In particular, some QSOs are observed to occur in interacting galaxy systems (e.g., Stockton 1990, and Bahcall, this volume), while such systems are generally observed in ultraluminous FIR galaxies (e.g., Sanders and Mirabel 1996). Since the latter are dominantly powered by starbursts, a starburst-AGN symbiosis in some form may be present in QSOs, if the above mentioned evolution model is correct.

We are engaged in a combined radio/optical/near-IR imaging project to investigate the nature of the radio emission in radio-quiet QSOs. We attempt to determine the level of starburst activity in radio-quiet QSO hosts, by comparing sub-arcsec resolution dual-frequency radio images with deep optical and near-IR images at the highest possible astrometric accuracy. The 25 objects in our equatorial PG QSO sample (to ensure visibility from ESO Chile) have all been detected at low radio flux density levels (selection criteria: $0.5\,\mathrm{mJy} \lesssim S(5\,\mathrm{GHz}) \lesssim 5\,\mathrm{mJy}$, $z \lesssim 0.45$, from Kellermann et al. (1989)), and their 5 GHz morphologies, at $\sim 0.5''$ resolution are known. This sample has average values of R and radio luminosity $P_{5\mathrm{GHz}}$ of 0.55 and $10^{22.7}$ W/Hz, respectively. Maximum values are 2.4 and $10^{24.4}$, so the sample is representative of the radio-quiet subpopulation. In fact, these QSOs are lying at the high end of the spiral galaxy radio luminosity function, and thus have the best chance of demonstrating a connection with ultraluminous far-infrared galaxies, if it exists.

We have made deep VLA observations of this sample, using the sensitive X-band (8.4GHz) system, at $0.25''$ resolution, and reaching map noise levels of $\lesssim 35\mu\mathrm{Jy}$, which ensures detection of radio sources of some tenths of mJy. Combination of the literature radio data with our 8.4GHz data indicates the presence of compact components, and allows detection of low surface brightness host galaxy emission. To complement these radio data, we have obtained sub-arcsecond resolution, optical and near-infrared images of the QSO host galaxies, which in several cases show distorted isophotes, double nuclei, extranuclear excess emission, etc. Using the Carlsberg Automatic Meridian Circle (La Palma) we are carrying out astrometric observations of the QSO fields, with accuracy of $\sim 0.3''(3\sigma)$. Overlay of radio and optical/nearIR images at this high accuracy, in combination with the radio spectral information will allow us to separate circumnuclear from nuclear radio emission and to possibly identify the former with blue starburst radiation. A preliminary account of this work was given in Gerritsen and Barthel (1995).

The astrometrically calibrated radio maps allow identification of compact nuclear components, which are commonly detected and found to have steep, flat, or inverted radio spectra. Two thirds of the sample QSOs have IRAS 25 and 60μm detections. Using these flux densities we have compiled FIR colors α_{25}^{60} and u-parameter values, where

$$u \equiv \log\left(S_{60\mu\mathrm{m}}/S_{6\mathrm{cm}}\right) \qquad (2)$$

The integrated u-values (FIR/total radio) range from ~ 1.0 to ~ 3.0. Hence several of the QSOs, while still having $R \sim 1$, are too radio-loud. Virgo spirals, the star forming disk in NGC 1068, and IRAS galaxies, including Arp 220, all have $u \approx 2.7$. Such star forming objects have steep FIR colors: $\alpha_{25}^{60} < -1.5$ (indeed $\alpha_{25}^{60} > -1.5$ is a powerful AGN selection technique – De Grijp et al. 1985). Among the steep α_{25}^{60} PG QSOs – which would by the way *not* be detected using the just mentioned De Grijp et al. technique – are some well studied star forming systems such as I Zw 1 (PG 0050+124) and Mk 1014 (PG 0157+001). Now the somewhat flatter α_{25}^{60} QSOs, where AGN dust dominates over star formation dust [2], indeed have an u-distribution which is too radio-loud. Subtracting the radio cores and determining the u-values for the non-AGN emission gives an improved u-distribution with average value closer to 2.7 and smaller dispersion. PG 0052+251, a $z=0.155$ QSO in a large spiral (Bahcall, Kirhakos, and Schneider (1996)) is a fine example. Its integrated u-value is 2.1, but subtraction of the compact, inverted spectrum radio core, leads to $u=2.5$ for the diffuse emission. It is however not impossible that substantial star formation occurs in (some of) the compact radio cores; brightness temperature and variability measurements are needed to address that issue.

A picture emerges where QSO host galaxies can have a substantial level of ongoing star formation. In that sense, an evolutionary connection to ultraluminous FIR galaxies can be envisaged. This star formation can be measured from FIR colors, high resolution radio imaging (e.g., the secondary, steep radio spectrum component in Mk 1014 coincides with blue extranuclear emission as seen on the HST image), and high resolution optical and NIR imaging. Detailed intercomparison of the imaging data, addressing the relative star formation magnitude is forthcoming. A search for molecular gas in QSOs with dominant cool dust will be of great interest.

It will be intriguing to study the relative importance of QSO host star formation with host type, environment, and radio-loudness, particularly since the classical picture of radio-loud QSOs living in elliptical hosts, while spiral hosts prefer radio-quiet QSOs is not valid any more (e.g., McLeod, this volume, and Bahcall, this volume). Such studies may eventually lead to an understanding of the evolving QSO luminosity functions.

Acknowledgements. The work of Barthel and Gerritsen is a collaborative effort with R. Sramek (NRAO VLA), D. Sanders (Univ. of Hawaii) and R. Argyle (RGO), to whom we are grateful for allowing us to present some of the results here. PB acknowledges travel support from the Leids Kerkhoven-Bosscha Foundation. JG acknowledges support by the Netherlands Foundation for Research in Astronomy (NFRA) with financial aid from the Netherlands organization for scientific research (NWO). The Carlsberg Automatic Meridian Circle is operated jointly by Copenhagen University Observatory, the Royal Greenwich Observatory and the Real Instituto y Observatorio de la Armada en San Fernando.

---

[2] The balancing effects of cool star formation dust and warm AGN dust is nicely illustrated by recent ISOPHOT data for three nearby Seyfert galaxies (Rodríguez-Espinosa et al. 1996)

References

Antonucci, R. (1993): ARAA 31, 473
Bahcall, J.N., Kirhakos, S., and Schneider, D.P. (1996): ApJ 457, 557
Barthel, P.D., Arnaud, K.A. (1996): MNRAS 283, L45
Barvainis, R., Lonsdale, C., Antonucci, R. (1996): AJ 111, 1431
Blundell, K.M., et al. (1996): ApJ 468, L91
Condon, J.J., Gower, A.C., Hutchings, J.B. (1987): AJ 92, 255
Crawford, T., et al. (1996): ApJ 460, 225
De Grijp, M.H.K., et al. (1985): Nat 314, 240
Falcke, H., Sherwood, W., Patnaik, A.R. (1996): ApJ 471, 106
Falcke, H., Patnaik, A.R., Sherwood, W. (1996): ApJ 473, L13
Foltz, C.B., et al. (1987): AJ 94, 1423
Gerritsen, J.P.E., Barthel, P.D. (1995): ESO Messenger 81, 12
Giuricin, G., Fadda, D., Mezzetti, M. (1996): ApJ 468, 475
Heckman, T.M. (1991), in 'Massive Stars in Starbursts', ed. Leitherer et al. (CUP), p. 289
Heckman, T.M. (1994), in 'Mass-Transfer-Induced Activity in Galaxies', ed. Shlosman (CUP), p. 234
Hes, R., Barthel, P.D., Hoekstra, H. (1995): AA 303, 8
Hooper, E., et al. (1995): ApJ 445, 62
Kellermann, K.I., et al. (1989): AJ 98, 1195
Kellermann, K.I., et al. (1994): AJ 108, 1163
Lonsdale, C.J., Smith, H.E., Lonsdale, C.J. (1993): ApJ 405, L9
Miller, L., Peacock, J.A., Mead, A.R.G. (1990): MNRAS 244, 207
Rodríguez Espinosa, J.M., et al. (1996): AA 315, L129
Rush, B., Malkan, M.A., Edelson, R.A. (1996): ApJ 473, 130
Sanders, D.B., et al. (1988): ApJ 325, 74
Sanders, D.B., Mirabel, I.F. (1996): ARAA 34, 749
Schmidt, M. (1969): ARAA 7, 527
Schmidt, M., et al. (1995): AJ 109, 473
Soifer, B.T., Houck, J.R., Neugebauer, G. (1987): ARAA 25, 187
Sopp, H.M., Alexander, P. (1991): MNRAS 251, 14P
Stockton, A. (1990): in 'Dynamics and Interactions of Galaxies', ed. Wielen (Springer, Berlin), p.440
Telesco, C.M. (1988): ARAA 26, 343
Visnovsky, K.L., et al. (1992): ApJ 391, 560
Weedman, D.W. (1986): in 'Quasar Astronomy' (CUP), p.157
Wilson, A.S. (1988): AA 206, 41

Adaptive Optics Observations of Quasar Hosts

M.N. Bremer[1,2], G. Miley[1], G. Monnet[3], M. Wieringa[4]

[1] Sterrewacht Leiden, Postbus 9513, 2300 RA Leiden, The Netherlands
[2] Institut d'Astrophysique de Paris, 98bis Bvd Arago, 75014 Paris, France
[3] ESO, Karl-Schwarzschild-Str. 2, D-85748 Garching bei München, Germany
[4] Australia Telescope National Facility, PO Box 76, Epping NSW 2121, Australia

Abstract. Adaptive Optics is an important technique that will be increasingly used to carry out ground-based studies of distant galaxies in general and distant quasar hosts in particular. It will allow ground-based studies at a similar spatial resolution to those currently only possible from space. The relative flexibility of observing from the ground, both in terms of available instruments and speed between applying and observing means that such techniques will complement space based observation. Here we present first results from a pilot project to study quasar hosts with adaptive optics, and discuss the need for better defined samples of distant AGN and galaxies suitable for adaptive optics study.

1 Introduction

The promise of Adaptive Optics is that it will provide (near) diffraction-limited images on 4 (and 8) meter class telescopes in the optical and near IR regimes. This will allow observations of distant galaxies, including quasar hosts, that have comparable or better resolution than that achieved in current space-based studies.

Although Adaptive Optics (AO) systems on large telescopes have been around for some years (*e.g.* see Beckers 1993 for a general review), they have only recently become sufficiently mature to allow the study of faint, distant galaxies as opposed to bright galactic objects.

The basic principles of AO systems are straightforward. Ground based seeing, as experienced in long (> 1 min) astronomical observations, is caused by the time averaging of variations in the incoming wavefront of light from an astronomical object, variations caused by turbulence in the atmosphere. AO systems remove these effects by sampling the wavefront at high enough spatial and temporal resolution so that the atmospheric distortions can be removed by introducing a continually varying opposite distortion in the wavefront. This counter-distortion is often introduced into the light beam by a deformable mirror.

For galactic (bright) objects, the kind of targets that have been observed with AO systems up until now, the usual technique has been to measure the distortions in the wavefront of the programme object directly. However, this has not been possible for objects at cosmological distances, as generally they are too faint for their incoming wavefronts to be sampled at the resolution necessary for a complete AO correction. For now, we have to select distant objects for their proximity on the sky to relatively bright foreground stars in order to practice this technique.

2 First Adaptive Optics Image of a Distant Quasar Host

As part of a pilot programme to prove the suitability of AO imaging for studying the hosts of distant AGN, we used ADONIS (Beuzit et al 1994) on the ESO 3.6m to image quasars drawn from the Véron-Cetty & Véron (1993) quasar catalogue. The quasars were selected to be close enough (within 30 arcsec) to a star that was bright enough ($v < 11$) to be used as a wavefront sensor. The distance criterion was chosen so that any quasar would be in the same isoplanatic patch (in the K-band) as the star, thereby allowing an adaptive correction of atmospheric effects to be made.

The technique for observing each quasar was as follows. The wavefront star was aquired by the wavefront sensor, and the high frequency correction for atmospheric variation was determined and applied. For bright galactic targets, the object itself would usually be used as its own wavefront reference source. Unfortunately the ADONIS wavefront sensors are not sensitive enough to correct on the quasars themselves (typically, they have $v > 15$). The field of the quasar was then imaged in K' using the SHARP (or SHARP II, Hofmann et. al. 1995) IR camera which serves as the detector for ADONIS.

After several minutes integration, the telescope was moved so that an image of a blank field near the quasar was obtained. With an ideal system, this step would not have been needed. However, ADONIS is an open optical bench system (betraying its experimental origins) and consequently suffers from dust on some in-focus surfaces. The effect of this dust can only be removed by imaging a blank field and subtracting the resulting image from the on-target exposure. The telescope then imaged a nearby star-star pair with a similar separation to the star-quasar pair. The same correction was applied to the stellar pair, with one star acting as the wavefront star and the other as the target. In this way an accurate representation of the telescope PSF was obtained for the quasar image. As a quasar can be up to 30 arcsec from its wavefront star, it will have a different PSF than that of the wavefront star. Simply recording an image of the wavefront star is not enough to estimate the telescope PSF at the quasar, the experiment *must* be repeated with a star-star pair.

All of the above was repeated several times over, depending upon the total integration time required on the quasar. However, the time available was usually limited by the long-term variation of atmospheric conditions, it is rare that perfect conditions (stable good intrinsic seeing, large isoplanatic patch) will last an entire night.

One of these quasars observed with this technique was Q0101-337. This quasar was discovered by Savage et al (1984) in an objective prism survey for quasars. The object was catalogued as a quasar with a possible redshift of z=2.21. Apart from this original reference, no further information existed in the literature on the object. Figure 1 shows a montage of the images of the quasar, the

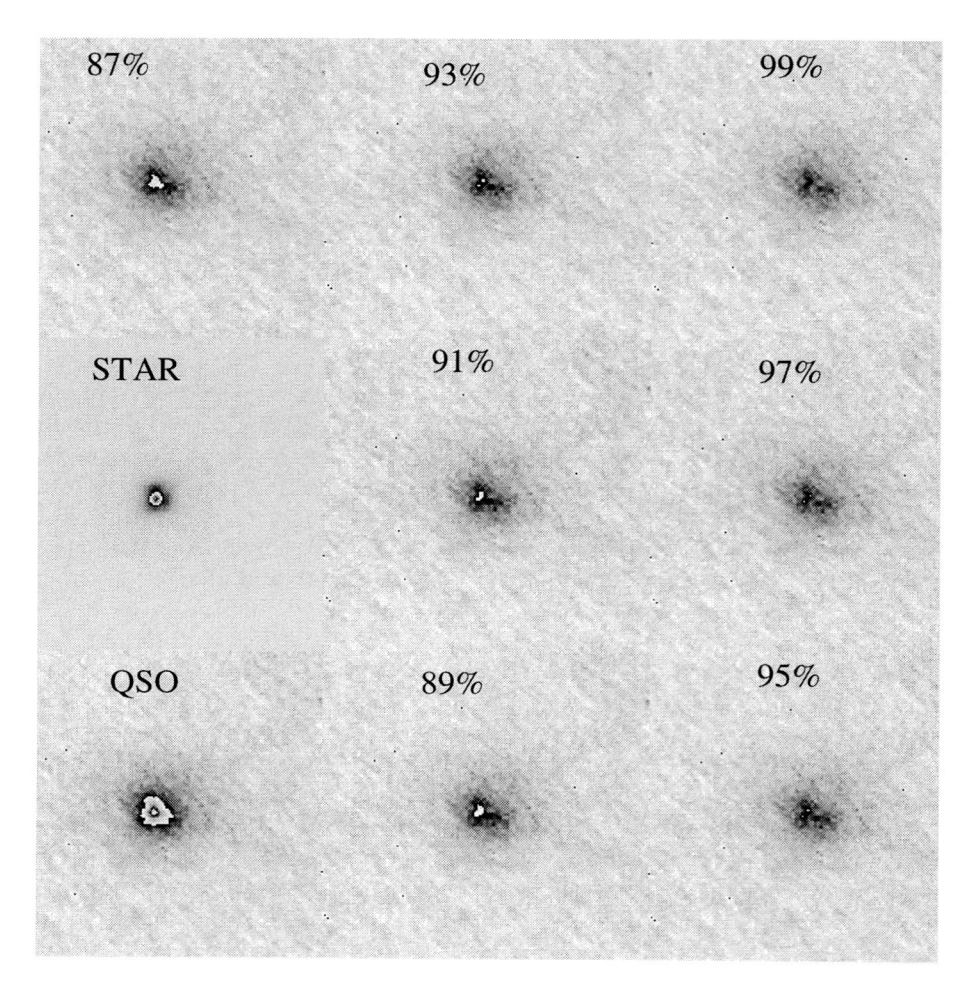

Fig. 1. K'-band image of the QSO 0101-337 and its model PSF-star. Apart from the two original images, those indicated with a percentage are the residual image having subtracted a scaled version of the PSF-star from the peak of the quasar emission. The percentage indicates the ratio of the peak-height of the PSF-star to the peak-height of the quasar image before subtraction. So for example, the image marked 95 % has had 95 per cent of the peak of the quasar emission removed by subtraction of the model-PSF star. The differences in the two original image immediately shows that the quasar is extended (star and quasar images are similarly scaled). Detail of emission due to the host galaxy is evident even after only 87 per cent of the nuclear emission has been removed.

model star (from the star-star pair) and the residual emission from the quasar host after scaling and subtracting the model star from the peak of the quasar emission. The percentages indicate the scaling of the star before subtraction, expressed as the ratio of the peak value in the scaled star to the peak value of the quasar emission. The structure in the host galaxy is clearly seen after only 87 per cent of the quasar light is removed. Figure 2 shows the residual emission from the host galaxy after 95 per cent of the nuclear light was subtracted. The residual high surface brightness structure, a possible dust lane (or double nucleus), is not a result of the over-subtraction of the nucleus; as can be seen from Figure 1 the peak of the point source emission is slightly to one side of the 'dust lane', and the sequence of subtractions shows no sign of over-subtraction or misalignment of the star with respect to the quasar nucleus. Figure 2 has been smoothed with a 0.2 arcsecond Gaussian, the original seeing with AO correction was 0.3 arcsec. For reference, 'raw' seeing of the ESO 3.6m at the time (*i.e.* without correction) was about 1.2-1.4 arcsec. One thing that is immediately apparent is that the host galaxy is quite bright ($K' \sim 17.5$) for a quasar at z=2.21. This prompted us to check the redshift, the quasar is infact at z=0.531. Thus, even with existing (non optimal) technology, we are able to image a distant quasar host, and therefore ordinary galaxies at similar distances with a spatial resolution equivalent to about 2 kpc, and signal to noise sufficient to resolve features such as dust lanes at the centres of the galaxies. All of this from the ground with a telescope that has had, up until now, notoriously bad seeing!

3 The Need for Better Samples

When we selected our sample of quasars by cross-correlating existing quasar catalogues with the HST Guide Star Catalogue, it became immediately apparent that the number of quasars suitable for AO studies was very low. Indeed, carrying out such cross-correlations is an excellent way of finding errors in catalogues, as it selects 'outliers'. For example, there are quasars selected for their proximity to bright galaxies (*e.g.* Q0032-086, Arp 1980). Unfortunately in some cases, these galaxies are misclassified as stars in the HST GSC by mistake. Such objects contaminate our cross-correlation. Even the quasar discussed above had an incorrectly catalogued redshift. This presumably came about because quasars near bright stars are generally not followed up after their discovery, as their fields are thought to make observations too difficult.

The situation will improve with better wavefront sensors and larger (8m) telescopes. Whereas the magnitude limit with ADONIS for a wavefront reference star is about $v = 11$ at 25 arcsec, and about $v \sim 13.5$ for correction on the object itself, newly commissioned and future AO systems will improve on this by 3-4 magnitudes, allowing fainter reference stars to be used. Laser guide stars will also improve the situation, providing the technical and political hurdles standing in the way of their use are overcome. Even then, for low-order, 'tip-tilt' correction to be applied, a $v < 17$ object must be within the isoplanatic patch of the object. If the highest resolutions are to be achieved on fainter quasars and other distant

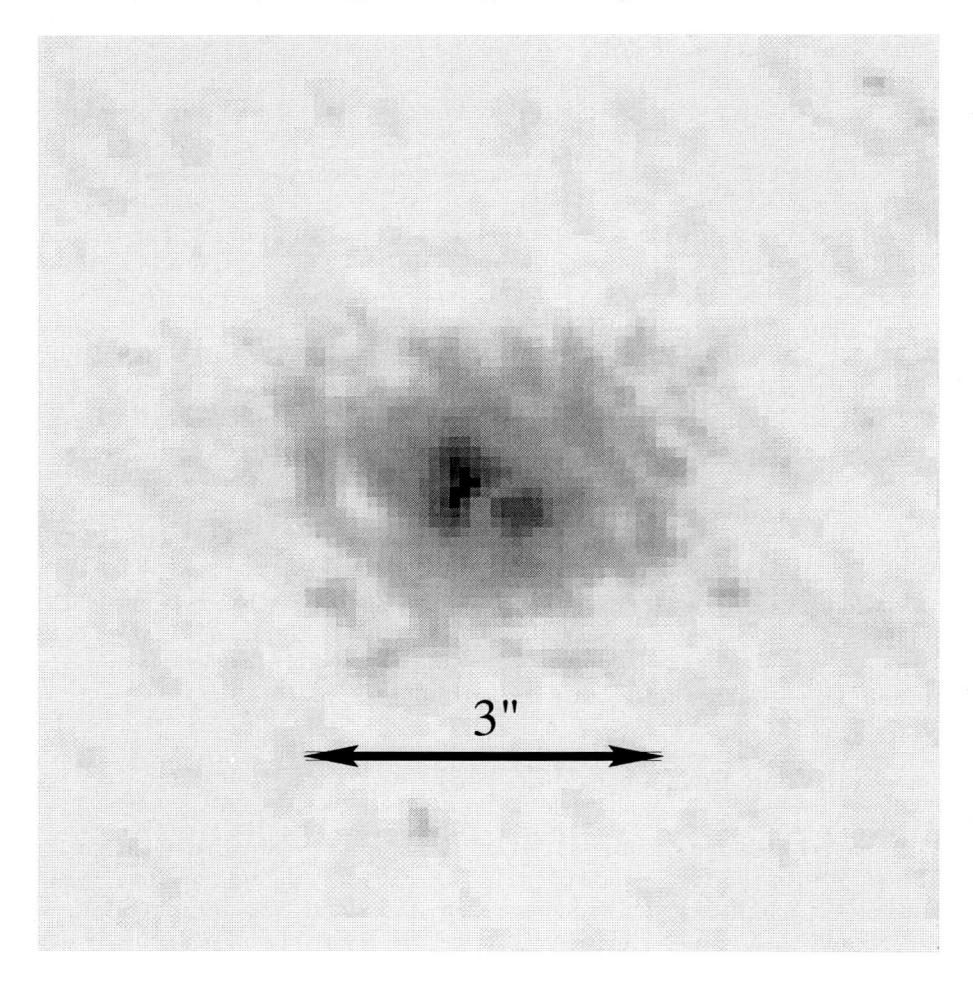

Fig. 2. K'-band image of the host of 0101-337 with 95 per cent of nuclear emission removed. The structure in the centre is real, not caused by over subtraction of the nuclear light. It is probably either a dust lane or double nucleus. The AO-corrected resolution in this image is about 0.3 arcsecs, equivalent to about 2 kpc at z=0.531.

AGN, a sample of distant AGN selected specifically for their proximity to bright stars must be available.

To this end we are carrying out a project to identify and obtain redshifts for radio sources close to stars in the HST GSC. In the southern hemisphere we are using the PMN (Griffith & Wright 1993) and Molonglo (Large et al 1981) catalogues to select our objects at high and low frequency respectively. The objects are selected for a catalogue position close to a bright ($v = 14$) star. These positions are improved upon by observations with the Australia Telescope Compact Array. Optical IDs are found by searching on the Digitised Sky Survey images, following up blank fields with imaging on ESO telescopes. Redshifts are now starting to be obtained at ESO for this sample.

A similar project is being carried out in the north, although in this case we select on radio spectral index rather than frequency. We select flat spectrum objects (mainly quasars) from the CLASS (Jackson et. al. 1995) and JVAS (Patnaik et. al 1992) surveys and steep-spectrum sources (mainly distant galaxies) from a combination of WENSS (Rengelink etal 1997) and higher frequency northern surveys.

The southern sample is ideally matched for the planned adaptive optics system on the VLT, whereas the northern sample will be immediately usable with the new CFHT Adaptive Optics Bonnette, described elsewhere in these proceedings.

4 Acknowledgements

MNB acknowledges support from an NWO programme subsidy. This work was supported in part by the Formation and Evolution of Galaxies network set up by the European Commission under contract ERB FMRX-CT96-086 of its TMR programme.

References

Arp, H., 1980, Proc. 9th Texas Symp, Ann. NYAS, 336, 94
Beckers, J.M. Ann. Rev. Astron. Astrophys, 1993, 31, 13
Beuzit J-L. et. al, 1994, SPIE, 2201, 955
Griffith M., Wright A., 1993, AJ, 105, 1666
Hofmann R., et. al. 1995, SPIE 2475, 192
Jackson N., et. al. 1995, MNRAS, 274, L25
Large M.I., et. al., 1981, MNRAS 194, 693
Patnaik et. al. MNRAS 254, 655
Savage A., et. al., 1984, MNRAS 207, 393
Rengelink et. al., 1997, A&A, in press.
Véron-Cetty M.-P, Véron P., 1993, ESO Scientific Report No. 13

HST Planetary Camera Images of Quasar Host Galaxies

P.J. Boyce[1], M.J. Disney[1], J.C. Blades[2], A. Boksenberg[3], P. Crane[4], J.M. Deharveng[5], F.D. Macchetto[2], C.D. Mackay[6] and W.B. Sparks[2]

[1] University of Wales College of Cardiff, P.O. Box 913, Cardiff, CF2 3YB, UK
[2] Space Telescope Science Institute, 3700 San Martin Drive, Baltimore, MD 21218, USA
[3] Royal Greenwich Observatory, Madingley Road, Cambridge, CB3 0EZ, UK
[4] European Southern Observatory, Karl Schwarzchild Strasse 2, D-85748 Garching, Germany
[5] Laboratoire d'Astronomie Spatiale du CNRS, Traverse du Siphon, Les Trois Lucs, F-13012 Marseille, France
[6] Institute of Astronomy, Madingley Road, Cambridge, CB3 0HA, UK

1 Introduction

We present here a summary of the main results from the full sample of quasars imaged with the Planetary Camera by the Faint Object Camera IDT during Cycles 5 and 6. The final observed sample consists of 14 objects. 11 of these were chosen from Table 1 of the catalogue of Veron-Cetty and Veron (1993) and are unambiguously quasars within the definition used by Veron-Cetty and Veron (i.e. $M_V < -23$, $H_o = 50$, $q_o = 0.5$). The other three were chosen from the lists of Low et al. (1988, 1989) who selected very luminous IRAS sources (i.e. $L_{IR} > 10^{12}$ $L_\odot$) which are also optical quasars based on the breadth of their emission lines and their appearance on the POSS. Two of these objects are included in Table 1 of Veron-Cetty and Veron. The third (IRAS 13218+0552) is slightly too faint and is listed by Veron-Cetty and Veron as a Seyfert Type I.

2 Observations and Data Reduction

The quasars were observed between 1994 and 1996 with the Planetary Camera (PC) within the WFPC2 (Traguer et al. 1994). The PC has a field of 800×800 CCD pixels of size $0.046''$ pixel^{-1}. All observations were taken through the F702W filter. All of the objects were observed for a total exposure time of 1800 seconds. For the first 6 objects observed this consisted of 3x600 second exposures. For the remaining objects four exposures were taken: two of 600 seconds and one each of 400 seconds and 200 seconds. Initial data processing involved the HST pipeline process and subsequent reductions were done on STARLINK, making use of STSDAS software packages supplied by the Space Telescope Science Institute. After bias–subtraction, flat–fielding and calibration the individual frames were median combined (to remove cosmic–ray events) to yield the "raw images" presented here.

3 Results

The first four objects observed were presented in Disney et al. (1995). All four objects appear to have luminous elliptical host galaxies (2→5 times brighter than $L^\star$). All four also show circumstantial evidence for interaction in that all have at least one very close companion object. Figure 1 reproduces Fig. 1 from this paper. It shows grey-scale plots of the four quasars after the subtraction of a best-fitting model PSF from the "raw images".

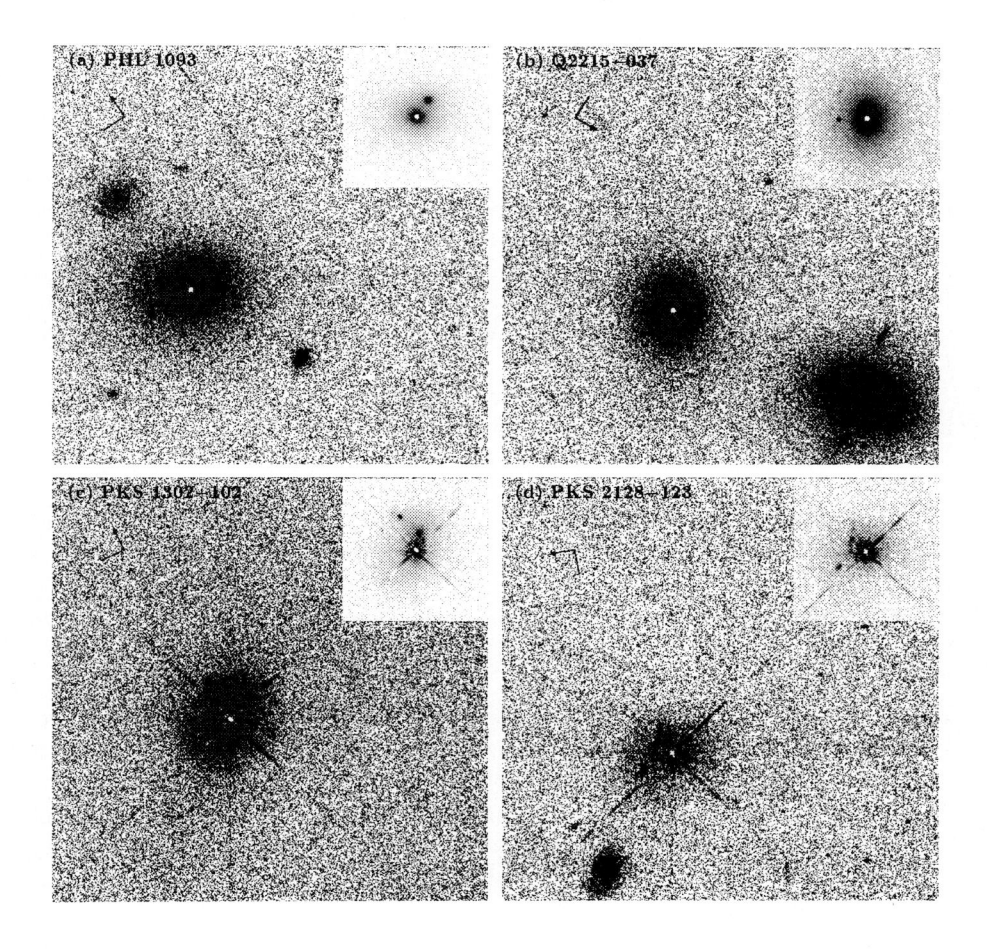

Fig. 1. Planetary Camera images of the 4 quasars described in Disney et al. (1995). The images show grey-scale plots after the subtraction of a model PSF from the "raw images". Two different intensity scales have been chosen; one to best show the host galaxies and the other (inset) to best show the companion objects. Each of the large panels is $27.6'' \times 27.6''$. North (arrow) and East are denoted.

In Boyce et al. (1996) we presented the results for the three luminous IRAS quasars included in the sample. Figure 2 reproduces Figs. 1-3 from this paper. All 3 quasars appear embedded in spectacular interactions between 2 or more luminous galaxies, probably spirals. We discussed the evolutionary connection, if any, between these 3 objects and the far more numerous Ultraluminous IR Galaxies. We argued that these 3 objects are probably young and therefore do not fit a scenario in which quasars emerge only in the later stages of an interaction when most of the dust has been blown away. It may be that we are simply viewing them from a fortuitous angle allowing a clear view into the cores.

In Figures 3 and 4 we present greyscale images of the remaining 7 quasars of the sample (Boyce et al., in prep). Each of these has had a best fitting PSF removed from it. The PSF was derived from a co-added 10x200 second exposure of a standard star. The scaling factor in each case was chosen so that, after subtraction, the residual "galaxy" image continued to increase monotonically to the smallest radius at which the image was not saturated on the shortest exposure. The diffraction spikes were also excluded from the fitting procedure since these are known to vary considerable across the PC frame. Although fairly crude, experiments show that this technique yields a probable error in the scaling factor of no more than ±20%. None of the images was saturated beyond a radius of 5 pixels in the shortest exposure and most do not suffer saturation outside the central 3 pixels.

In Figs. 3 and 4, the inner few pixels in each image (where there was significant saturation) have been replaced by null values as have the pixels in the regions of the diffraction spikes. This was necessary in order to determine a scaling factor for the subtraction of the PSF. However, it also means that all the data displayed in each frame have had the PSF removed with reasonable accuracy. Brief comments on each of the objects follow (all quoted values assume $H_o=75$ and $q_o=0$):

3C351: The subtraction process reveals a host galaxy with a luminosity profile well fitted by an r-qtr law and a luminosity several tenths of a magnitude brighter than $L^\star$. There is a close companion at a separation of 28 kpc.

PKS0312-77: An elliptical host galaxy is clearly seen in this case. It is the brightest host in the sample and has a similar luminosity to a brightest cluster member. An apparent companion is seen at a projected separation of 23 kpc.

MS 07546+3928: The host galaxy in this case appears to have a core with an r-qtr law luminosity profile with a large ring structure outside this. There is an apparent companion at a projected separation of 19 kpc.

PG 1216+069: This object has the brightest nucleus of any in the sample. After the removal of a best-fitting PSF there is still much residual non-galaxy light left. This is the so-called "scattered light", nuclear light which has been scattered off the back of the CCDs over a wide-angle of the image. This is the only object is which this has caused a major problem and we consider it imprudent to draw any serious conclusions from the subtracted image. The apparent companion seen in this image is actually a filter ghost.

PG 1358+04: The subtraction process leaves only a little residual light in the inner few pixels. If the host galaxy is an elliptical, the upper limit to its magni-

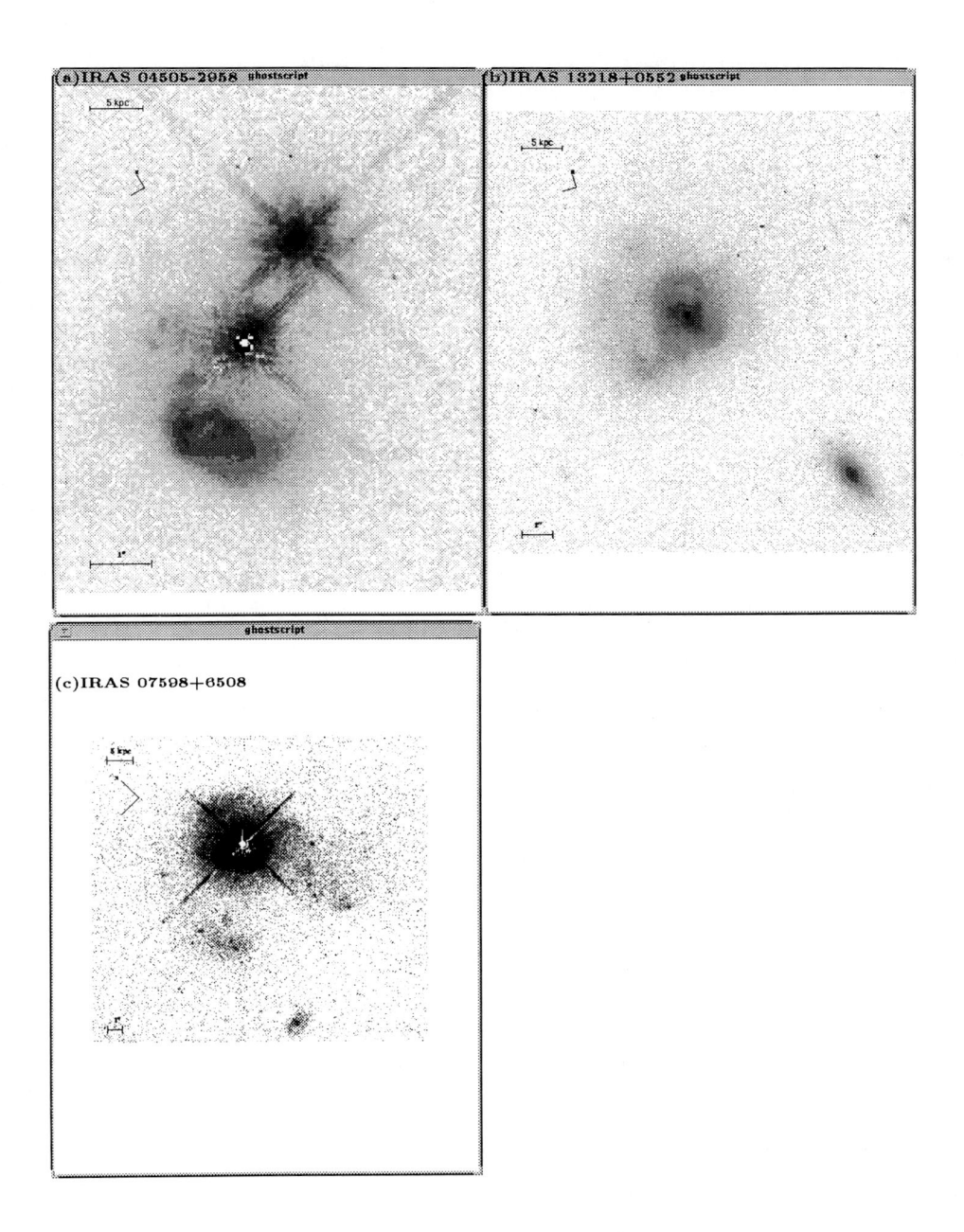

Fig. 2. Planetary Camera images of the 3 IRAS quasars described in Boyce et al. (1996). The upper scale bar represents 5 kpc ($H_o=75$, $q_o=0$) and the lower scale bar represents $1''$. North (N) and East are also denoted.

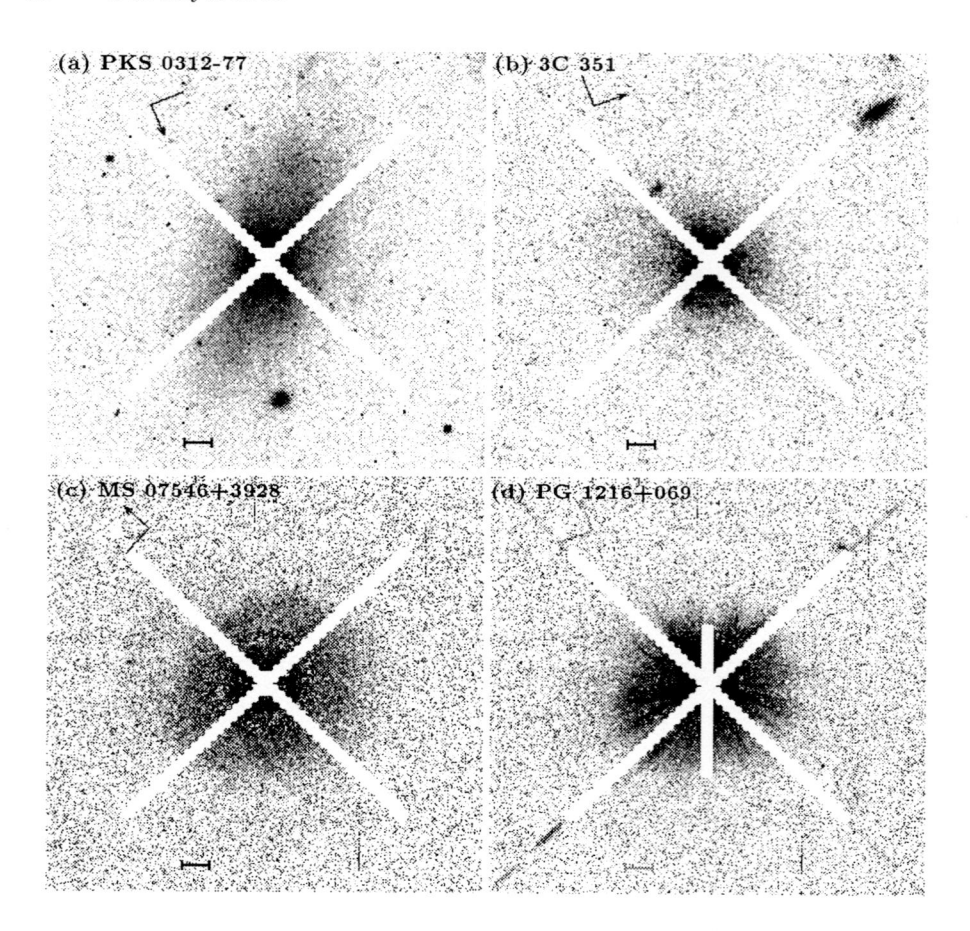

Fig. 3. Planetary Camera images of quasars following subtraction of a scaled PSF as described in the text. North (arrow) and East are denoted. The scale bar represents $1''$.

tude is $M_V = -20.6$. There is no observed companion objects within 100 kpc.

PKS 0202-76: The subtraction process leaves only a little residual light in the inner few pixels. If the host galaxy is an elliptical, the upper limit to its magnitude is $M_V = -20.9$. There is no observed companion objects within 100 kpc.

PG 0043+039: The subtraction process leaves only a little residual light in the inner few pixels. If the host galaxy is an elliptical, the upper limit to its magnitude is $M_V = -20.3$. The nearest companion objects has a projected separation of 83 kpc.

In Table 1 we summarise the properties of the whole sample of quasars. All values assume $H_o = 75$ and $q_o = 0$. Radio Loud Quasars are denoted by a $\star$. The values of M_{QSO} and M_{Host} refer to the F702W filter system of the PC. The error of $\pm 20\%$ in the PSF subtraction factor implies an error of ± 0.2 mag in

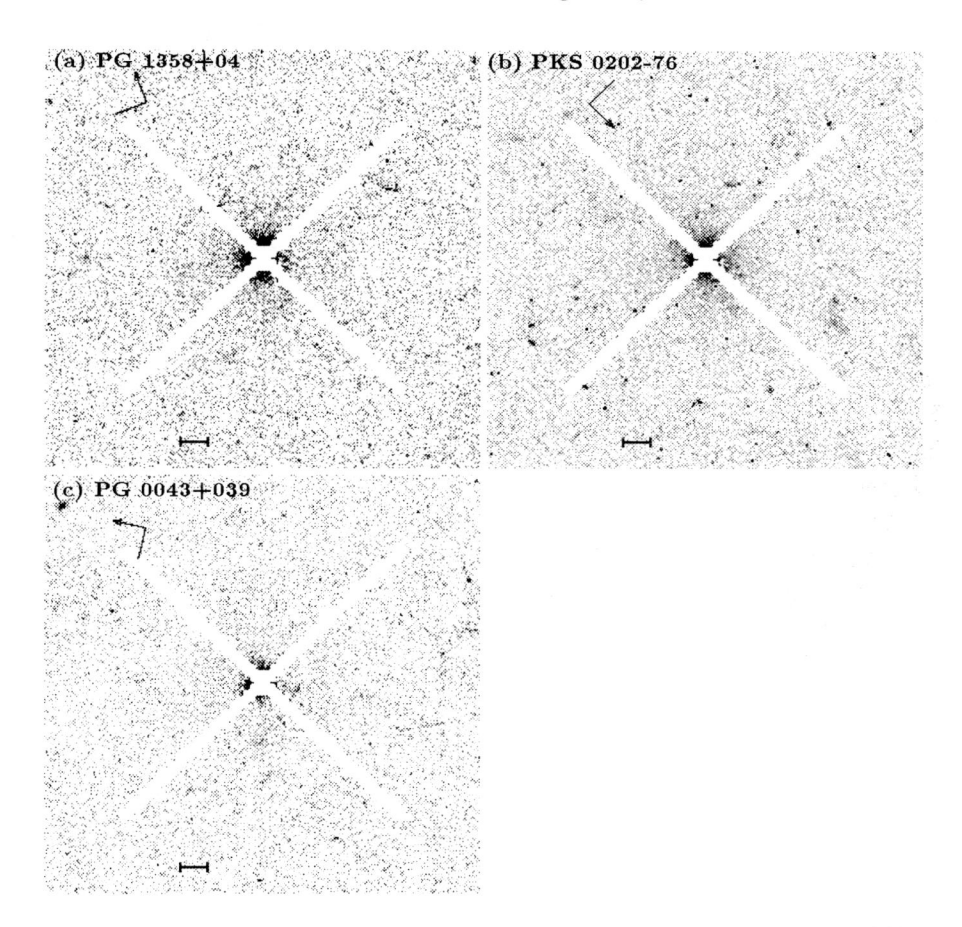

Fig. 4. Planetary Camera images of quasars following subtraction of a scaled PSF as described in the text. North (arrow) and East are denoted. The scale bar represents $1''$.

these values. The values for $M_{Host}(V)$ were obtained by transforming the measured F702W aperture magnitudes to V by applying the k-corrections calculated by Fukugita et al. (1995). In the cases of PG 1358+04, PKS 0202-76 and PG 0043+039 the $M_{Host}(V)$ value quoted is an upper limit calculated by assuming that the underlying host is a typical elliptical. The values listed for PG 1216+069 should be treated with extreme caution due to the severe scattered light problem in this case. The "companion" column lists the apparent separation to the nearest companion if this is less than 30 kpc. The three IRAS quasars are listed as "Inter." since they are clearly interacting systems although one cannot clearly separate the components involved.

Table 1. Derived results for the whole sample of quasars

Object	m$_V$	redshift	M$_{QSO}$	M$_{Host}$	M$_{Host}$(V)	Companions
PG 1358+04	16.3	0.427	−24.4	−22.2	>−20.6	>30 kpc
PKS 0202-76*	16.9	0.389	−23.5	−22.4	>−20.9	>30 kpc
PG 0043+039	16.0	0.385	−23.7	−21.8	>−20.3	>30 kpc
IRAS 04505-2958	16.0	0.286	−22.6	−22.0	−20.8	Inter.
IRAS 07598+6508	15.5	0.148	−21.9	−20.2	−19.5	Inter.
IRAS 13218+0552	17.1	0.201	−18.2	−19.9	−19.1	Inter.
PHL 1093*	17.1	0.258	−22.0	−22.5	−21.3	6 kpc
MS 22152-0347	17.2	0.241	−21.6	−21.9	−20.8	8 kpc
PKS 1302-102*	15.2	0.286	−24.6	−22.6	−21.4	7, 15 kpc
PKS 2128-123*	16.1	0.501	−25.7	−22.8	−21.0	25 kpc
3C 351*	15.3	0.371	−24.5	−23.3	−21.9	28 kpc
PKS 0312-77*	16.1	0.223	−22.6	−24.4	−23.3	23 kpc
MS 07546+3928*	14.4	0.096	−22.9	−21.4	−20.7	19 kpc
PG 1216+069*	15.7	0.334	−26.3	−25.2	−23.8	>30 kpc

4 Discussion

Of the 14 quasars in the sample, we have unambiguously detected the host galaxy in 10 cases. The morphologies of the hosts in the three IRAS quasars are not clear due to the highly disturbed nature of these systems. In the other 7 cases the hosts appear to be elliptical. The elliptical hosts detected cover a wide range of M$_V$ from −20.8 to −23.3 (slightly fainter than L* to brighter than brightest cluster member). In this small sample there is no apparent relationship between M$_{QSO}$ and M$_{Host}$. It should be noted that we find 1 example of a Radio Quiet Quasar with an elliptical host galaxy. A popular idea before the advent of HST observations was that Radio Quiet Quasars would lie in spiral hosts.

Of the four galaxies for which we cannot unambiguously detect a host galaxy, one (PG 1216+069) suffers badly from scattered light and has the brightest nuclear source in the sample. The other three, however, have relatively modest central sources and do not suffer from scattered light to any measurable extent. The host galaxies of these objects must be at least several tenths of a magnitude fainter than L* if they are elliptical objects. Alternatively they could be normal surface brightness spirals or low surface brightness galaxies or dwarfs.

Seven of the quasars have companion objects within 30 kpc of the nucleus, suggestive of interactions. In addition the three IRAS quasars are also clearly interacting. However, in four cases there is no close companion objects detected nor any obvious evidence of ongoing interaction. One of these objects (PG 1216+069) has a very bright nucleus and suffers greatly from scattered light. A close companion object would be very hard to detect in this object. The other three are those presented in Fig. 4 for which we could not detect a host galaxy to limits fainter than L*. If one assumes that interaction is the trigger to the nuclear activity in all quasars, then this is a slightly puzzling result. If the three objects have fainter host galaxies but are, nonetheless, still involved in interactions, then why can we not detect close companion objects ? Just because the host galaxy is faint would not automatically imply that an interacting companion is also faint. Two objects in the sample (3C 351 and PKS 2128-123) have higher redshifts but we still clearly detect companion objects in these cases.

5 Concluding Remarks

The results of our work are in broad agreement with those found by Bahcall and co-workers from their WFC imaging of a sample of 20 quasars (Bahcall et al. 1994, 1995a, 1995b, 1996). They found examples of quasars in diverse environments including ellipticals (with luminosity from L* to that of brightest cluster members), normal spirals, and complex systems of gravitationally interacting components. They found 7 examples of Radio Quiet Quasars with elliptical host galaxies although they also found 3 Radio Quiet Quasars with spiral hosts. Of their 20 objects, they found 5 objects in which there is no evidence of a recent interaction.

Taken together, the two samples illustrate the ability of the HST to study the host galaxies of quasars and their environs with a vastly improved resolution than previous work. The results have not as yet, however, led to a single coherent pattern of quasar behaviour. This may emerge as larger numbers of quasars are observed. We can also expect even better data from the NICMOS camera on HST with its ability to perform non-destructive read-outs at very short time intervals. This is vital for studying the bright inner regions close to the nucleus.

References

Bahcall, J.N., Kirhakos, S., Schneider, D.F. (1994): ApJ **435**, L11
Bahcall, J.N., Kirhakos, S., Schneider, D.F. (1995a): ApJ **447**, L1
Bahcall, J.N., Kirhakos, S., Schneider, D.F. (1995b): ApJ **450**, 486
Bahcall, J.N., Kirhakos, S., Schneider, D.F. (1996): preprint
Boyce, P.J. et al. (1996): ApJ **473**, 760
Disney, M.J. et al. (1995): Nat **376**, 150
Fukugita, M., Shimasaku, K., Ichikawa, T. (1995): PASP **107**, 945
Low, F.J., Huchra, J.P., Kleinman, S.G., Cutri, R.M. (1988): ApJ **327**, L41
Low, F.J., Cutri, R.M., Kleinman, S.G., Huchra, J.P. (1989): ApJ **340**, L1
Trauger, J.T. et al. (1994): ApJ **435**, L3
Veron-Cetty, M.P., Veron, P. (1993): A Catalogue of Quasars and Active Nuclei (Sixth Edition), ESO Scientific Report, No. 13

The Hosts of $z = 2$ QSOs

Itziar Aretxaga[1], B.J. Boyle[2] and Roberto J. Terlevich[3]

[1] Max-Planck-Institut für Astrophysik, Karl-Schwarzschild-Str. 1, Postfach 1523, 85740 Garching, Germany
[2] Anglo-Australian Observatory, PO Box 296, Epping, NSW 2121 Australia
[3] Royal Greenwich Observatory, Madingley Road, Cambridge CB3 0EZ, UK

Abstract. We present results on the hosts of four high-redshift ($z \approx 2$) and high luminosity ($M_B \lesssim -28$ mag) QSOs, three radio-quiet one radio-loud, imaged in R and K bands. The extensions to the nuclear unresolved source are most likely due to the host galaxies of these QSOs, with luminosities at rest-frame 2300Å of at least $3-7$ per cent of the QSO luminosity, and most likely around $6-18$ per cent of the QSO luminosity. Our observations show that, if the extensions we have detected are indeed galaxies, extraordinary big and luminous host galaxies are not only a characteristic of radio-loud objects, but of QSOs as an entire class.

1 Introduction

The study of high redshift ($z \approx 2$) QSOs offers a unique opportunity to investigate conditions in the early universe. In the currently favoured cold dark matter cosmogony, this epoch corresponds to the period when normal galaxies formed through hierarchical coalescence (Carlberg 1990), thereby giving rise to enormous concentrations of gas in the center of the galaxies, which could feed a central black hole (Haehnelt & Rees 1993) or provoke a massive starburst episode (Terlevich & Boyle 1993). As such, this picture is consistent with the observation that the QSO population peaks at these redshifts (Schmidt et al. 1991).

Searches for galaxies hosting high-redshift QSOs were first carried out in radio-loud objects, with spectacular results: luminosities several magnitudes brighter than the most luminous galaxies in the nearby Universe were observed (Lehnert et al. 1992). However, radio-loud quasars are only a small fraction ($< 1\%$) of all QSOs, and many of the conclusions drawn from radio-loud objects might be unrepresentative of the conditions in the early universe. It is, therefore, important to look into the properties of radio-quiet QSOs, as they may be better indicators of the general properties of galaxies at high redshifts. We present here deep multicolor images of one high-redshift radio-loud and three high-redshift radio-quiet QSOs.

2 Data Analysis

The four QSOs were observed in the Harris R passband with the auxiliary port of the 4.2m William Herschel Telescope (WHT) at the Observatorio de Roque de los Muchachos in La Palma, and in K band with the Cassegrain focus of the 3.5m at the Observatorio de Calar Alto in Almería. The QSOs were selected so as to have bright stars in the field (20 $\lesssim \theta \lesssim$ 50 arcsec), which enabled us to define the point spread function (PSF) of each observation accurately (see Aretxaga et al. 1995,1997 for a detailed description of data and analysis).

For each QSO field, we defined a 2-dimensional PSF using the brightest of the closest stellar companion to the QSO. We then scaled the PSF to match the luminosity of the QSO and other nearby stars over the same region, and subtracted the scaled PSF from them. The remaining residuals in the non-PSF stars provided an accurate check of the validity of the subtraction process. We accepted the PSF subtraction if the residuals in the non-PSF stars accounted for less than 1σ of the Poisson noise expected from the subtraction technique. We have detected R-band residuals in excess of 3σ in the following QSOs: 1630.5+3749 (4σ), PKS 2134+008 (3σ) and Q 2244−0105 (3.7σ). All the residuals show a 'doughnut' shape with a well of negative counts in the centre. This indicates that there is a flatter component below the PSF in the centres of the QSOs, from which the nuclear (PSF) contribution has been over-subtracted. As an example, Fig.1 shows the R-band residuals for the radio-quiet QSOs 1630.5+3749, after subtracting a luminosity-scaled PSF.

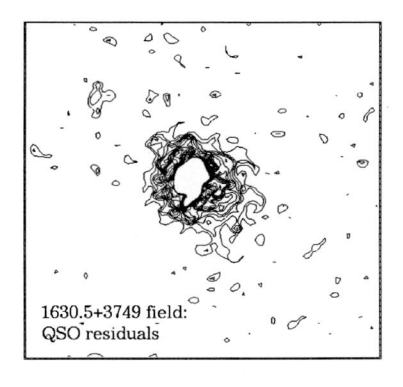

Fig. 1. R-band residuals of the QSO 1630.5+3749 after PSF subtraction (Aretxaga et al. 1995). The size of the frame is 6″x6″ .

To estimate the true luminosity of these systems, we subtracted smaller amounts of the PSF in order either a) to produce zero counts in the center of the residuals or b) to achieve a flat-top profile with no depression in the center. We regard these quantities as lower limits (3–7% of the QSO luminosity) and best estimates (6–18% of the QSO luminosity), respectively, of the total luminosity of these extended components (Table 1). In all cases, the FWHM of the flat-top residual profiles are significantly larger than the FWHM of the stars in each field: 1″.05 vs. 0″.7 for 1630.5+3749, 0″.8 vs. 0″.7 for PKS 2134+008 and 0″.84 vs. 0″.7 for Q 2244−0105.

The K-band images of the same hosts show no significant extensions to the stellar profiles (see Fig.2), over a 1σ limiting magnitude $\mu_K \approx 23$ mag/arcsec^2 (Aretxaga et al. 1997). These limits are consistent with previous non-detections of $z \approx 2$ radio-quiet hosts in K band (Lowenthal et al. 1995). If there are no colour gradients, from the $r \sim 2 - 3''$ non-detection limits, the colours of the hosts are $R - K \lesssim 3.3$ mag.

Table 1. Magnitudes of the QSOs and their extensions (Aretxaga et al. 1995)

Name	M_R^* (QSO)	M_R^1	R^1	M_R^2	R^2	FWHM^2 (arcsec)
1630.5+3749	−28.7	−24.8	21.7	−25.9	20.9	1.05
PKS 2134+008	−30.1	−25.9	20.4	−26.6	19.8	0.80
Q 2244−0105	−29.0	−25.6	20.9	−26.7	19.9	0.84

* Total QSO M_R absolute magnitude, including host ($H_0 = 50$ Km s^{-1} Mpc^{-1}, $q_0 = 0.5$).
[1] Properties for hosts with zero counts in the center after the PSF subtraction.
[2] Properties for hosts with flat-topped profile after the PSF subtraction.

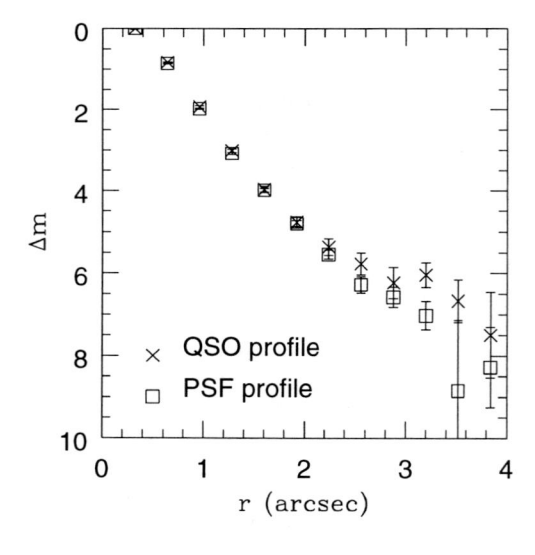

Fig. 2. Comparison between the QSO (1630.5+3749) radial profile and the PSF stellar radial profile in K-band, where crosses indicate QSO and squares stellar profiles

3 Discussion: the Origin of the 'Fuzz'

From a sample of one radio-loud and three radio-quiet QSOs with suitable PSF stars, we have detected R-band extensions in three cases (Aretxaga et al. 1995) and no extensions in K-band (Aretxaga et al. 1997). The best estimates for R-band (2300Å rest-frame) luminosities of these systems lie between 6–18% of the total QSO luminosity.

3.1 Scattering

For the lobe-dominated radio-loud QSO sample of Lehnert et al. (1992), the red colours of the hosts favour a host galaxy origin of the excess light, rather than scattering by dust or electrons in the halo of the QSOs. Our colour limits for the excess light, $R - K \lesssim 3.3$ mag, include colours as blue as those of the QSOs themselves ($R - K \approx 2.3$ mag), and are therefore consistent with the colours expected from the optically thin scattering case and, also, with those of a young stellar population. They, however, exclude the colours of passively evolved bulge populations (eg. Bressan et al. 1993).

However, most scattering models proposed to date require the presence of a powerful transverse radio-jet (e.g. Fabian 1989) which is unlikely to be present in either the radio-quiet QSOs, or the core-dominated radio-loud QSO around which we have detected extensions.

3.2 Nebular Continuum

An alternative origin for the hosts could be extended nebular continuum, seen to be a major contribution to the UV continuum in three powerful radio galaxies (Dikson et al. 1995). However, if our hosts are due to nebular luminosities of $M_R \sim -26.5$ mag, the predicted narrow Hβ luminosities would be about 3×10^{44} erg s^{-1}. From the PSF light we derive that the luminosity of the broad component is about double that. If this is so, the QSOs would exhibit prominent narrow lines with central peak intensities of more than 3 times those of the broad lines. The spectra of our QSOs do not show such prominent narrow lines .

3.3 Host Galaxies

There is some circumstantial evidence that the extensions we have detected are most probably the galaxies which host these QSOs:

a) The radial profile of the R-band hosts, derived from the flat-top solutions, falls approximately as an $r^{1/4}$-law for radii $r \gtrsim 0.6$ arcsec (Fig. 3). Profiles derived for radii smaller than the FWHM of the observations are usually unreliably recovered by flat-top subtractions, as confirmed by our the numerical simulations of galaxy+PSF. However, total luminosities and sizes are parameters which can be recovered well if the galaxy contribution exceeds 3% of the QSO+galaxy system.

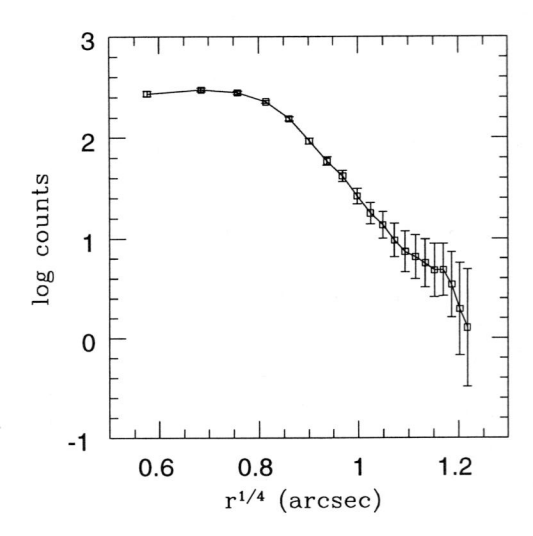

Fig. 3. Radial profile of the R-band host of the QSO 1630.5+3749, in a log counts vs. $r^{1/4}$ diagram: an $r^{1/4}$ profile would appear as a straight line.

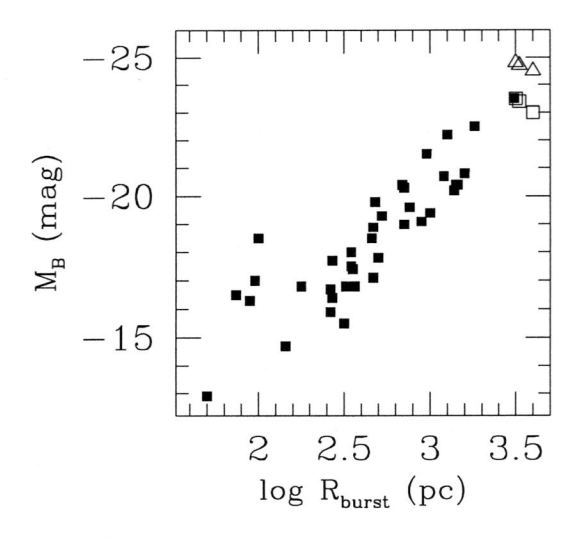

Fig. 4. Luminosity–size relationship for nearby H II galaxies (Telles 1995). H II galaxies are marked with filled squares. The three QSOs studied here lie in this relation if their SEDs are $f_\nu \propto \nu^0$ (open triangles) to $f_\nu \propto \nu^{0.5}$ (open squares). These SEDs are typical of young H II galaxies

b) The luminosities and radii of our hosts lie in the luminosity-radius relation of local young H II galaxies (Fig.4). We converted the UV luminosities of the hosts (observed R-band) to rest-frame B-band using the spectral energy distribution (SED) of local H II galaxies ($f_\nu \propto \nu^\alpha$, with $0 \lesssim \alpha \lesssim 0.5$). This is equivalent to converting the B-band luminosities of the H II galaxies to rest-frame 2300 Å and then comparing the UV luminosity-radius relationship of H II galaxies with that of the hosts. Notice that there is at least one H II galaxy which is as big and luminous as our hosts. At $z \approx 2$, an unevolved $L_\star$ galaxy with SED typical of an H II galaxy would appear to be about 3 mag fainter than the hosts we have detected. The star formation rates involved would be of the order of a few hundreds of solar masses per year.

Galaxies as luminous as the extensions detected here have already been found in the imaging survey of lobe-dominated radio-loud QSOs carried out by Lehnert et al. (1992). Four of the objects of their sample, with similar redshifts to those in our sample, show 'fuzz' around the PSF of the nucleus. In the observed R frame the absolute magnitude of this 'fuzz' ranges from -25.6 to -26.9 mag, as derived from the B and K colours they report, which compares to the -25.9 to -26.7 mag we found in our study of two radio-quiet and one core-dominated radio-loud QSO. Our observations show that, if the hosts we have detected are indeed galaxies, young massive and luminous galaxies are not only a characteristic of radio-loud QSOs, but of QSOs as an entire class. Indeed, the only radio-loud QSO studied in this sample does not exhibit a significantly larger or a more luminous extension than those of radio-quiet QSOs. One of the radio-quiet QSOs exhibits no significant evidence for any extension in any band.

Acknowledgments: This work was supported in part by the 'Formation and Evolution of Galaxies' network set up by the European Commission under contract ERB FMRX-CT96-086 of its TMR programme. IA also has been partly supported by the EEC HCM fellowship ERBCHBICT941023.

References

Aretxaga I., Boyle B.J., Terlevich R.J. 1995, MNRAS, 275, L27.
Aretxaga I., Boyle B.J., Terlevich R.J. 1997, MNRAS, in preparation.
Carlberg R., 1990, ApJ, 350, 505.
Bressan et al. 1994, ApJS,94,63.
Dickson R., Tadhunter C., Shaw M., Clark, N & Morganti R., 1995, MNRAS, 273, L29.
Fabian, A.C. 1989, MNRAS, 238, 41P.
Haehnelt M.G. & Rees, M., 1993, MNRAS, 263,168.
Lehnert M.D, Heckman T., Chambers K.C. & Miley G.K., 1992, ApJ, 393, 68.
Lowenthal J.D., Heckman T.M., Lehnert M.D. & Elias J.H., 1995, ApJ, 439, 588.
Schmidt M., Schneider D.P. & Gunn J.E., 1991, in ASP Conf. Ser. 21, p109.
Telles E. 1995, Ph.D. Thesis, Cambridge Univ.
Terlevich R.J. & Boyle, B.J., 1993, MNRAS, 262, 491.

Observing the Galaxy Environment of QSOs

Klaus Jäger[1], Klaus J. Fricke[1], and Jochen Heidt[2]

[1] Universitäts-Sternwarte, Geismarlandstraße 11, D-37083 Göttingen, Germany
[2] Landessternwarte Heidelberg, Königstuhl, 69117 Heidelberg, Germany

Abstract. We outline our recently started program to investigate the galaxy environment of QSOs, in particular of radio–quiet objects at intermediate redshifts.

1 Studies of QSO Environments

Galaxy environment studies of QSOs and of their host galaxies are important since tidal interaction with neighbouring galaxies is thought to be one fundamental mechanism for triggering AGN activity (e.g. Osterbrock (1993)). The search for galaxy density enhancements (clusters) and/or for the presence of close companions near QSOs in support of this scenario is only part of the story. In addition, one wants to look for qualitative differences in the environments associated with intrinsic QSO properties (e.g. host galaxy types, radio–loud/quiet), and for the evolution with redshift. E.g. previous studies (cf. Ellingson and Yee (1994), Yee and Ellingson (1993)) for $z < 0.6$ show basically that radio–loud QSOs tend to lie in galaxy clusters while radio–quiet QSOs are found in poorer environments. On the contrary, a few recent observations at $z > 1$ (Hutchings et al. (1995), Hutchings (1995)) show hints for both radio–loud and radio–quiet QSOs to reside in compact groups of galaxies. This seems to imply an evolution in density and composition of the cluster environment of radio–quiet QSOs at the intermediate redshifts. A careful comparison and interpretation of several studies of this kind seems necessary as principle difficulties exist for the observation and analysis of QSO environment data, e.g. the interpretation of different statistical methods for detecting and measuring galaxy clustering.

2 A Research at Intermediate Redshifts

There are nearly no data of QSO environments within $0.6 < z < 1$. We have started a program on the observation of fields around radio–quiet QSOs at these redshifts. This research is part of an extensive QSO environment study over a large redshift range. Our main goals are the detection of galaxy groups or clusters around the QSOs, an investigation into the density evolution of the environments, into the role of the host galaxy types and of gravitational interactions with companions for the development of QSO activity. The first step of our program is a deep imaging survey (multicolor broad/narrow–band) of fields around a large sample of QSOs to detect host galaxies, close companion- and cluster candidates. Beforehand all fields were checked for contamination by known foreground galaxy clusters, radio sources, bright foreground galaxies or stars. As a second step spectroscopy of selected targets is intended to verify the physical association of environment galaxies with the QSOs and to assess their state of activity. A preliminary analysis of our obtained data holds already some promising results. A comparison of R– and I– band observations of fields around the QSOs reveals evidence for the physical association of companion galaxies with some QSOs. While the galaxies are easily detected in the I–band the comparable R–band image shows these objects to be more diffuse and fainter. This is to be expected at redshifts $z > 0.75$ due to the restframe wavelength ranges covered by the selected broad–band filters (eg. in the R–band below the 4000 Å break). Fig. 1 shows e.g. the compact group of objects near the QSO 2249–0154 at $z = 0.83$ (I–band image, 50 min., Calar Alto 2.2m telescope). 11 objects within 20 arcsec or within a projected distance of about 300 kpc around the QSO can be seen (one being a likely foreground star). There are some other cases where we found close–companion galaxies or even cluster candidates revealing the typical shape and size of a cluster to be expected at this redshift. In summary, the first inspection of our data supports the presumption that radio-quiet QSOs reside in a variety of environments at intermediate redshifts.

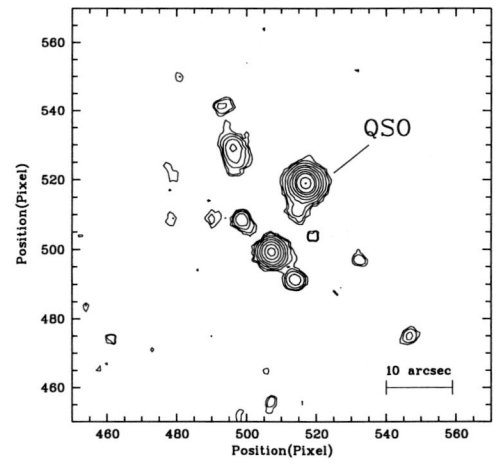

Fig. 1. The close environment of the radio–quiet QSO 2249–0154.(cf. text)

References

Ellingson, E., Yee, H.K.C. (1993): ApJ Suppl. **92**, 33
Hutchings, J.B., Crampton, D., Johnson, A. (1995): AJ **109**, 73
Hutchings, J.B., (1995): AJ **109**, 928
Osterbrock, D.E. (1990): ApJ 404, 551
Yee, H.K.C., Ellingson, E. (1993): ApJ **411**, 43

Limitations of Differential CCD Photometry Due to Weather Conditions

Dimitris Sinachopoulos[1], Alice Devillers[1], and Michael Geffert[2]

[1] Observatoire Royal de Belgique, Avenue Circulaire 3, B-1180 Bruxelles, Belgium
[2] Sternwarte der Universität Bonn, Auf dem Hügel 71, D-53121 Bonn, Germany

Abstract. We discuss the influence of weather conditions on the photometric accuracy of CCD observations. The accuracy seems to be inferior to the generally expected accuracy. We conclude that the minimum size of the emulated aperture and the method of the sky background determination have to be chosen very carefully.

1 Introduction

Since 1995 we have been performing a photometric monitoring of the gravitational lens QSO 0957+561 (twin QSO) (Sinachopoulos et al. 1996) mainly using the 1m Cassegrain telescope of the Hoher List Observatory (Germany) in combination with the 2kx2k CCD camera. The project aims mainly at the improvement of the determination of the time delay $\Delta t(AB)$. Since a monitoring programm requires a high coverage of observations over weeks to months it is for observations in central Europe important to test the use of clear but not always photometric nights.

In this paper we discuss the limitations of the accuracy of the differential CCD photometry caused by not optimal weather conditions.

2 Observations and Photometric Reduction

Observations of two nights, 28 April (1995) and 5 January (1996), were used for the present investigation. Details of the instrumentation can be found in Reif et al. (1994) and Sinachopoulos et al. (1996).

Observations of the first night were performed with a focal reducer and one pixel corresponds to 0.8 arcsec. In this dark, non photometric night, some faint clouds reflected the weak light pollution, creating a rather high and slightly inhomogeneous sky background. In addition, the illumination of the villages in the region of the Observatory create a permanent weak light pollution at Hoher List. Six exposures of five minutes each, taken in R Bessel filter, were added to one CCD frame. Both components of the gravitational lens (GL) were well exposed in this night, since we collected around 40000 ADUs per component.

Observations of the second night were taken directly in the cassegrain focus and have a pixel size of 0.4 arcsec. In this humid, full moon night the same filter as above was used. Ten exposures of three minutes were added.

The seeing in both nights was around 2.8 arcsec. The magnitudes of the lens components were R = 16.4 mags in the first and 17.1 mags in the second night.

For each of the local standard stars (Schild and Weekes, 1984) we collected a few hundred thousands ADUs, since they are much brighter than the GL.

For performing aperture photometry on the CCD frame the command MAGNITUDE/CIRCLE of ESO-MIDAS version NOV 93 (ESO Manual, 1994) has been used. The photometric values given below correspond to instrumental values corrected for zeropoint. The command MAGNITUDE/CIRCLE computes the magnitude of the specified object by integrating over the central area defined by a circular aperture. The local background as integrated over the given background area was subtracted. The command uses three concentric circles defined by the corresponding parameters $F_{siz}, N_{siz}, B_{siz}$ given in pixels. The innermost (flux) circle has a radius of $F_{siz}/2$ pixels. Flux from star plus sky is measured there. The sky background is measured in the area $\pi \cdot ((0.5 \cdot F_{siz} + N_{siz} + B_{siz})^2 - (0.5 \cdot F_{siz} + N_{siz})^2)$.

3 Sky Background Determination

In the following the influence of the sky background on the determination of the instrumental magnitude and differential photometry is discussed.

3.1 Sky Background Around a Bright Star

We selected one well exposed bright star in such a way that no other stars are present in its neighborhood on the CCD frame. Aperture photometry of this star using different diaphragm diameters was performed. We chose $N_{siz}=0$ and $B_{siz}=8$ arcseconds. A rms. of the magnitudes of 0.004 mags for diaphragms between 25 and 88 arcseconds was found. For diaphagms smaller than 25 arcseconds, light is lost from the star and added to the sky, which makes the star appear 0.04 mags fainter for a diaphragm of 16 arcsecs. Due to faint stars in the area of the star diaphragms larger than 88 arcseconds could not be considered.

On the frame of the first night, which was a dark one, we selected another well exposed bright star using the same criteria. We performed aperture photometry of this star using $N_{siz} = 0$, $B_{siz} = 0$ and an aperture with a diameter $F_{siz} = 25$ arcseconds. The magnitude of the star was determined five times by using the sky at several locations around the star. The rms. of the magnitude of this well exposed star was 0.007 mags using different sky measurements with a separation from the star up to 45 arcsec.

On the frame of the second night (full moon!) we repeated the same procedure for three well exposed stars. The rms. of the magnitudes of the stars fell to 0.03 mags. This improves slightly to 0.025 mags for taking sky backgrounds up to distances of only 27 arcseconds from the stars.

3.2 Sky Background Around the Gravitational Lens

We repeated the previous procedure for the GL on the CCD frame of the first night. This time, the sky was measured at nine different positions around the GL, at distances always smaller than 45 arcsecs. As already mentioned above the GL was well exposed, since we collected around 80000 ADUs for both components. Nevertheless, the rms. of the GL magnitudes was 0.017 mags per sky measurement. One explanation for this may be the inhomogenenity of the background around the GL. As HST deep sky images show, there are many faint galaxies in the field close and around this twin QSO. Although they are not visible on our frames, it is possible that they contribute to a less accurate sky determination.

We repeated the previous procedure for sky background determinations at larger distances from the GL using the CCD frame of the first night. Rough distances (x,y in arcmin) of these fields from the GL are given together with the corresponding angular separation and the magnitude determined for the complete GL are listed as well.

Table 1. Instrumental magnitudes depending on different sky backgrounds

(x,y) [arcminutes]	ρ	R [magn.]	(x,y) [arcminutes]	ρ	R [magn.]
(-1.0, 0)	1	15.55	(3.5,10)	11	15.68
(-1.0, 5)	5	15.41	(5.0, 0)	5	16.02
(-1.0,10)	10	15.48	(5.0, 5)	7	15.65
(1.5, 0)	1	15.62	(5.0,10)	11	15.83
(1.5, 5)	5	15.53	(7.0, 0)	7	16.03
(1.5,10)	10	15.85	(7.0, 5)	9	15.98
(3.5, 0)	3	16.01	(7.0,10)	12	16.54
(3.5, 5)	6	15.59			

A magnitude variation of more than 1 mags with a rms. of 0.3 mags is seen in our table! Even within a distance of 3 arcmin from the GL we found strong background variations causing differences of magnitude estimations up to 0.46 mags!

4 Differential Photometry

Differential CCD photometry of two star pairs from Schild and Chofin (1986) was performed using the ten frames of the second night with exposures of 3 minutes each. The accuracy of their local standard star photometry is 0.03 mags. We used a diaphagm of $F_{siz} = 39$, $N_{siz} = 0$, $B_{siz} = 4$ arcseconds.

For the stars No. 5 and No. 3 with $\Delta R = 0.24$ mags and angular separation of 70 arcseconds we obtained a magnitude difference $\Delta R = 0.22 \pm 0.05$ mags per exposure, or ± 0.01 mags for the combined frame. For star No. 5 and a brighter one at a distance of 130 arcseconds a value of $\Delta R = 0.54 \pm 0.04$ mags per exposure, or ± 0.01 mags for the combined frame was found.

The instrumental magnitude determination of each star on each exposure had an rms. of 0.03 mags corresponding to 0.01 mags in ten exposures. This indicates the photometric quality of the night as well.

5 Conclusion

- For a slightly light polluted sky, we found an optimal diaphragm diameter of about 25 arcseconds.
- A photometric accuracy of 0.007 mags in a non photometric night may be achieved for a well exposed star, if the sky background is measured at a distance smaller than 50 arcseconds from the star. Sky background determinations at higher angular separations lead to higher uncertainties.
- For faint stars, even not visible objects like nebulae and galaxies in the area may produce systematic irregularities in the sky background, decreasing the photometric accuracy and introducing systematic errors.
- Magnitude differences of well exposed stars with an angular separation of up to two arcminutes can be estimated in medium quality nights with an accuracy of 0.01 mags or better.

Acknowledgements.

This research was mainly carried out in the framework of the project "Service Centres and Research Networks", initiated and financed by the "Belgian Federal Scientific Services (DWTC/SSTC)".

References

ESO-MIDAS (1994): Users Guide, Vol. A, European Southern Observatory
Reif, K. et al. (1994): AG Abstr. Ser. 10, 243
Schild, R.E., Weekes, T. (1984): ApJ **277**, 481
Schild, R.E., Cholfin, B. (1986): ApJ **300**, 209
Sinachopoulos, D. et al. (1996): IAU Symp. **No. 173**, 53-54
Kippenhahn, R., Weigert, A. (1990): *Stellar Structure and Evolution* (Springer, Berlin, Heidelberg), 68–76

The Host Galaxy of HE 1029−1401

Lutz Wisotzki

Hamburger Sternwarte, Gojenbergsweg 112, D-21029 Hamburg, Germany
email: lwisotzki@hs.uni-hamburg.de

1 Introduction

I present observations of the low-redshift QSO HE 1029−1401, discovered 1991 in the course of the Hamburg/ESO survey (Wisotzki et al. 1991). With $V = 13.9$ it is one of the brightest QSOs in the sky, and with $z = 0.085$, i.e. $M_V = -25$, the most luminous known QSO at $z < 0.1$.

2 Image Analysis

The data shown here were collected with EFOSC1 at the ESO 3.6 m telescope. A 150 sec exposure R band direct image was taken at $1\rlap{.}''3$ seeing. The host galaxy of HE 1029−1401 is so prominent that no image processing is needed to unambiguously detect it (Fig. 1). The azimuthally averaged surface brightness profile is well fitted by a de-Vaucouleur spheroidal model with a half-light radius of 6.9 kpc and an integrated total magnitude of $R = 16.0$, resulting in a luminosity of $M_R = -22.9$, including Galactic extinction but neglecting K

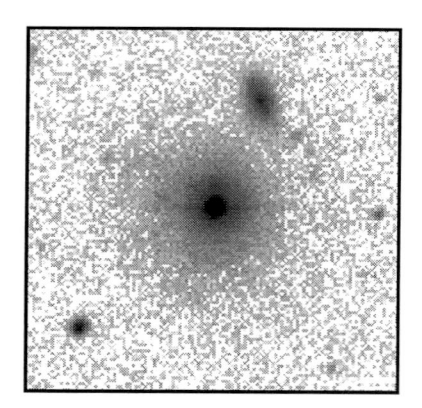

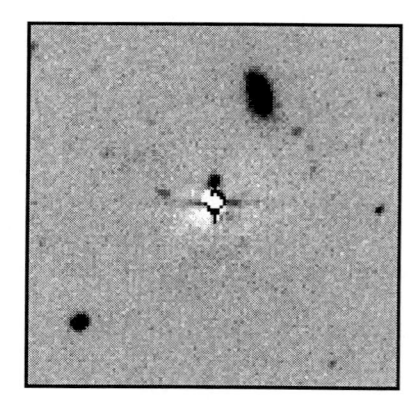

Fig. 1. Left: R band image in logarithmic gray scale representation. The image size is $84'' \times 84''$. The nearby bright galaxy is a background object. Right: Same image after subtraction of point source and model host galaxy, in linear gray scale.

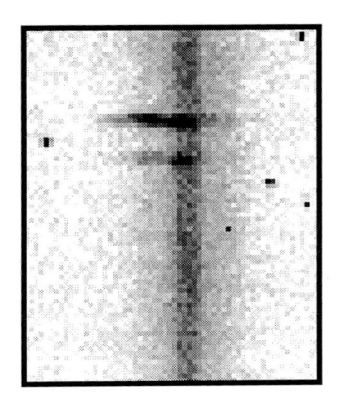

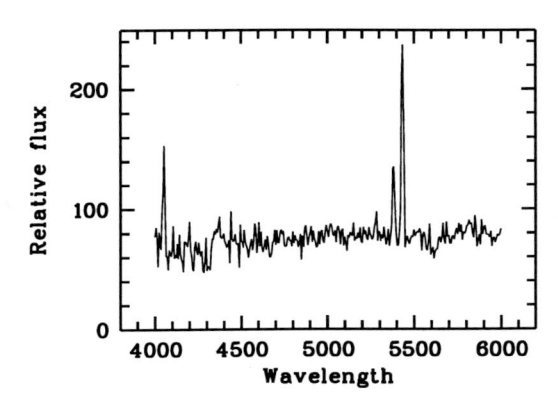

Fig. 2. Left: Section of the off-nuclear long slit spectrum around Hβ and [O III] after subtraction of model for scattered nuclear light. Right: Extracted and calibrated host galaxy spectrum.

corrections. In a second step, two-dimensional models of both the scaled PSF and the elliptical galaxy were subtracted from the image. (see Fig. 1). Three features require attention: The compact companion 4″ north of the nucleus; the asymmetric residuals, indicating deviations of the host galaxy structure from azimuthal symmetry; and a strange weak bow-shaped feature about 10″ south of the centre.

3 Off-Nuclear Spectroscopy

Spectra were also obtained with EFOSC1; a 1 min 'snapshot' of the QSO nucleus was followed by a 40 min exposure with the slit offset by 3″ to the north. Resolution was ~ 20 Å. To construct a long-slit spectrum free from nuclear contribution, a model image of the scattered nuclear light was subtracted (Fig. 2). In the extracted one-dimensional spectra no broad emission lines are present, indicating that the deblending was successful. Several stellar absorption features at the expected redshifted wavelengths are clearly visible: Ca II H+K, G-band, Mg $\lambda 5175$, Na D. The continuum seems to be much bluer than expected for a late-type stellar population. The strong decentred [O III] emission blob is probably excited by the QSO nucleus, as the line ratio [O III]/Hβ is ~ 8 while [O II] $\lambda 3737$ is still very strong, typical for a non-stellar exciting continuum. I conclude that while the overall appearance of this host galaxy is extremely smooth and regular, a closer investigation reveals many indications of perturbation, possibly caused by a recent merger event.

References

Wisotzki L., Wamsteker W., Reimers D., 1991, A&A 247, L17

Mount Teide and strange rocks

Part 3

STAR FORMATION AND THE ISM IN QUASAR HOSTS

Molecular Gas in Quasar Hosts

Richard Barvainis

MIT Haystack Observatory, Westford MA USA 01886

Abstract. The study of molecular gas in quasar host galaxies addresses a number of interesting questions pertaining to the hosts' ISM, to unified schemes relating quasars and IR galaxies, and to the processes fueling nuclear activity. In this contribution I review observations of molecular gas in quasar hosts from $z = 0.06$ to $z = 4.7$. The Cloverleaf quasar at $z = 2.5$ is featured as a case where there are now enough detected transitions (four in CO, and one each in CI and HCN) to allow detailed modeling of physical conditions in the molecular ISM. We find that the CO-emitting gas is warmer, denser, and less optically thick than that found in typical Galactic molecular clouds. These differences are probably due to the presence of the luminous quasar in the nucleus of the Cloverleaf's host galaxy.

1 Introduction

Prior to the flight of the IRAS satellite in the mid-1980s, it was assumed by most workers in the field that quasar infrared emission was nonthermal in character, for both radio quiet and radio loud quasars. The broad-band view of the IR provided by IRAS showed that the continua were not necessarily well-described by a simple power law, although in some instances such a description could suffice. The question of whether the IR might rather be dust emission was then brought to the fore, with many implications for the IR itself, and also for the nature of the energetically dominant optical/UV Big Blue Bump. One potential discriminant between the nonthermal (synchrotron) and dust models would be the presence of CO emission commensurate with the far-IR strength, and this test was what first led some of us to look for CO from quasars back in the late '80s.

CO was in fact detected in amounts expected from the dust model, and this provided one avenue of evidence among many leading to the current widely accepted view that emission from dust is the dominant IR process in most radio quiet and normal (i.e., non-blazar) radio loud objects (see Barvainis 1992 for a review). This having been established, there are a number of other interesting and important questions that can be addressed by the study of molecular gas in quasars. They include:

 - How is the host ISM affected by the presence of the AGN?
 - Do quasar hosts have enhanced ISM as a result of interactions and mergers?
 - How are quasars and luminous IR galaxies related, and how are they fueled?
 - What are the properties of star formation at high redshift?
 - And, speculatively, can we use CO to measure redshifts of obscured quasars at ultra-high z?

Here I review observations to date of molecular gas in quasar hosts and discuss some of what has been learned, with a focus on the Cloverleaf quasar where we now have enough transitions to perform detailed modeling of the molecular gas within the inner few hundred parsecs of the active galactic nucleus.

2 Overview of Observations

2.1 Low Redshifts

The molecular species with the strongest lines at radio wavelengths is CO, a diatomic molecule with rotational transitions spaced by 115 GHz. Initial searches focussed on the $J = 1 \to 0$ transition in low redshift quasars detected at 60 and/or 100 μm by IRAS. CO had already been seen in abundance in luminous IR galaxies (now believed by many to harbor hidden quasar nuclei that have been both activated and obscured by processes accompanying galactic interactions and mergers - see Sanders and Mirabel 1996 for a review).

Sanders, Scoville, & Soifer (1988) reported detection of CO(1–0) emission from the infrared-excess quasar Mrk 1014 at $z = 0.16$, with a total molecular gas mass $M_{H_2} \sim 4 \times 10^{10} M_\odot$ (using a conversion factor from CO luminosity to H_2 mass appropriate for Galactic molecular clouds; however, this value may be too large for IR galaxies and quasars – see below). Barvainis, Alloin, & Antonucci (1989) detected strong lines in both CO(1–0) and CO(2–1) from the $z = 0.06$ quasar I Zw 1, which showed double-horned profiles very similar to the HI profile from this system. Sanders et al (1989) detected CO(1–0) from the IR-discovered quasar IRAS 07598+6508. CO detections from several other low-z quasars were reported by Alloin et al (1992), lending credence to the dust model for quasar IR emission. In contrast to the objects mentioned so far, which are all radio quiet, Scoville et al (1993) detected CO from 3C48, a radio loud quasar at $z = 0.3$ with a massive elliptical host showing signs of a recent merger (Stockton and Ridgway 1991; Barvainis 1993).

2.2 High Redshifts

The first claimed detection of molecular gas at high redshifts was the CO(3–2) line from IRAS F10214+4724, an ultraluminous IR galaxy at $z = 2.28$, by Brown and Vanden Bout (1991). Although the original report turned out to be incorrect (Radford et al 1996), a follow-up study by Brown and Vanden Bout (1992) did correctly measure a (weaker) line in CO(3–2), and also a line in CO(4–3); the CO(3–2) line was confirmed at many telescopes world-wide (see Radford et al for references). CO(6–5) has been detected as well, by Solomon, Radford, & Downes (1992), but a claimed detection of CO(1–0) by Tsuboi & Nakai (1994) was not confirmed by Barvainis (1995).

Today we know two important facts about F10214+4724 which were not known in 1992: It is gravitationally lensed (e.g., Broadhurst and Lehar 1995), and it contains a hidden quasar in its nucleus (Goodrich et al 1996). This makes it similar in several ways to the second high redshift object to be discovered in CO, the Cloverleaf quasar at $z = 2.56$ (Barvainis et al 1994), to which we turn in the next section for a more detailed look. After accounting for lensing magnification, current estimates for the molecular gas mass in F10214+4724 have been scaled down considerably relative to earlier reports, but are still fairly large: $M_{H_2} \sim 2 \times 10^{10} M_\odot$ [using the Galactic value of $M_{H_2}/L'_{CO} = 4.0 M_\odot$ (K km s^{-1} pc^2)$^{-1}$]. This estimate is based on interferometric observations by Downes, Solomon, & Radford (1995), who also find that the bulk of the CO is confined to the inner part of the galactic nucleus ($r_{CO} \leq 400 - 800$ pc). These authors conclude that this object is not very different from luminous IR galaxies in the local Universe.

At the very highest redshifts known, a very recent and remarkable discovery has been made of CO(5–4) emission from the quasar BR1202–0725, at $z = 4.7$. Simultaneous papers in Nature by Ohta et al (1996) and Omont et al (1996) confirm the reality of the CO(5–4) detection, while Omont et al also show a probable detection of CO(7–6). The Omont et al observations, with $2''$ resolution, were able to resolve the CO source into two spatially distinct components, separated by $4''$. One component is precisely coincident with the optical quasar, but the second component has no optical counterpart on deep HST images. This second component is either a gravitationally lensed image in which the optical part is obscured by passage through a dust cloud in the (as-yet-undetected) lensing galaxy, or it could be a real, optically-weak, molecular companion. The latter interpretation is by far the more interesting one, for reasons I will touch upon in the final section of this review.

2.3 No Redshifts

All of the quasars detected in CO so far have been selected for having either far-IR detections from IRAS, or submillimeter detections from ground-based observations in the 400–1300 μm region. The submm dust spectrum is rising so steeply toward shorter wavelengths that at high redshifts ($z \gtrsim 1$) the favorable K-correction keeps up with geometrical dilution, so that more distant sources are no harder to detect in the submm continuum than nearer ones. Detectable far-IR/submm flux (using current instruments) seems to be a necessary but not sufficient condition for CO detections, and is the only known selection criterion for CO that does in fact work at least part of the time.

Searches based on other criteria have not been successful. For example, Barvainis & Antonucci (1995) selected most of the known $z \approx 4$ quasars for which the 115 GHz CO(1–0) line frequency is shifted into the VLA 22–24 GHz band. They failed to obtain detections at typical levels of $L'_{CO} \lesssim 10^{11}$ K km s^{-1} pc^2 for 10 objects. Evans et al (1996) derived limits up to an order of magnitude lower in some cases for a sample of 11 powerful radio galaxies lying in the redshift range $1 < z < 4$, only one of which had a thermal far-IR/submm detection. Brown & Vanden Bout (1993) reported detection of CO(3–2) emission at $z = 2.14$ from a damped Lyman–α absorption system towards the quasar Q0528–250, and Frayer et al (1994) reported CO(1–0) and CO(3–2) from a DLA at $z = 3.137$. Both of the CO(3–2) DLA results were refuted by more sensitive observations, in which a number of other DLA systems were also searched for CO and not detected (Wiklind & Combes 1994; Braine, Downes, & Guilloteau 1996).

We should not leave this section on nondetections without mentioning that the redshift to be used for tuning one's receiver for molecular line observations of quasar hosts should be based on LOW IONIZATION broad optical emission lines, e.g. Hα, Mg II, etc., or narrow forbidden lines like [OIII]. 'High ionization' lines such as Lyman–α or CIV can be blueshifted with respect to systemic by as much as several thousand km s^{-1}, and tunings based on such lines will often result in incorrect placement of narrow mm-wavelength spectrometers.

3 The Nuclear ISM in the Cloverleaf Quasar

The Cloverleaf is a broad absorption line quasar at $z = 2.56$, gravitationally lensed into four spots in the optical with separations of about $1''$. The total magnification factor is estimated to be ~ 10.

Our initial search for molecular gas in the Cloverleaf was motivated by a detection of strong submm flux during a survey of BALQs using the JCMT (Barvainis, Antonucci, & Coleman 1992). After two missed attempts at CO(3–2) using the IRAM 30m telescope, caused first by a bad redshift (based on Lyman–α, which in this object is blueshifted by ≈ 900 km s^{-1} relative to systemic), and then by a mistuned receiver, a strong detection was finally obtained at the IRAM Plateau de Bure Interferometer and confirmed using the IRAM 30m telescope (Barvainis et al 1994). From follow-up observations at the 30m we now have three additional CO transitions ($J = 4 \to 3$, $5 \to 4$, and $7 \to 6$), a detection of the neutral carbon fine-structure transition CI($^3P_1 - {}^3P_0$), and a probable detection of HCN(4–3).

Given the four observed CO lines, it is possible to proceed with detailed modeling of the CO-emitting gas based on the line ratios and other constraints. The line brightness temperature ratios relative to the strongest line [CO(4–3)] are: $(3-2)/(4-3) = 0.83\pm0.16$, $(5-4)/(4-3) = 0.73\pm0.16$, and $(7-6)/(4-3) = 0.68 \pm 0.13$. The roughly constant T_B from (3–2) to (4–3), followed by a falloff in the higher-J transitions, suggests that the optical depths in the CO lines are only modest ($\tau_{4-3} \lesssim 3$), and that the gas is relatively warm ($T \gtrsim 100$ K) and dense ($n_{H_2} \gtrsim 3 \times 10^3$ cm^{-3}). The gas may be heated by X-rays from the quasar, and the optical depths in each transition are lowered because of the distribution of population over many states at high temperature. These conclusions are based on escape probability modeling by Phil Maloney, and are presented along with the line measurements in Barvainis et al (1997).

Because of the high temperature and low optical depth of the gas, it has a much higher emissivity per unit mass than is typical of Galactic molecular clouds. This means that application of the standard Galactic L'_{CO} to M_{H_2} conversion factor [4 $M_\odot$ (K km s^{-1} pc^2)$^{-1}$; e.g., Radford, Solomon, & Downes 1991] would overestimate the total molecular mass, according to our model, by about an order of magnitude. Our derived mass for the Cloverleaf is $M_{H_2} \approx 2 \times 10^{10} m^{-1} h^{-2} M_\odot$, where m is the lensing magnification factor and h is Hubble's constant in units of 100 km s^{-1} Mpc^{-1}. For $m = 10$ and $h = 0.7$ this represents a relatively modest mass of roughly $4 \times 10^9 M_\odot$, based on the *observed* CO emission (it is possible that there is significant mass outside the X-ray heated zone that would escape detection because the gas is cool and optically thick).

A conflict has been noted in luminous IR galaxies between the molecular mass derived from mm-wavelength CO emission line observations, and the dynamical mass derived from central stellar velocity dispersions (obtained from near-IR CO absorption bands). Shier, Rieke, & Rieke (1994) find that there must be 4–10 times less molecular mass in NGC 1614 and IC 694 than estimated using the Galactic conversion factor. A factor of 10 below the Galactic value is, perhaps not coincidentally, the conversion we derived for the Cloverleaf.

Is the molecular mass in the Cloverleaf consistent with the dynamical mass? We have recently obtained a high resolution (0.5″) image of CO(7–6) and find an upper limit to the diameter of image C of the quad of $< 0.25''$ (Alloin et al 1997, A&A, in press), yielding an intrinsic source radius of $< 2.4 m^{-1} h^{-1}$ kpc ($q_0 = 0.5$). For $m = 10$ and $h = 0.7$, this gives $r < 0.34$ kpc. Combining this source size with the CO line width of 375 km s^{-1} yields

$$M_{\rm dyn} \approx \frac{r \Delta V_{\rm FWHM}^2}{G \sin^2 i} \lesssim 2 \times 10^{10} M_\odot,$$

where the inclination i has been taken to be 45°. This is comfortably above the derived H$_2$ mass of $4 \times 10^9 M_\odot$ (for the same m and h).

To summarize this section, the detectability of the Cloverleaf in molecular and neutral atomic transitions is largely due to high emissivity of the gas and magnification from gravitational lensing, rather than an extremely large mass of gas.

4 The Connection Between Quasars and ULIRGs

The idea that the impressive infrared luminosities ($L_{\rm IR} \gtrsim 10^{12} L_\odot$) of ultra-luminous IR galaxies (ULIRGs) might be supplied, or at least augmented, by a hidden quasar goes back to shortly after their discovery by IRAS (e.g., Antonucci & Olszewski 1985). Sanders et al (1988) expanded this idea to encompass an evolutionary link between ULIRGs and quasars, with an interaction or merger between two galaxies driving a large amount of gas and dust into the galactic nucleus (e.g. Barnes & Hernquist 1996), creating a quasar by providing fuel for a resident supermassive black hole. The quasar would be hidden at first, and the object would appear as a ULIRG. The IR luminosity might be supplied by accompanying starbursts and reprocessing of the AGN UV/X-ray emission via dust. Later, as the gas and dust were consumed or dispersed, a classical optical quasar would emerge.

The gas and dust phase, lasting from before the emergence of the quasar until sometime after, should produce strong molecular line emission. Indeed, ULIRGs do of course show copious CO emission (see Sanders & Mirabel 1996), as do quasars with strong far-IR emission (see above). In this context, IRAS F10214+4724 and the Cloverleaf make a very interesting pair for comparison. Their redshifts are similar, their IR spectra are almost identical, and their CO emission properties are very similar. The two objects differ primarily in the optical/UV region, where the Cloverleaf shows a fairly typical (perhaps slightly reddened) big blue bump, and broad emission lines, whereas F10214+4724 has a weak and steep optical continuum and a Seyfert 2 spectrum in unpolarized light. However, in *polarized* (i.e., scattered) light, F10214 shows broad emission lines (Goodrich et al 1996), confirming that a hidden quasar lies in the nucleus (and can probably be seen unobscured from some other vantage point).

In comparing these properties we (Barvainis et al 1995) have suggested that F10214 and the Cloverleaf may in fact be the same type of object. Both would be emerging from the obscured phase, with a thick and extensive disk or torus of material still present in the nucleus. The differences in their observed properties are likely due to orientation effects: we see the Cloverleaf from above the torus and F10214 from the side. The nuclear light in F10214 is probably scattered by electrons, or perhaps dust clouds, located in the opening of the torus, and is then reddened by passage through the upper parts of the torus. The observed molecular line emission originates in this still large reservoir of dust and gas, evident via other means in both objects. In this general picture of ULIRGs and quasars, molecular gas plays a critical role as the primary fuel for the nuclear activity, both for starbursts and for the AGN.

5 Future Prospects

High redshift molecular line observations strain current instrumentation to its limits, but the recent spectacular detection of CO at $z = 4.7$ shows that continued searches are certainly worth pursuing. Of the three known CO-emitting objects at $z > 0.5$, two (F10214+4724 and the Cloverleaf) are definitely lensed, while the third (BR1202–0725) has a double CO source, with both CO components having submm counterparts but only one having an optical counterpart. Determining whether this object is lensed will be difficult but critical to evaluating the amount of molecular mass in high redshift systems, and the future potential for observing CO and other molecules at high ($z = 1 - 5$) and possibly ultrahigh ($z > 5$) redshifts. Perhaps molecular studies can currently only be done on magnified systems – which will limit the number of candidates, but will not alter the method of selection by submm continuum, since that component will be magnified too.

If, however, BR1202–0725 is not lensed, it may represent a population of extremely gas rich systems in the early Universe. The companion CO source is particularly intriguing in this regard because it has no optical signature, raising the possibility that an abundant population of dusty molecular objects could have heretofore escaped detection. Is the sharp falloff in the optical quasar population for $z \gtrsim 3$ a real decline, or are many early quasars simply obscured by the products of massive, galaxy-forming starbursts? Such objects might appear observationally very much like the ultraluminous IR galaxies in the local Universe, many of which are known to host hidden quasars (e.g., Wills et al 1992; Hines et al 1995).

One argument that has been made against the obscured quasar hypothesis is based on the near-complete optical identification of radio-selected samples (e.g., Shaver et al 1996), but this conclusion applies mainly to obscuration by intervening absorbers along the line-of-sight, and does not appear to strongly constrain internal absorption. Furthermore, we know that radio-loud objects are found in elliptical hosts (as was frequently confirmed at this meeting), whereas local ULIRGs are predominantly radio quiet objects in heavily disturbed systems. The proposed hidden quasars at high z might well be similar to the latter, and such systems may simply present the wrong type of environment for the production of powerful radio sources. Therefore arguments based on identification of radio loud samples might not apply. It should be noted that Webster et al (1995) have asserted, based on a wide range in B–K colors found in a radio-selected sample, that up to 80% of radio loud and radio quiet quasars may be reddened enough to be missed by the usual optical/UV search methods. However, this result has been disputed as being due to effects other than reddening by dust (Sergeant & Rawlings 1996; Boyle & di Matteo 1995; Srianand 1997, this volume; Gonzalez–Serrano 1997, this volume), and it appears the jury is still out on this point.

If there *are* dusty molecular optically-obscured quasars at ultra-high redshift, how might we find them? Random-field surveys for submm sources are being planned for the new SCUBA camera on the JCMT, and these surveys may turn up a candidate population based on crude spectral energy distributions (3–4 points). But without optical emission lines, it will be difficult to obtain accurate redshifts. An alternative might be to determine redshifts via CO observations. At high redshifts the CO rotation ladder is compressed in frequency by the factor $1/(1+z)$, so that for say $z = 8$ there would be a CO line every 12 GHz. Identifying two such lines would uniquely provide the object's redshift. Such an approach, however, would be impractical using current instruments. The next significant advance in CO-detection technology will be the 100m Green Bank Telescope (GBT), scheduled for completion in 1998. In the 35–50 GHz range it will be able to go several times deeper than the IRAM 30m telescope or the PdBI and OVRO interferometers can at 3mm (and if, at some point in the future, the GBT is able to operate at 80–100 GHz as planned, it will be even more sensitive). Additionally, the GBT will have a wide instantaneous bandwidth of at least 2 GHz, so that stepping through frequency in a search for lines will be more efficient than possible now.

The combination of SCUBA and the GBT should prove powerful for molecular studies of known high redshift quasars as well. SCUBA detections in the submm are likely to provide a good selection of candidates with known redshifts for deep integrations by the GBT. A new 50m telescope for 3mm and 1mm operation, the Large Millimeter Telescope (LMT), is being built by a US/Mexico collaboration, and should be an impressive high-z CO machine when finished in

~ 5 years. In the longer term (~ 10 years) the next step in sensitivity will come with the new millimeter arrays being planned by the US and Japan (especially if combined), and by Europe, which, by observing at 2mm or 1mm, will beat the GBT by factors of several. This progress should be just about enough to match the lensing advantage enjoyed by objects like F10214+4724 and the Cloverleaf, thereby allowing molecular studies of high redshift quasars and galaxies having luminosities comparable to those of 'ordinary' ultraluminous IR objects found in the local universe.

References

Antonucci, R., & Olszewski, E.W. (1985): AJ, 90, 2203.
Alloin, D., Barvainis, R., Gordon, M.A., & Antonucci, R. (1992): A&A, 265, 429.
Barnes, J.E., & Hernquist, L. (1996): ApJ, 471, 115.
Barvainis, R., Alloin, D., & Antonucci, R. (1989): ApJ, 337, L69.
Barvainis, R., Antonucci, R., & Coleman, P. (1992): ApJ, 399, L19.
Barvainis, R. (1992): In *Testing the AGN Paradigm*, ed. S.S. Holt, S.G. Neff, & C.M. Urry, p. 129. New York: AIP Conference Proceedings.
Barvainis, R. (1993): Nature, 366, 298.
Barvainis, R., Tacconi, L., Alloin, D., Antonucci, R., & Coleman, P. (1994): Nature, 371, 586.
Barvainis, R. (1995): AJ, 110, 1573.
Barvainis, R., & Antonucci, R. (1995): PASP, 108 187.
Barvainis, R., Antonucci, R., Hurt, T., Coleman, P., & Reuter, H.-P. (1995): ApJ, 451, L9.
Barvainis, R., Maloney, P., Alloin, D., & Antonucci, R. (1997): ApJ, in press.
Boyle, B.J., & di Matteo, T. (1995): MNRAS, 277, L63.
Braine, J., Downes, D., & Guilloteau, S. (1996): A&A, 309, L43.
Broadhurst, T., & Lehar, J. (1995): ApJ, 450, L41.
Brown, R.L., & Vanden Bout, P.A. (1991): AJ, 102, 1956.
Brown, R.L., & Vanden Bout, P.A. (1992): ApJ, 397, L19.
Brown, R.L., & Vanden Bout, P.A. (1993): ApJ, 412, L21.
Downes, D., Solomon, P.M., & Radford, S.J.E. (1995): ApJ, 453, L65.
Evans, A.S., et al (1996): ApJ, 457, 658.
Frayer, D.T., Brown, R.L., & Vanden Bout, P.A. (1994): ApJ, 433, L5.
Goodrich, R., et al (1996): ApJ, 456, L9.
Hines, D., et al (1995): ApJ, 450, L1.
Ohta, K., et al (1996): Nature, 382, 426.
Omont, A., et al (1996): Nature, 382, 428.
Radford, S.J.E., Solomon, P.M., & Downes, D. (1991): ApJ, 369, L15.
Radford, S.J.E., Downes, D., Solomon, P.M., Barret, J., & Sage, L.J. (1996): AJ, 111, 1021.
Sanders, D.B., Scoville, N.Z., & Soifer, B.T. (1988): ApJ, 335, L1.
Sanders, D.B., et al (1988): ApJ, 325, 74.
Sanders, D.B., et al (1989): A&A, 213, L5.
Sanders, D.B., & Mirabel, I.F. (1996): ARAA, 34, xx.
Scoville, N.A., Padin, S., Sanders, D.B., Soifer, B.T., & Yun, M.S. (1993): ApJ, 415, L75.

Serjeant, S., & Rawlings, S. (1996): Nature, 379, 304.
Shaver, P.A., et al (1996): Nature, 384, 439.
Shier, L.M., Rieke, M.J., & Rieke, G.H. (1994): ApJ, 433, L9.
Solomon, P.M., Downes, D., & Radford, S.J.E. (1993): ApJ, 398, L29.
Stockton, A., & Ridgway, S.E. (1991): AJ, 102, 488.
Tsuboi, M., & Nakai, N. (1994): PASJ, 46, L179.
Webster, R.L., Francis, P.J., Peterson, B.A., Drinkwater, M.J., & Masci, F.J. (1995): Nature, 375, 469.
Wiklind, T., & Combes, F. (1994): A&A, 228, L41.
Wills, B., et al (1992): ApJ, 400, 96.

CO Emission Lines from a Quasar at z=4.7

T. Yamada[1], K. Ohta[2], R. Kawabe[3], K. Kohno[3], K. Nakanishi[2], and M. Akiyama[2]

[1] Astronomical Institute, Tohoku University, Aoba-ku, Sendai, 980-70, Japan
[2] Department of Astronomy, Faculty of Science, Kyoto University, Kyoto, 606-01, Japan
[3] Nobeyama Radio Observatory, National Astronomical Observatory, Minamimaki, Minamisaku, Nagano, 384-13, Japan

Abstract. We have detected a CO (J=5-4) emission line from the quasar BR1202-0725 at z=4.7 with the Nobeyama Millimeter Arrays. The estimated mass of the CO emitter is as large as 10^{11} $M_\odot$ (without considering unknown gravitational-lensing effects). It becomes clear that at least some quasars are closely related to the galaxy-formation events at high redshift. Here we review our results, comparison of our results to those by Omont et al. by IRAM arrays with higher spatial resolution, and discuss the possible history of the CO emitting objects.

1 CO Luminosity of Nearby Quasars

Before going to high redshift quasars, let us summarize the CO properties of nearby quasars. CO emission lines from several nearby quasars have been successfully detected (Barvainis and Antonucci 1989; Alloin et al. 1992; Scoville et al. 1993). CO luminosity of these quasars are very large, and probably, in the local Universe, the most luminous objects in CO emission lines are quasars. On the other hand, it is not clear whether quasars are always luminous in CO. Those quasars detected in CO are biased to those detected in far-infrared wavelength, although it does not have to mean that the sample is biased to those objects with FIR-excess; quasars are luminous in FIR in general although there is some significant scatter in their SED distribution (Elvis et al. 1994).

Yamada (1994) compiled the CO data for nearby quasars and Seyfert 1 galaxies which had been observed both in CO and X-ray wavelength (Figure 1a). X-ray emissions represent powers of active nucleus and they are little contaminated by star-formation components. CO emissions represent overall (dense) gas contents in the host galaxy, and, in general, star formation activity. They are not contaminated by AGN components. Thus, by studying the correlation between CO and X-ray luminosities, we can study the relation between star-formation activities and AGN activities. Note that those luminosities do not have to correlate with galaxy mass scale. It is known that nuclear luminosity does not correlate with host galaxy luminosity (McLeoad, this volume). Probably nuclear luminosity represents the accretion rate onto the massive black hole. It is directly related to only very small scale structure ($\lesssim$ 1pc) and small mass. Nevertheless, in Figure 1a, we see strong correlation between these quantities over the large luminosity range. Since the correlation is also seen for flux (Figure 1b), it must not be due to the detection-limit effect. It shows that the more powerful AGN are hosted in more CO luminous objects. To explain this we can consider indirect connections between accretion rate onto MBH and large-scale gas contents or star formation activities (Yamada 1994). The important aspect in Figure 1 is that the quasars

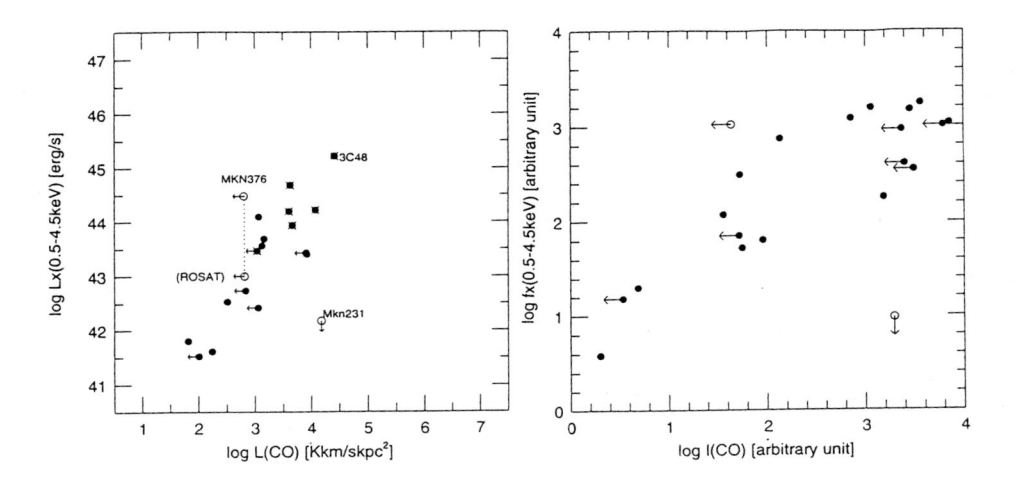

Fig. 1. Correlation between X-ray and CO luminosities (flux) of nearby Seyfert 1 galaxies and quasars.

as luminous as 10^{45} erg/s in X-ray has 10^{10} K km/s pc^2 in CO which corresponds to $\sim 5 \times 10$ $M_\odot$ in hydrogen molecule with conversion factor of 4.5 ($M_\odot$/[K km/s ps^2]). If the correlation holds in general and for high-reshift luminous quasars, the most luminous quasars may have host gas mass of $\gtrsim 5 \times 10^{12}$ $M_\odot$, which is comparable with the stellar mass of the most massive giant elliptical galaxies in the local universe.

2 Comparison Between NMA and IRAM Result

We observed the quasar BR1202-0725 at z=4.7 with Nobeyama Millimeter Arrays (NMA) and detected CO (J=5-4) emission line with 7 σ level. The details have already been published in Ohta et al. (1996). Omont et al. (1996) also detected CO (5-4) line independently with higher spatial resolution. See Omont et al. (1996) for the detail of their observations and results. Here we compare the two results and describe how they are consistent to each other.

In our NMA observation, a 10 mJy peak is detected at 101.19 GHz with 220 km/s width in FWHM. Besides this peak, 2σ (in 150km/s bin) emission is

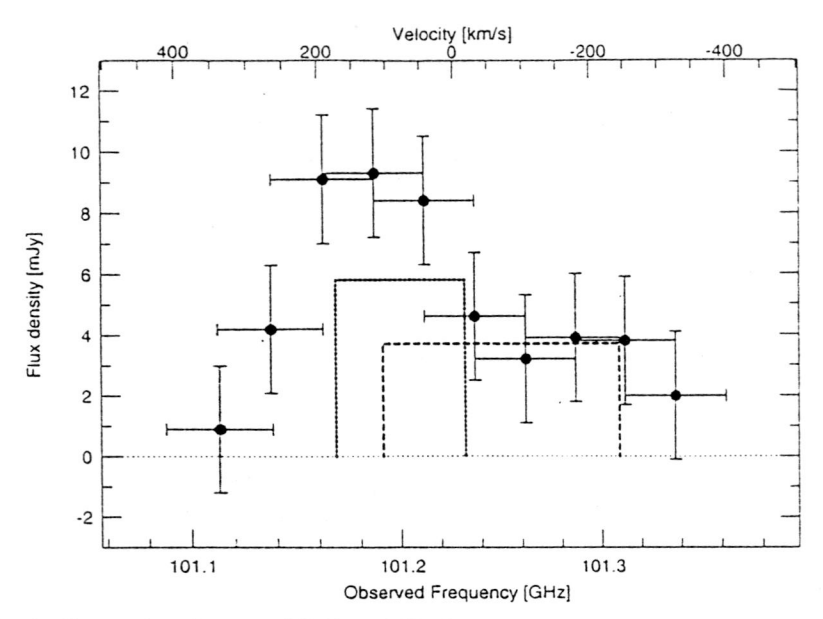

Fig. 2. Comparison between NMA and IRAM results. Solid histogram shows the NMA results and dotted and dashed line shows the IRAM two components schematically.

seen at higher frequency, up to 101.3 GHz (Figure 2). The emission is not resolved significantly, but the channel map suggest that this broad high-frequency component is marginally shifted to south-east of the peak at 101.19 GHz. We integrate the spectrum in frequency ranges centered at 101.19 GHz and found the maximum signal to noise ratio with 370 km/s velocity width. The integrated flux density is 2.7±0.4 Jy km/s.

Omont et al. (1996) resolved two peaks: one coincides with the optical position of the quasar and the other is located 5 arcsec northwest from the quasar (denoted by NW component). The quasar component is peaked at 101.20 GHz with peak flux density of 6 mJy and 190 km/s width (FWHM). Total flux of this component is 1.1 Jy km/s. The NW component is at 101.25 GHz and relatively broad but less bright, with 3 mJy, 350 km/s, and 1.3 Jy km/s.

Thus, the results of two groups are broadly consistent, especially; the components peaked at 101.2 GHz coincide if in the NMA observations we see the sum of the two components resolved with IRAM arrays at the frequency. However, there is also a small inconsistency. Although NMA results show the broad component with $I_{peak} \sim 4$ mJy up to higher frequency, too, the location on the sky seems different from the NW component in the IRAM Arrays' results; this might be due to the low S/N ratio of the component.

3 Physical Properties of CO Emitter Associated with BR1202-0725

Here we summarise the physical quantities of the CO emitters of BR1202-0725. The hydrogen molecular gas mass evaluated from the CO luminosity is, M(H$_2$) = 8.4 × 10^{10} h^{-2} (q_0 = 0.5) with the CO to H$_2$ conversion factor of 4.5, the value obtained for the CO (J=1-0) transition and for the giant molecular clouds in the Galaxy. The value of the conversion factor is highly uncertain. It may depends, at least, density, tempelature, and metallicity. If the CO emission is only partially optically thin (Barvainis, this volume) the value may be as small as ~ 1, but if the metallicity is lower the value becomes larger. Note that with q_0 = 0.5, the reduced Hubble constant should be as small as 0.5 to obtain the age of the universe compatible with globular-cluster ages, and it makes the mass larger. Even if the conversion factor is as small as 1, it is compensated by using h=0.5. Thus, we believe the molecular hydrogen mass is at least ~ 10^{11} $M_\odot$, nearly equal to the stellar mass of present-day luminous galaxies.

The CO emitter is not resolved with NMA. From the beam size, we evaluated the upper-limit of the size of the object as $R < 9h^{-1}$ kpc in radius assuming a spherical object. Each of the two component resolved in the IRAM observation also seems unresolved, and it constrains $R < 4h^{-1}$ kpc.

From this, with the evaluated gas mass of ~ 10^{11} $M_\odot$, we can evaluate the average gas density. From the NMA observations, the averaged gas density is constrained as $\rho > 3 \times 10^{-3} h$ $M_\odot$ pc^{-3}. From the IRAM observation, for each component, $\rho > 0.5 h$ $M_\odot$ pc^{-3}. This is compared with average density of virialized objects in the universe at z = 4.7, ρ_{vir} ~ 9 × 10^{-3} (δ_{vir} / 170) Ωh^2 $M_\odot$ pc^{-3}, or average density at maximum expansion, $\rho_{max-exp}$ ~ 3 × 10^{-4} ($\delta_{max-exp}$ / 5.6) Ωh^2 $M_\odot$. The obtained average density of the CO emitter of BR1202-0725 is higher than these values, which means the object had collapsed at higher redshift (z ~ 20), or it experienced contraction at least factor of several by cooling. Collapsing of such a massive object at z ~ 20 seems unexceptional, and we suggest that this may be the direct evidence of the contraction of the gas by cooling at high redshift.

We now consider how this object is typical or atypical in the universe. The estimated mass of 10^{11} $M_\odot$ is only that of the molecular gas component. Contribution of atomic gas and stellar mass (already formed) may increase a factor larger. Moreover, if we consider massive dark matter associated with gas component, the total mass of the object may be as large as 10^{12} $M_\odot$. These are compared with the collapsing redshift of halos with various mass scale and peak hight in the standard Cold Dark Matter model (e.g., Figure 2 of Haehnelt and Rees 1993). With some uncertainty of the normalization of the initial density fluctuation, the object collapsed at z=4.7 with 10^{12} $M_\odot$ corresponds to a $\gtrsim 4\sigma$ peak. If the collapsing redshift of the CO emitter is larger than 4.7, it is a further rarer object in the universe. If the mass of the host object of other quasars at high reshift are obtained in the future, we can thus constrain the galaxy-formation models by comparing the number densities of quasars and corresponding density peaks predicted by models.

4 Chemical Enrichment

Since we detected CO emission line, there must be substantial amount of metals in the objects. We consider about the chemical evolutional history of the object.

Pei and Fall (1995) studied the model of cosmic chemical evolution based on the metallicity informations of quasar absorption lines. If we extrapolate their model to z = 4.7, the averaged metallicity of the universe is at most 0.01 solar, strongly depending on the cosmological parameters. Wampler et al. (1996) studied the metallicity of absorption line systems toward BR1202-0725. There are a few systems whose redshifts are very close to the quasar. For the z=4.672 system, they reported that oxygen abundance could be larger than solar while carbon to oxygen abundance ratio is less than solar. For the Dampled Lyman-alpha system at z=4.4 the metallicity is estimated as 0.01 of the solar.

It is difficult to estimate the abundance of these metals only from the CO line observation. Since CO (5-4) line may be optically thick, we cannot in principle evaluate the total mass of carbon or oxygen in CO molecule. Also, since the conversion factor changes with metallicity, the estimated hydrogen mass cannot use to determine the metallicity. Nevertheless we are tempted to speculate the possible metal abundance and consider about the possible star-formation history. Assuming the optically *thin* case and thermal equilibrium at 80 K, we calculate the lower bound of the carbon mass as $\sim 10^6$ $M_\odot$. The true value may be 10-100 times of this value, namely $\sim 5 \times 10^7$ [F/50] $M_\odot$, where F is the unknown factor for the correction. The dynamical mass is $\sim 10^{11}$ $M_\odot$, so we tentatively adopted this value for the hydrogen molecular gas mass. Then the carbon abundance could be estimated as ~ 0.1 [F/50] [M(H)/10^{11} $M_\odot$]$^{-1}$ of the solar value. We can compare this crude estimation with chemical evolution models of galaxies. At z = 4.7, the age of the universe is ~ 1 Gyr (H_0 = 50 km/s/Mpc, q_0 = 0.5). For the so-called "solar-neighborhood model", namely those with steep IMF with long (~ 10 Gyr) time scale, the carbon abundance is at most ~ 0.1 solar at the age of 1 Gyr (Matteucci et al. 1992). On the other hand, for the "giant elliptical model", with flat IMF and short ($\lesssim 1$ Gyr) time scale it is easy to achieve 0.3 solar carbon abundance at age of 0.1 Gyr. Thus, although the argument may be too speculative, the star formation history like "giant elliptical models" seems to be favored for the CO emitter of BR1202-0725 considering the very young age of the universe.

5 Quasar Hosts and Forming Galaxies

The detection of luminous CO emission line from BR1202-0725 suggest the quasar activity is strongly related with galaxy formation phenomena. In Figure 4, we plotted the high redshift quasars observed in CO on the Lx-L$_{CO}$ diagram. BR1202-0725 is, roughly, located at the extension of the correlation senn for the low-z quasars and Seyfert 1s. The Cloverleaf also lie on the sequence although both the luminosities are amplified by the gravitational lens probably with different factor. If the correlation holds in general, quasars are as gas rich as expected for forming galaxies.

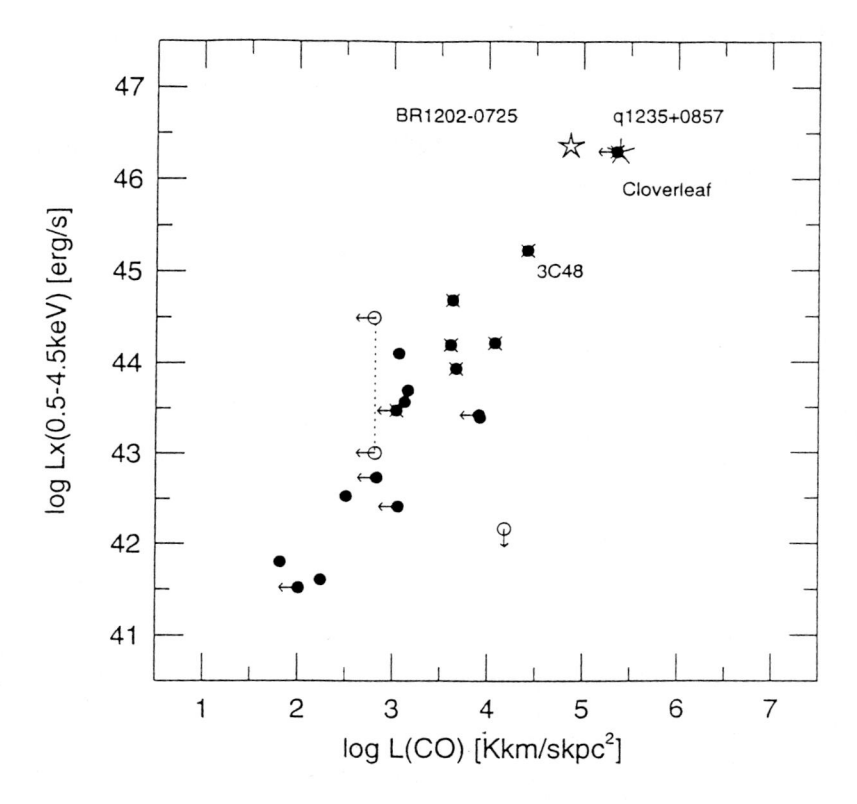

Fig. 3. L_x-L_{CO} correlation including high-redshift quasars.

References

Alloin, D. et al. (1992):Astron. Astrophys., **265**, 429.
Barvinis, R. and Antonucci, R. (1989):Astrophys. J. Suppl., **70**, 257.
Elvis, B. et al. (1996):Astrophys. J. Suppl., **95**, 1.
Haehnelt, M. and Rees, M. (1993):Mon. Not. Roy. astr. Soc., **263**, 168.
Ohta, K. et al. (1996):Nature, **382**, 426.
Omont, A. et al. (1996):NAture, **382**, 428.
Pei, Y.C. and Fall, S.M. (1995):Astrophys. J., **454**, 69.
Scoville, N.Z. et al. (1993):Astrophys. J., **415**, L75.
Wampler, J. et al. (1992):Astron. Astrophys., **316**, 33.
Yamada, T. (1994):Astrophys. J., **423**, L27.

Molecular Gas and Star Formation in I Zw 1

E. Schinnerer[1], A. Eckart[1], L.J. Tacconi[1], N. Thatte[1], N. Nakai[2], S.K. Okumura[2]

[1] Max Planck Institut für extraterrestrische Physik, Germany
[2] Nobeyama Radio Observatory, Japan

Abstract. We have investigated the ISM of the I Zw 1 QSO host galaxy. We have detected a circum-nuclear gas ring of diameter $\sim 1.5''$ (1.5 kpc) in its CO line emission and mapped the disk and the spiral arms of the host galaxy in the ^{12}CO(1-0) line at 115 GHz as well as in the H-band (1.65 μm). We derive an inclination for the disk of i = (38±5°) using the CO rotation curve and the R-band image (Hutchings & Crampton 1990). Knowing the inclination we obtain a dynamical mass of $(3.1\pm1.2)\times10^{10}$ M$_\odot$ and a cold molecular gas mass of $(5.5\pm1.1)\times10^9$ M$_\odot$. Together with an estimate of the stellar contribution to the mass and light from NIR spectroscopy we derive the $\frac{N_{H_2}}{I_{CO}}$-conversion factor to be comparable to $2\times10^{20}\,\frac{cm^{-2}}{K km s^{-1}}$ found for molecular gas in our galaxy and many nearby external galaxies.

1 Introduction and Observations

I Zw 1 at a redshift of z=0.0611 (Condon et al. 1985) is thought to be the closest QSO because of its high optical nuclear luminosity (M_B=-23.45^{mag} Schmidt & Green 1983) and because it shows spectral properties of high-redshift extragalactic nuclei (e.g. CIV (λ 1549$\mathring{A}$) is blueshifted by 1350 km/s (Buson & Ulrich 1990)). The host galaxy disk has been detected in the V, R and H-band (Bothun et al. 1984, Hutchings & Crampton 1990, McLeod & Rieke 1995) and also in the ^{12}CO(1-0) line (Barvainis et al. 1989). The two objects near the edges of the disk are a foreground star in the north and a companion galaxy in the west (Hutchings & Crampton 1990). The QSO nucleus is located in a gas rich host galaxy disk, which makes I Zw 1 an ideal candidate to study the properties of QSO hosts.

A first detection of compact CO emission was done with the NRO millimeter array in 1994. We now mapped the ^{12}CO J=1-0 line emission in I Zw 1 in Jan./Feb. 1995 with the IRAM millimeter interferometer on the Plateau de Bure (Guilloteau et al. 1993) at 2'' resolution. The K band (2.20 μm) observations were done in Jan. 1995 with the MPE 3D camera (Weitzel et al. 1996) at the 3.6m telescope in Calar Alto. The H-band spectrum (1.65 μm) was observed in Dec. 1995 at the 4.2m William Herschel Telescope (WHT) in La Palma. An image of I Zw 1 in the H-band (1.65 μm) was taken with the ESO IRAC2 camera on the 2.2m telescope in La Silla, Chile, in July 1995.

2 Results

The Distribution of the Molecular Gas in I Zw 1. As seen in Fig. 1 I Zw 1 has extended ^{12}CO (1-0) emission. We compared the IRAM 30m single dish spectrum taken by Barvainis et al. (1989) with a total spectrum of our IRAM PdB Interferometer observations to derive the flux emitted by the disk of host galaxy. Both spectra show a double horned profile. From our high resolution ^{12}CO (1-0) map we measure a FWHM source size of the nucleus of about 3.3″. Using correction factors for the beam filling we decomposed the 0.078 Jy flux measured by the 30m telescope into a nuclear (0.056 Jy) and a disk (0.022 Jy) component. Using the measured flux and calculated beam filling factors we get a disk diameter of 9″ (12″) or higher for a Gaussian (flat) flux distribution. With velocity widths of 380 km/s for the nucleus and 310 km/s (distance horn-to-horn) for the disk and an $\frac{N_{H_2}}{I_{CO}}$-conversion factor of $2.0\times10^{20}\frac{cm^{-2}}{Kkms^{-1}}$ we get a molecular gas mass in the nucleus of $5.5\times10^9 M_\odot$ and for the disk a gas mass of $1.9(0.5)\times10^9 M_\odot$. From the rotation curve we derive a dynamical mass for the nucleus (3.3″) of $(3.1\pm1.2)10^{10} M_\odot$ assuming solid body rotation and a rotation velocity of (290 ± 60) km/s.

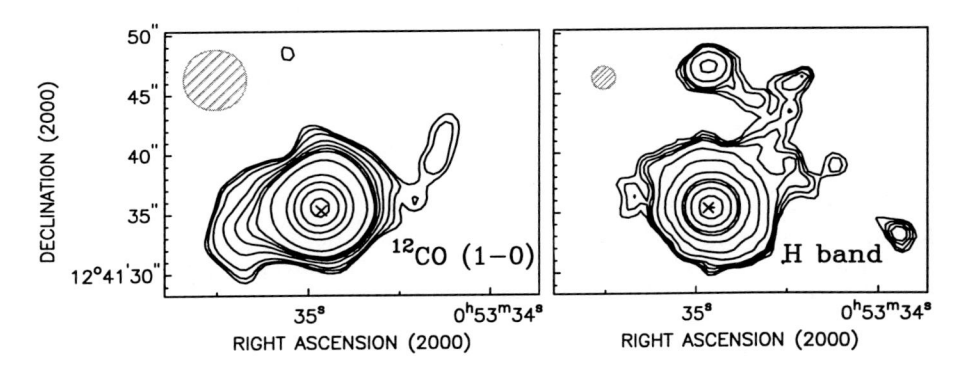

Fig. 1. Maps of I Zw 1 in the ^{12}CO(1-0) line (left) and the (1.65 μm) continuum.

The Dynamics of the Molecular Gas in I Zw 1. To get limits on the inclination of I Zw 1 we assumed a typical rotation velocity of 230 km/s for the flat part of the rotation curve. From this we derived a value for the inclination of $(38\pm5)°$. This value and the position angle of the kinematic major axis of 45° that we get from the CO velocity field is in agreement with the observed ellipticity of the R-band image (Hutchings & Crampton 1990) assuming an intrinsically circular disk. In Fig.2 we show two position-velocity diagrams taken along the kinematic major axis, along with calculated model pv-diagrams. The pv-diagram made at a high angular resolution of 1.9″ has two flux peaks at velocities of ±180 km/s, indicating a nuclear ring with a diameter of about 1.6″. The modeled pv-diagram matches the measured one very well except for a single emission peak at a velocity of 220 km/s and approximately 2″ E and 2″ S of the nucleus – probably a giant molecular cloud near or in the nuclear ring. The disk emission peaks at about ±140 km/s in the second pv-diagram having an extend of about 30 kpc and showing a low velocity dispersion of about ~ 17 km/s. The nuclear source exhibits high velocity dispersion (isothermal case) of 350 km/s. The model flux distribution is plotted in Fig. 3.

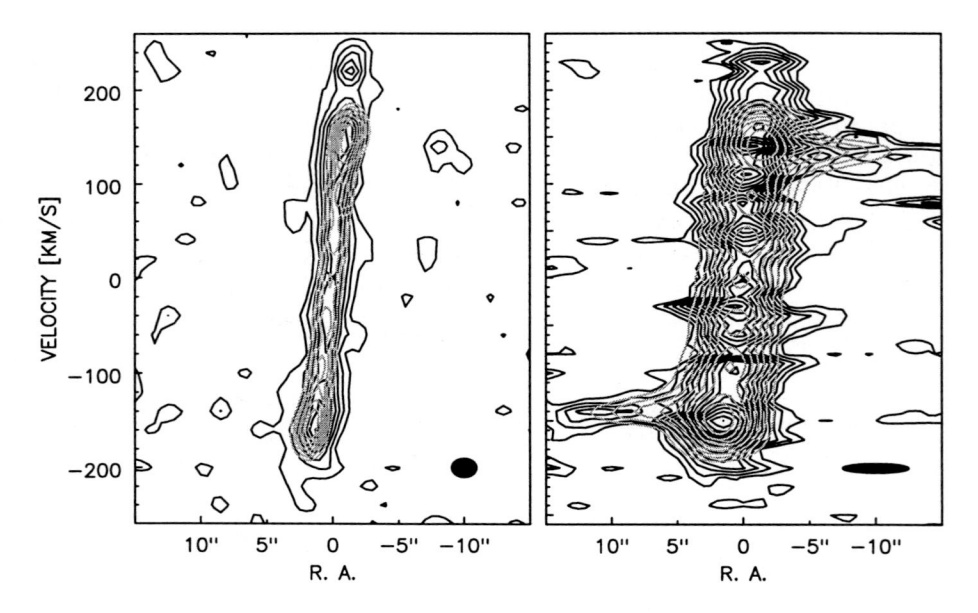

Fig. 2. pv-diagram of the ^{12}CO(1-0) line along the kinematic major axis for the high angular resolution map (1.9″ FWHM) (left) and low angular resolution map (5″ FWHM) (right). The modelled pv-diagrams are overlayed in gray.

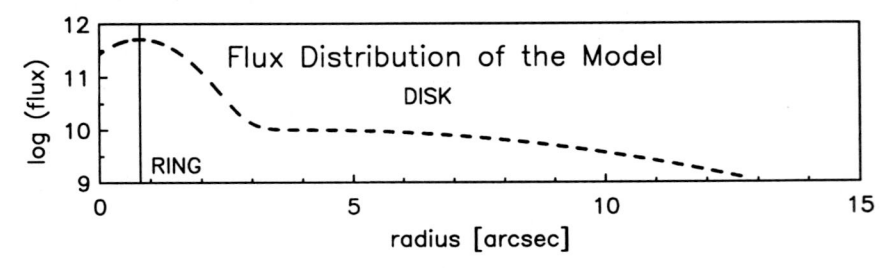

Fig. 3. Flux distribution as a function of radius for the assumed model. A nuclear ring at 0.8″ is superposed on an underlying and much less luminous disk is used.

Star Formation in the Disk. The host galaxy is clearly detected in the optical, the NIR and in its molecular line emission (see Fig. 1). There are two spiral arms starting east and west from the nucleus. The western spiral arm is detected in the V, R and H band (Bothun et al. 1984, McLeod & Rieke 1995) and in ^{12}CO (1-0) line emission. This indicates that there are star forming regions in this spiral arm as the V, R and H band traces young to late type stars. The material for this star formation is concentrated on the spiral arms. On the other hand the eastern spiral arm is only detected in the molecular line emission. From that we conclude that the star formation in the western spiral arm is possibly enhanced due to an interaction with the companion in the west of I Zw 1.

Stellar Light in the Nucleus of I Zw 1. We present an H and K band (1.58 - 2.40 μm) NIR-spectrum of the nuclear region (3.0″ aperture) with a spectral resolution of about R$\sim$ 1000 - 750 (see Fig. 4). The most prominent line is the Paα line. Its maximum flux density is 2/3 the continuum flux density. Brγ and Brδ are two other H-recombination lines detected in the nucleus. The hydrogen recombination line ratios indicate little or no extinction towards the nucleus of the QSO. The other emission line is due to [SiVI] ($\lambda 1.962\mu$m, coronal) and is, like the [CIV] ($\lambda 1549 \text{Å}$), blueshifted by 1350 km/s. In the H-band spectrum the ^{12}CO (6-3) overtone bandhead is clearly detected indicating a prominent stellar contribution from the continuum light of the nucleus. We used the depth of this line to estimate the fraction of stellar flux in the H-band due to late type giants and supergiants. Consistent with the result of Barvainis (1990) and Eckart et al. (1994) we find that about 20% (30%) of the K (H) band flux are of stellar origin and extincted by an A$_V$ $\sim 10^{mag}$. So we detected a stellar nuclear bulge in the central 3.0″.

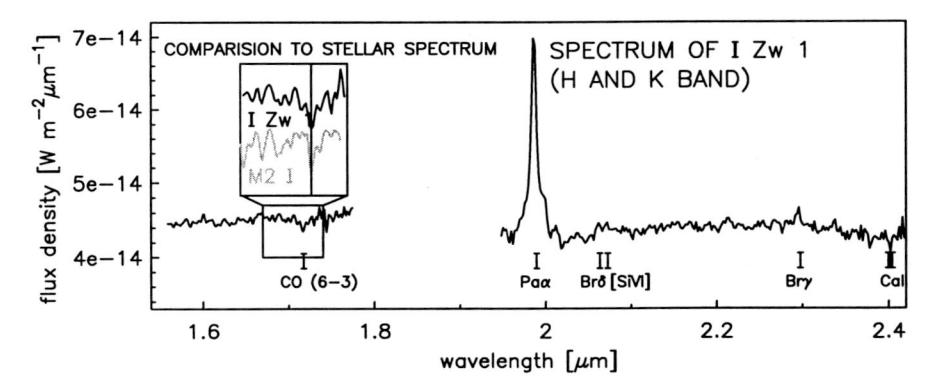

Fig. 4. NIR-spectrum of I Zw 1 in the H and K band (1.58 - 2.40 μm). The spectrum represents the inner 3″ of the nucleus. For the CO(6-3) overtone a comparison to a stellar spectrum of a M2 I star (Dallier et al. 1996) is shown.

A Starburst in the Circum-Nuclear Ring? Starburst rings are observed in a variety of galaxies. To investigate the properties of the molecular ring in the nucleus of I Zw 1 which we detected in the ^{12}CO (1-0) line emission, we have applied a starburst model (Krabbe et al. 1994) to the circum-nuclear region. For the model, the K band luminosity L_K, the bolometric luminosity L_{bol}, the Lyman continuum luminosity L_{Lyc} and the supernova rate ν_{SN} are used as input parameters to be fitted. Under the assumption that 20% (see above) of the K-band flux is stellar and suffers from an extinction A$_V \sim 10^{mag}$ we get a reasonable solution assuming that only $\sim$ 0.5 % of the Lyman continuum luminosity and therefore about $\sim$ 0.5 % of the Brγ line flux (used to derive L_{Lyc}) is due to the starburst, and that this flux is highly extincted with respect to the rest of the emission. Our result is a decaying starburst (decay-time: 5×10^6 yrs) which started about 4×10^7 yrs ago with a Salpeter IMF and an upper mass cut-off of ≈ 50 M$_\odot$. The present day stellar mass we get from our model calculations of this starburst is 9.1×10^9M$_\odot$.

Comparison to other Starburst Rings. The properties of the starburst ring in the nucleus of I Zw 1 are similar to those found for the Seyfert1 galaxy NGC 7469 (Genzel et al. 1995) and the HII-galaxy NGC 7552 (Schinnerer et al 1997). The rings all have comparable diameters of about 1 kpc. Any differences could be due to the slightly different age of the individual rings. It is suggestive that some fraction of the luminosity observed for QSO and Seyfert's is due to a starburst in the nuclear regions of their host galaxies.

	NGC 7552	NGC 7469	I Zw 1
Type	LINER/H II	Seyfert 1	QSO/Sey 1
Distance	20 Mpc	66 Mpc	180 Mpc
Size	17 kpc	20 kpc	$\sim$ 25 kpc
Ring Size	1 kpc	1 kpc	1.5 kpc
L_{bol} [10^{10}L$_\odot$]	3	10	7
L_K [10^9L$_\odot$]	0.7	3	9
L_{Lyc} [10^9L$_\odot$]	2.8	9.5	0.1
ν_{SN} [yr^{-1}]	0.1	0.4	0.6
$\frac{L_{bol}}{L_{Lyc}}$	10	28 - 38	515
$\frac{L_K}{L_{Lyc}}$	0.24	$\sim$0.3	63
$\frac{10^9 \nu_{SN}}{L_{Lyc}}$	0.034	0.049	4.753
m_u [M$_\odot$]	$\approx$ 100	$\leq$ 100	$\approx$ 50
t_{burst} [yr]	$\sim 1.5\times10^7$	$\sim 1.5\times10^7$	$\sim 4\times10^7$
type	dec.	dec. (?)	dec.

Test of the $\frac{N_{H_2}}{I_{CO}}$-Conversion Factor. The standard Galactic conversion factor results in a molecular gas mass of M$_{H_2}$=(5.5$\pm$1.1)$\times10^9$M$_\odot$. We used the dynamical mass of M$_{DYN}$=(3.1$\pm$1.2)$\times10^{10}$M$_\odot$ and the stellar mass of the starburst of M$_{ST}$=(9.1$\pm$2.9)$\times10^9$M$_\odot$ to test the conversion factor. Correcting for helium (36 % of the molecular gas mass) we can now test the conversion factor via

$$\frac{M_{DYN} - M_{ST} - 0.36 M_{H_2}}{M_{H_2}} = \frac{M_{DYN} - M_{ST}}{M_{H_2}} - 0.36 = 3.5 \pm 3.0 \quad (1)$$

We find the $\frac{N_{H_2}}{I_{CO}}$-conversion factor to agree within the errors with the canonical value of $2\times10^{20} \frac{cm^{-2}}{Kkms^{-1}}$ (Strong et al. 1987). Remaining differences can easily be explained by uncertainties in our assumptions of the temperatures of the gas, the geometry assumed in calculating the dynamical mass, a higher fraction of warm molecular gas and also by an old bulge population of late type dwarfs which contributes $\sim 2\%$ to the K-band flux. However, our result on I Zw 1 and the findings for other lower redshift galaxies indicate that the average $\frac{N_{H_2}}{I_{CO}}$-conversion factor is valid in these extragalactic objects as well.

3 Summary and Conclusions

Our study of the host galaxy of I Zw 1 spans a broad range in wavelength. We mapped the structure (including spiral arms) of a QSO host in its emission of the $^{12}CO(1-0)$ molecular line. We have detected a circum-nuclear starburst ring similar to the ones observed in a variety of other active galaxies. From the comparison to other starburst rings we find that the properties of such rings are similar. Therefore it seems likely that this QSO host has properties like 'normal' galaxies. This may apply for QSO hosts in general. The enhanced starformation in the western spiral arm of the host may indicate tidal interaction with a companion. A test of the $\frac{N_{H_2}}{I_{CO}}$-conversion factor shows that this factor can probably also be applied to QSO hosts galaxies.

References

Barvainis, R., 1990, *Ap. J.*, 353, 419.
Barvainis, R., Alloin, D., Antonucci, R., 1989, *Ap. J. (Letters)*, 337, L69.
Buson, L.M., and Ulrich, M. H., 1990, *Astron. Astrophys.*, 240, 247.
Bothun, G. D., Heckman, T. M., Schommer, R. A., Balick, B., 1984, *Astron. Journ.*, 89, 1293.
Condon, J.J., Hutchings, J.B., Gower, A.C., 1985, *Astron. Journ.*, 90, 1642.
Dallier, R., Boisson, C., Joly, M., 1996, *Astron. Astrophys. Suppl.*, 116, 139.
Eckart, A., van der Werf, P.P., Hofmann, R., Harris, A.I., 1994, *Ap. J.*, 424, 627.
Genzel, R., Weitzel, L., Tacconi-Garman, L.E., Blietz, M., Krabbe, A., Lutz, D., Sternberg, A., 1995, *Ap. J.*, 444, 129.
Guilloteau, S., et al. 1992, å262624)
Halpern, J.P., and Oke, J.B., 1987, *Ap. J.*, 312, 91.
Hutchings, J.B., Crampton, D., 1990, *Astron. Journ.*, 99, 37.
Krabbe, A., Sternberg, A., and Genzel, R., 1994, *Ap. J.*, 425, 72.
Kruper, J.S., Urry, C.M., and Canizares, C.R., 1990, *Ap. J. Supp.*, 74, 347.
McLeod, K. K., Rieke, G. H., 1995, *Ap. J.*, 441, 96.
Schmidt, M., Green, R.F., 1983, *Ap. J.*, 269, 352.
Strong, A. W., et al. 1987, Proc. 20th Intern. Cosmic Ray Conf., I, 125
Thronson, H.A. Jr., Telesco, C.M., 1986, *Ap. J.*, 311, 98.
Weitzel, L., Krabbe, A., Kroker, H., Thatte, N., Tacconi-Garman, L.E., Cameron, M., Genzel, R., 1996, *Astron. Astrophys. Suppl.*, 119, 531.

The Epoch of Major Star Formation in High-z Quasar Hosts

Yoshiaki Taniguchi[1,2], Nobuo Arimoto[3,4], Takashi Murayama[1], Aaron S. Evans[5,6], David B. Sanders[5], and Kimiaki Kawara[7,8]

[1] Astronomical Institute, Tohoku University, Aoba, Sendai 980-77, Japan
[2] Royal Greenwich Observatory, Madingley Road, Cambridge CB3 0EZ, UK
[3] Institute of Astronomy, University of Tokyo, Mitaka 181, Japan
[4] Physics Department, University of Durham, South Road, Durham, DH1 3LE, UK
[5] Institute for Astronomy, University of Hawaii, Honolulu HI 96822, U.S.A.
[6] Astronomy Department, Caltech, Pasadena, CA 91125, U.S.A.
[7] ISAS, Yoshianodai, Sagamihara 227, Japan
[8] ISO Science Operations Centre, Villafranca, 28080 Madrid, Spain

Abstract. We present the results of our observing program on NIR spectroscopy of high-redshift (z) quasars which have been undertaken both at Kitt Peak National Observatory and at Mauna Kea Observatory, University of Hawaii. These data are utilized for studying the epoch of major star formation in high-z quasar hosts.

1 Introduction

The major epoch of star formation in galaxies is one of the most important topics in modern astrophysics, because it is significantly related to the formation of galaxies and quasars as well as to cosmology. Massive stars formed in the first episode of star formation have a lifetime of 10^6 to 10^7 years and then release Type II supernova (SNII) products (primarily the α-elements such as O, Ne, Mg, Si, etc., but comparatively little iron). It takes a much longer time for Type Ia supernovae (SNIa) to release iron. The different nucleosynthesis yields and timescales of SNIa's and SNII's thus make the abundance ratio [α/Fe] a potentially useful cosmological clock with which one can identify the epoch of first star formation in galaxies. It is therefore important to study chemical properties of high-redshift (z) objects.

Since it is considered that the heavy elements in the broad line regions (BLRs) come from stars in a host galaxy, systematic study of chemical properties of BLRs of quasars at high redshift is of particular interest (Hamann & Ferland 1993). Rest-frame optical emission lines, which are usually used to study chemical properties of nearby objects, are redshifted to the near-infrared (NIR) in these quasars. Recent NIR spectroscopy of high-z quasars has shown that the rest-frame optical spectra are dominated by singly ionized iron (FeII) emission as well as hydrogen recombination lines (Hill, Thompson, & Elston 1993; Elston, Thompson, & Hill 1994) suggesting long-lasting star formation in the nuclear regions of the quasar hosts (~ 1 Gyr).

In order to study the major epoch of star formation in high-z quasar hosts, we present new results of our NIR spectroscopy of high-z ($z > 3$) quasars; 1) B1422+231 ($z = 3.62$; Patnaik et al. 1992), 2) PKS 1937−101 ($z = 3.79$; Lanzetta et al. 1991), and 3) S4 0636+68 ($z = 3.2$; Stickel & Kuhr 1994).

2 Observational Results

The two quasars, B1422+231 and PKS 1937−101, were observed by using the long-slit Cryogenic Spectrometer (CRSP) with a 256×256 InSb detector array at the f/15 focus of the Kitt Peak National Observatory (KPNO) 4 meter telescope while the other quasar, S4 0636+68, was observed by using the KSPEC at the Cassegrain focus of the UH 2.2 m telescope. The details of the observations and the data reduction are given elsewhere (Kawara et al. 1996; Murayama et al. 1997). The spectra of the three quasars are shown in Fig. 1. We describe their important observational properties below.

2.1 B1422+231

The spectrum in Fig. 1 shows the emission lines, MgIIλ2798, Hγ, Hβ, and [OIII] λ5007 as well as a marginal detection of CIII] λ2326. Note that this is the first detection of [OIII]λ5007 in a quasar beyond $z = 3$ (Kawara et al. 1996). [OIII]λ5007 relative to Hβ is smaller in B1422+231 than the LBQS composite (Francis et al. 1991). The broad feature of optical Fe II emission lines is present. The feature shortward of MgII emission line is due to UV Fe II emission lines. The flux ratio Fe II(UV + opt)/Mg II of B1422+231 is comparable to that of the LBQS composite spectrum: 12.2 ± 3.9 for B1422+231 and 8.9 for the LBQS composite. Note that Fe II(UV) and Fe II(opt) denote Fe II emission in 2000 − 3000 Å and 3500 − 6000 Å in the rest-frame, respectively. Wills, Netzer, & Wills (1985) give a mean ratio of 7.8 ± 2.6 for nine low-z quasars with $z = 0.15-0.63$. It is thus suggested that the major iron enrichment has already been done in this quasar host.

2.2 PKS 1937−101

The [OIII]/Hβ ratio is similar to that of LBQS composite quasar spectrum. Since the observed K-band spectrum can be fit well solely by the emission lines of [OIII]λ4959,5007, Hβ, Hγ, and the linear continuum, there seems little optical FeII emission which is ubiquitously observed in either high-z quasars (Hill et al. 1993; Elston et al. 1994; Kawara et al. 1996) or most low-z quasars (Boroson & Green 1992). The J-band spectrum shows also little evidence for UV FeII emission feature, either. We fit the continuum emission with a power law of $F_\nu \propto \nu^{-0.50}$, which is almost consistent with the average continuum spectrum of quasars, where the power-law index ranges from -0.3 (Francis et al. 1991) to -0.7 (Sargent et al. 1989). The UV spectra of most quasars, regardless of radio loudness (Bergeron & Kunth 1984), are dominated by the FeII features as well as the power-law continuum emission. Therefore, both the lower flux and the featureless property of the J band spectrum are explained by the absence of UV FeII emission features in PKS 1937−101. In the red edge of the J-band spectrum, a blue part of MgIIλ2798 emission can be seen.

2.3 S4 0636+68

The NIR spectrum of this quasar was first reported by Elston et al. (1994) who showed that its rest-frame optical spectrum is significantly dominated by FeII emission lines, suggesting an iron overabundance than in the solar neighbourhood. Although our new measurement has confirmed the presence of FeIIλ5169 emission, its intensity relative to that of Hβ emission is significantly weaker than that of Elston et al. (1994). The intensity ratio of FeII(opt)/Hβ is estimated to be 3.5±1.1. This value is slightly larger than those of low-z quasars; 1.63±0.88 (Wills et al. 1985). However, we cannot conclude that S4 0636+68 belongs to a class of strong iron quasars (cf. Lípari, Terlevich, & Macchetto 1993).

3 Discussion

We discuss the nature of high-z quasars in viewed from their rest-frame optical spectra. There is a tendency that the quasars with $z < 3.5$ show strong FeII emission (Hill et al. 1993; Elston et al. 1994) while those with $z > 3.5$ show strong [OIII] emission. It should be, however, mentioned that the strong optical FeII emission of S4 0636+68 reported by Elston et al. (1994) is not confirmed in this study. One interesting spectroscopic property known for low-z quasars is the anticorrelation between the strength of optical FeII and [OIII] emission lines, although its physical mechanism is not fully understood (Boroson & Green 1992). We examine if the high-z quasars follow the same anticorrelation. In Fig. 2, we show the relationship of the equivalent width ratios between ([OIII]λ4959+λ5007)/Hβ and FeIIλ4434-4684/Hβ. The low-z quasars studied by Boroson & Green (1992) show a loose, but statistically significant anticorrelation. It is also known that the radio-loud quasars tend to be located in the lower portion of this diagram (i.e., weak FeII emitters). PKS 1937−101, B1422+231 (Kawara et al. 1996), and the radio-quiet, high-z quasars studied by Hill et al. (1993) share the same property as those of low-z quasars. On the other hand, the radio-loud quasars studied by Elston et al. (1994) and Hill et al. (1993) do not follow the same as low-z quasars although our new measurement of S4 0636+68 shows that the ratio is consistent with those of low-z quasars. If there would be many strong iron radio-loud quasars at high redshifts, we would have to introduce a new class of quasars which has not yet been observed at low redshifts.

Finally, we discuss the epoch of major star formation in the host galaxies of B1422+231 and PKS 1937−101.

1) B1422+231: We show that the ratio of Fe II/Mg II, including UV Fe II emission lines, in the broad-line gas of some quasars at $z = 3.6$ is almost identical to those at the low-redshift quasars. This may imply that the Fe/Mg

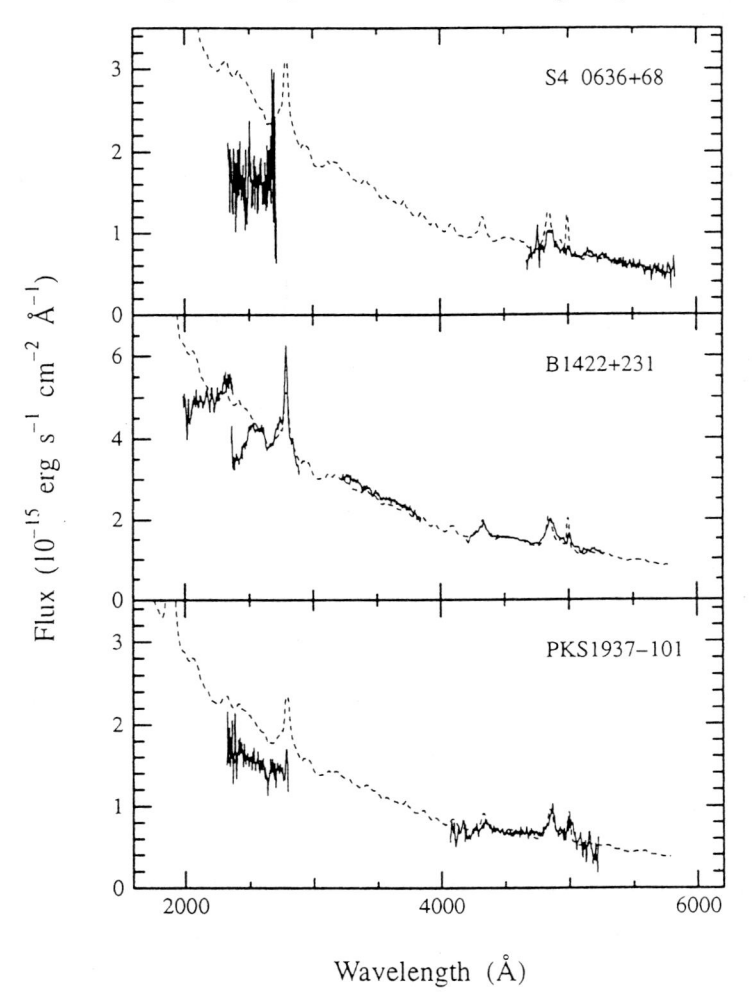

Fig. 1. The rest-frame optical spectra of the three high-z quasars; S4 0636+68, B1422+231, and PKS 1937−101. The dashed spectrum in each panel is the mean spectrum of LBQS quasars taken from Francis et al. (1991).

abundance at the center of some quasar host galaxies did not change after $z = 3.6$. It is generally considered that Mg is preferentially produced in massive star supernovae (SNe II, Ib, and Ic) on short time scales (2–10 Myr), while Fe is mainly created by accreting white dwarf supernovae (SNe Ia) in much longer time scales (1–2 Gyr). The Fe/Mg abundance ratio should be 1/4–1/2 of the solar value until SNe Ia start to produce significant amount of Fe. When SNe Ia dominate the Fe production, the Fe/Mg abundance increases up to the values in low-redshift quasars and is kept nearly constant since then. Although it is not straightforward to derive the Fe/Mg abundance from the present data, the

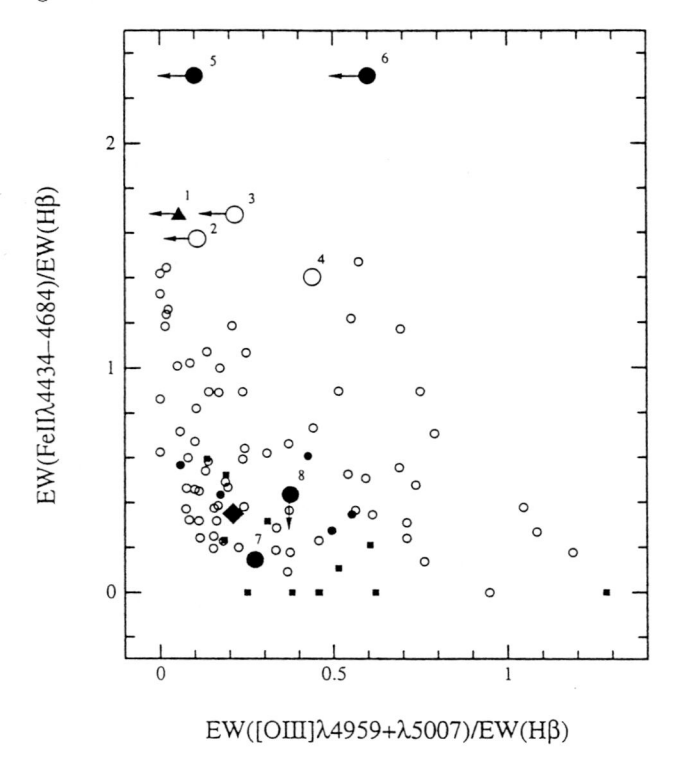

Fig. 2. Diagram between ([OIII]λ5007+λ4959)/Hβ equivalent width ratio and FeIIλ4434-4684/Hβ one for low-z (small symbols; Boroson & Green 1992) and high-z ($z > 2$) quasars (large symbols; Hill et al. 1993; Elston et al. 1994; Kawara et al. 1996; this study). Radio-quiet, radio-loud with flat spectrum, and radio-loud with steep spectrum are shown by open circles, filled circles, and filled squares, respectively. B2 1225+317 is shown by the filled triangle because its radio spectrum is unknown. The numbers given for the high-z quasars correspond to; 1. B2 1225+317, 2. Q1246−057, 3. Q0933+733, 4. Q1413+117, 5. S4 0636+68, 6. Q0014+813, 7. B1422+231, and 8. PKS 1937−101. The filled diamond shows our result for S4 0636+68.

similarity in Fe II/Mg II between B1422+231 and low-redshift quasars (and the LBQS composite spectrum) suggests that the host galaxy of B1422+231 had already been in the late evolutionary phase of the Fe enrichment at $z = 3.6$. Yoshii, Tsujimoto, & Nomoto (1996) derived $\sim$ 1.5 Gyr for the lifetime of SN Ia progenitors from the analysis of the O/Fe and Fe/H abundances in solar neighbourhood stars. If the Fe enrichment started at 1.5 Gyr after the onset of the first star formation, the host galaxy of B1422+231 would have formed at $z \geq 15$ for $q_0 = 0.0$ and $H_0 = 100$ km s^{-1} Mpc^{-1} while at $z \geq 6$ for $q_0 = 0.0$ and $H_0 = 50$ km s^{-1} Mpc^{-1}.

2) PKS 1937−101: The little evidence for Fe emission lines suggests that the major epoch of star formation in this quasar host is different from that in

B1422+231. The α elements, such as O and Mg, come from SNII's of massive star origin and thus are quickly expelled into the interstellar space after the major episode of star formation (within a few 10^6 to 10^7 years). It is considered that the N enrichment is delayed ($\sim 10^8$ years) because it is partly a secondary element formed by CNO burning in stellar envelopes (Hamann & Ferland 1993). The rest-frame ultraviolet spectra of PKS 1937−101 taken by Lanzetta et al. (1991) and Fang & Crotts (1995) show evidence for NVλ1240 emission. Therefore, the nuclear gas has already been polluted with N, implying that the elapsed time from the major star formation is longer than $\sim 10^8$ years (Hamann & Ferland 1993). However, our observation has shown that the major Fe enrichment has not yet been made in PKS 1937−101. The bulk of iron comes from SNIa's whose progenitors' lifetime is very likely to cluster around ~ 1.5 Gyr (Yoshii et al. 1996). Therefore, the Fe enrichment may start at 1.5 Gyr after the onset of the first, major star formation in quasar host galaxies. These arguments, therefore, specify the epoch of major star formation in PKS 1937−101; $\sim 10^8$ - 1.5×10^9 years before redshift 3.787. Namely, the initial star formation would occur at $3.9 < z < 6.7$ for $H_0 = 50$ km s^{-1} Mpc^{-1} and $q_0 = 0$, while at $4.0 < z < 17$ for $H_0 = 100$ km s^{-1} Mpc^{-1} and $q_0 = 0$. Recent theoretical prescription on the star formation at high-z suggests that the major epoch of star formation may occur $z < 5$ although subgalactic structures may exist even at $z > 10$ (Rees 1996). Provided that the smaller H_0 is more preferable, the present observation is consistent with this prescription.

References

Bergeron, J., & Kunth, D. 1984, MNRAS, 207, 263.
Boroson, T. A., & Green, R. F. 1992, ApJS, 80, 109.
Elston, R., Thompson, K. L., & Hill, G.J. 1994 Nature, 367, 250.
Francis, J. P., Hewett, P. C., Foltz, C. B., Chaffee, F. H., Weymann, R. J., & Morris, S. L. 1991, ApJ, 373, 465.
Hamann, F. & Ferland, G. 1993, ApJ, 418, 11.
Hill, G. J., Thompson, K. L., & Elston, R. 1993, ApJ, 414, L1.
Kawara, K., Murayama, T., Taniguchi, Y., & Arimoto, N. 1996, ApJ, 470, L85.
Lanzetta, K. M., Wolfe, A. M., Turnshek, D. A., Lu, L., McMahon, R. G., & Hazard, C. 1991, ApJS, 77, 1.
Lípari, S., Terlevich, R., & Macchetto, F. 1993, ApJ, 406, 451
Murayama, T., Taniguchi, Y., Evans, A. S., Sanders, D. B., Ohyama, Y., Kawara, K., & Arimoto, N. 1997, in preparation.
Patnaik, A. R., Browne, I. W. A., Walsh, D., Chaffee, F. H., & Foltz, C.B. 1992, MNRAS, 259, 1p.
Rees, M. 1996, in the proceedings of 37th Herstmonceux conference on "HST and the High Redshift Universe", edited by N. Tanvir, A. Aragón-Salamanca, & J. V. Wall, in press.
Sargent, W. L. W., Steidel, C. C., & Boksenberg, A., 1989, ApJS, 69, 703.
Stickel, M., & Kuhr, H. 1994, A & AS, 103, 349
Wills, B. J., Netzer, H., & Wills, D. 1985, ApJ, 288, 94.
Yoshii, Y., Tsujimoto, T., & Nomoto, K. 1996, ApJ, 462, 266.

The Nuclear Stellar Cluster in NGC 1068

Niranjan Thatte[1], Roberto Maiolino[1], Reinhard Genzel[1], Alfred Krabbe[1], Harald Kroker[1]

Max-Planck-Institut für extraterrestrische Physik (MPE), Giessenbachstraße, D-85748 Garching, Germany

Abstract. We present new near-infrared integral field spectroscopy and adaptive optics imaging of the nucleus of NGC 1068. Using the stellar CO absorption features in the H and K bands, we have identified a moderately extincted stellar cluster centered on the nuclear position and of intrinsic size ~50 pc. We show that this nuclear star cluster is probably $5-13 \times 10^8$ years in age and contributes at least 10% of the total nuclear luminosity of $\sim 1 \times 10^{11}$ $L_\odot$.

1 Introduction

It is not known how much of the nuclear luminosity in Seyfert galaxies originates from stars, as opposed to accretion disks. Terlevich (1991) and collaborators (Cid Fernandes and Terlevich 1991, Cid Fernandes and Terlevich 1995) propose that Seyfert nuclei are composed of an extreme compact nuclear starburst occuring in a dense, metal-rich environment. In their model, Wolf-Rayet stars and type II supernovae in a dense medium account for the *entire* energy output of the Seyfert nucleus. At the other extreme lie models where active galactic nuclei are powered by massive black holes, via the release of gravitational energy (see Rees (1984) for a review). The presence of stars in the nucleus of NGC 1068 has been inferred from spectroscopic observations at visible wavelengths (Koski 1978, Malkan and Filippenko 1983). However, little is known about the geometry, age or luminosity of the stellar population. Using the stellar CO absorption features in the near-infrared, which are much less affected by extinction, we have observed a compact nuclear stellar cluster in the nucleus of NGC 1068, and constrained its age and luminosity. Our observations show that at least 10% of the nuclear luminosity of 1×10^{11} $L_\odot$ of NGC 1068 (assumed distance of 14 Mpc) has a stellar origin.

2 Observations and Data Reduction

The nucleus of NGC 1068 was observed with the MPE 3D near-IR imaging spectrometer (Weitzel et al. 1996) in conjunction with the tip-tilt adaptive optics system ROGUE II (Thatte et al. 1995). 3D is an integral field spectrometer which obtains *simultaneous* spectra for each of 256 spatial pixels covering a square region of sky with over 95% fill factor. The spectral range may be chosen "on-the-fly" to cover the H or K near infrared windows. The spatial pixel scale may be chosen (via ROGUE II) to be 0$''$3 or 0$''$5 per pixel, depending on the atmospheric seeing conditions. Further details on the two instruments and the data reduction are presented in Weitzel et al. (1996) and Thatte et al. (1995).

The observations were carried out at the 4.2 meter WHT[1] on La Palma in December 1995 and January 1996. Median tip-tilt corrected seeing for the K band observations (0$''$3 per pixel, 140 second integrations, total of 5240 seconds on-source) was 0$''$94, while that for the H band observations was 1$''$24 (0$''$3 per pixel, 60 second integrations, total of 1680 seconds on-source). An equal amount of time was spent on blank sky 3$'$ EW of the object, for sky background subtraction. The resolving power ($R \equiv \lambda/\Delta\lambda$) was 1000, Nyquist sampled using two settings of a piezo-driven flat mirror. However, for the purposes of this analysis, the data were convolved to an effective resolution of 700. The data reduction was carried out using the 3D data analysis package, which is based on the GIPSY [2] (van der Hulst et al. 1992) package. We performed wavelength calibration, spectral and spatial flatfielding, dead and hot pixel correction and division by a reference stellar spectrum obtained during the observations. Data cubes from individual exposures were recentered using the centroid of the broad band continuum and then coadded. Observations of H and K photometric standard stars were used to establish the absolute flux scale of the spectra.

Maps covering three stellar absorption features of interest were extracted from the resulting data cube. The Si feature ($4s^1P^0 \rightarrow 4p^1P$) at 1.59 μm, the CO $v = 3 \rightarrow 6$ bandhead feature at 1.62 μm and the CO $v = 0 \rightarrow 2$ bandhead feature at 2.29 μm all arise exclusively from late type stars. A linear fit to the line free regions of the spectrum in the vicinity of each feature was used to establish the continuum level for each spatial pixel. Only channels shortward of the absorption bandhead were used to fit the continuum for the 2.29 μm feature. The equivalent width of each feature was measured using the wavelength intervals specified by Origlia, Moorwood and Oliva (1993), to enable direct comparison with their data. We also made maps of the absorption flux in each feature (shown in fig. 1 for the CO 2.29 μm feature), as well as maps of the line free continua in the vicinity of the absorption lines.

3 Properties of the Nuclear Stellar Cluster

3.1 Morphology

The dominant component of the K band nuclear light from NGC 1068 is thermal emission from hot dust (Quirrenbach et al. 1997). The dust emission fills in the stellar CO absorption features longward of 2.29 μm, dramatically reducing their equivalent width (EW). However, a map of the absorption flux in the 2.29 μm CO feature, shown in figure 1, represents the spatial distribution of light from late type stars, even in the presence of dilution. If we assume a homogeneous

---

[1] The William Herschel Telescope is operated by the Royal Greenwich Observatory, Cambridge, U.K.
[2] the Groningen Image Processing System

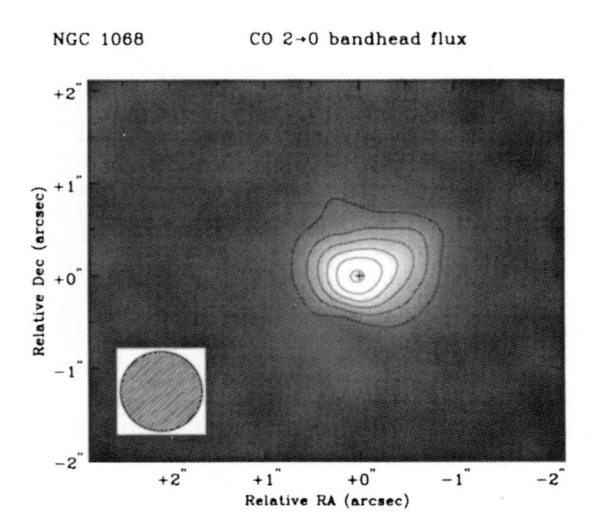

Fig. 1. Map of the distribution of the CO 0→2 bandhead absorption flux toward the nuclear region of NGC 1068. Contours are in increments of 13% of the peak absorption flux, starting at 30%. The location of the K band emission peak is indicated by a +. The instrumental point spread function is shown as a beam in the lower left corner of the figure.

stellar population (which would show a constant CO EW) figure 1 maps the number density of stars in the nucleus. The nuclear stellar cluster is spatially resolved, with an intrinsic FWHM of $0''.66 \pm 0''.04$ (~45 pc), computed from the observed FWHM of $1''.15 \pm 0''.03$ for the CO flux map and $0''.94 \pm 0''.02$ for the line free continuum. The H band flux maps for the Si 1.59 μm feature and the CO 1.62 μm feature also yield the same value, although the H band seeing was somewhat worse than for the K band observations. The stellar light distribution is centered on the K band continuum peak (marked by a + in figure 1) and appears axisymmetric. Due to a small amount of astigmatism present in the ROGUE II interface, the instrumental point spread function was not exactly circular.

3.2 Measuring the Nuclear Stellar Light

Origlia et al. (1993) have shown that the ratio of the equivalent width of the 1.62 μm 3→6 bandhead CO feature to that of the 1.59 μm Si feature forms a very good temperature indicator for late type stars. Their work has been further extended by Dallier, Boisson and Joly (1996), who confirm the previous result. Toward NGC 1068 and other Seyfert galaxy nuclei, these late-type stellar features are substantially diluted by thermal dust emission. However, since the two features are very closely spaced in wavelength, the ratio of the two EWs is insensitive to dilution. Oliva et al. (1995) observe a very good correlation for ellipticals, spirals and starburst galaxies between the EW of the CO 3→6 bandhead and the ratio EW(CO 1.62 μm)/EW(Si 1.59 μm), which serves as a dilution insensitive temperature indicator. For Seyfert galaxies, they use this tight correlation to compute the continuum flux dilution at 1.6 μm. The dilution fraction, D, is computed from the measured EW and the instrinsic EW of the CO 3→6 feature (obtained from the above correlation) using the equation

$$1 - D = \frac{EW_{obs}}{EW_{intr}} = \frac{F_{late-type\ stars}}{F_{total}} \quad (1)$$

where F denotes the flux at the wavelength of the stellar feature and EW_{obs} and EW_{intr} refer to the observed and intrinsic stellar equivalent widths, respectively.

In the K band, the dilution cannot be obtained in the same manner, due to the absence of two strong, closely spaced stellar absorption features. Instead, we used the stellar temperature measured from the H band data (via the dilution insensitive ratio EW(CO 1.62 μm)/EW(Si 1.59 μm)) to estimate the intrinsic EW of the CO 2.29 μm feature. We then calculated the dilution fraction from the observed EW, using equation 1.

Table 1. Dilution of starlight in the nuclear region of NGC 1068

Radius range (″)	H band dilution (%)	K band dilution (%)	K band starlight flux (mJy)
0 – 0.5	71 ± 5	94 ± 5	13.4
0.5 – 1	64 ± 5	88 ± 5	21.0
1 – 1.5	53 ± 5	75 ± 5	18.0
1.5 – 2	46 ± 5	66 ± 5	17.0
2 – 2.5	47 ± 5	54 ± 5	21.2

Table 1 shows the dilution at 1.62 and 2.29 μm for the central 5″ of NGC 1068. We chose five annuli with radii ranging from 0.″5 to 2.″5, which cover the entire spatial extent of the 3D field of view. Our choice of annuli centered on the K band continuum peak is justified by the almost circular symmetry seen in the CO flux map shown in figure 1. The stellar temperatures derived from the ratio EW(CO 1.62 μm)/EW(Si 1.59 μm) correspond to spectral types in the range K5 to M2. This is consistent with the observations of Oliva et al. (1995) for a range of galaxy types.

3.3 Stellar Light Profile

The *stellar* light profile within a radius of 10″ from the nucleus of NGC 1068 was plotted by combining the 3D measurements with K band images from the MPE

SHARP camera. Close to the nucleus, where the contribution from hot dust emission is substantial, the starlight flux as a function of radius is computed from the measured total flux and the dilution fraction, using 3D spectroscopy. At radii larger than the 3D field of view, we obtained the starlight flux by subtracting the central compact dust source from the SHARP K band speckle image. At radii greater than 2″, we expect that starlight contributes most of the K band light. Figure 2 shows the *stellar* light profile within a radius of 10″ in NGC 1068. The agreement of the profiles measured by 3D and SHARP verifies that the contribution from hot dust at large radii in the SHARP map is indeed very small.

The nuclear star cluster is clearly revealed as a separate component contributing substantially to the stellar K light in the central few arc seconds, superposed on the galaxy bulge light profile which shows the characteristic turnover at a radius of ∼5″. For comparison, figure 2 shows light profiles of two other galaxies, one for M87 (Kormendy and Richstone 1995), which does not contain a central core, and the other for M31 (Faber et al. 1997), where a separate nuclear core is distinctly visible. Both galaxy profiles have been scaled in distance and intensity for ease of comparison.

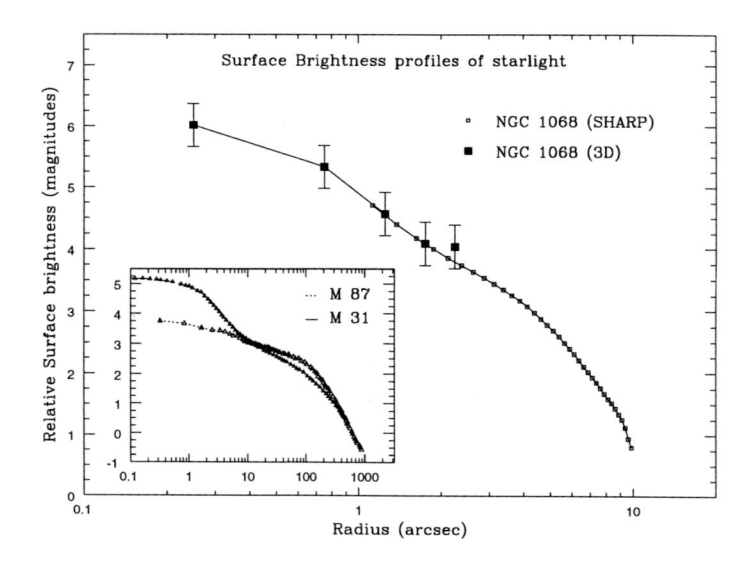

Fig. 2. Radial surface brightness profile of starlight within the central 10″ of NGC 1068. The profile is a composite of data from the MPE speckle camera SHARP at large radii (small open squares) and the MPE 3D spectrometer (large filled squares) at small radii. The SHARP points were obtained by subtracting the compact dusty source from the K band speckle image. For comparison, radial light profiles of M87 (Kormendy and Richstone 1995) and M31 (Faber et al. 1997) are also shown, shifted by arbitrary amounts along the two axes (but *not* scaled).

3.4 Extinction Toward the Nuclear Cluster

The continuum flux at 1.6 and 2.3 μm, combined with the dilution fractions computed above, allow us to determine the flux from the nuclear regions due to late type stars. To calculate the true luminosity of the stellar cluster, we need to correct for extinction effects. Antonucci and Miller (1985) estimated that starlight contributes $\sim$82% of the continuum flux at 5400 Å within a 2″.8 aperture, based on a comparison with the spectrum of M32. Their measurement corresponds to a V magnitude of 13.53 for starlight within the 2″.8 aperture. Integrating over the same aperture in our K band data, we obtain a K magnitude of 10.13. Comparing the observed V−K color with the value predicted for clusters older than 10^8 years (section 3.5) by Leitherer and Heckman (1995), we obtain a color excess of 1.8 magnitudes, which corresponds to an A_V of 2 magnitudes, assuming a screen model for the extinction.

3.5 Luminosity of the Central Cluster

The fraction of nuclear K band flux due to starlight, corrected for the extinction toward the nuclear cluster, yields a total K band luminosity of the central cluster of 1.7×10^8 L$_\odot$, within a radius of 2″.5, and 6.5×10^7 L$_\odot$ within a radius of 1″. The spectral resolution of our data is not sufficient to measure the velocity dispersion of stars in the central few arc seconds, so we have used velocity dispersion measurements made by Terlevich, Diaz and Terlevich (1990), Oliva et al. (1995) and Dressler (1984). Terlevich et al. (1990) measure a velocity dispersion of 153 ± 15 km s^{-1} in a 2″.8 aperture, Oliva et al. (1995) measure 161 ± 20 km s^{-1} in a 4″.4 × 4″.4 aperture, and Dressler (1984) measures 143 ± 5 km s^{-1} in a 2″.0 × 1″.7 aperture. All three measurements are fully consistent with each other within the errors, although a different stellar feature was used in each case. Although there is no significant trend with decreasing aperture, the measurement by Dressler (1984) covers the smallest aperture, and can be directly used to compute a dynamical mass estimate relevant to our observations.

We assume that the nuclear cluster is a virialized, isotropic, isothermal distribution and use the virial theorem to estimate a dynamical mass. Using

$$M_{\rm dyn} = \frac{2\sigma_*^2 R}{G} \qquad (2)$$

where σ_* is the projected stellar velocity dispersion over an aperture of radius R, we obtain a dynamical mass of 6.5×10^8 M$_\odot$ within a radius of 1″. The dynamical mass represents an *upper limit* on the stellar mass since the gas mass as well as the mass of the Seyfert nucleus may significantly contribute to the dynamical mass estimate. The corresponding lower limit to the intrinsic L_K/M ratio of the nuclear stellar cluster is 0.1 L$_\odot$/M$_\odot$, where we have used L$_\odot$ = 3.83×10^{33} ergs s^{-1}.

To estimate the age of the nuclear star cluster, we have computed the L_K/M_* ratio for an evolving stellar cluster using the Sternberg and Kovo (1996) code (see also Genzel et al. 1995). This method uses the stellar tracks of Meynet et al. (1994) for solar metallicity stars with a Salpeter initial mass function ($dN(M)/dM \sim M^{-2.35}$) down to 2 $M_\odot$ and an upper mass cutoff of 100 $M_\odot$. For masses lower than 2 $M_\odot$ we modeled the IMF as a broken power law (Miller and Scalo 1979). The exact lower mass cutoff point is not critical, as the IMF shows a turnover at masses below 0.18 $M_\odot$. King et al. (1996) and De Marchi and Paresce (1996) get similar factors for mass contributions from stars less than 2 $M_\odot$ using HST observations of globular clusters. For the star formation history, we assumed an essentially constant star formation rate ($\sim \exp(-t/t_{scl})$, with a time constant (t_{scl}) of 1×10^9 years).

Comparing with the observed values, we obtain an *upper limit* to the cluster age of 1.3×10^9 years, and a *lower limit* ratio of total bolometric to K band luminosity of 60. The total bolometric luminosity due to stars within a radius of $2''\!.5$ is then *at least* 1×10^{10} $L_\odot$. Consequently, the stars contribute at least 10% of the total nuclear bolometric luminosity of 1×10^{11} $L_\odot$ (Telesco et al. 1984). It is very likely that the nuclear region of NGC 1068 also contains stars older than 1 Gyr. Taking the mass contained within $\sim$70 pc in our own galaxy (4.5 $\times$ 10^8 $M_\odot$, Genzel, Hollenbach and Townes 1994) or in the nearby Seyfert galaxy Circinus (3×10^8 $M_\odot$, Maiolino et al. 1997) as a measure for the underlying old stellar cluster in NGC 1068, the L_K/M ratio of the younger component sampled in the near infrared would be 2.4 times larger ($L_K/M \sim 0.24$) and its age would be 4.6×10^8 years. This moderate age component would then contribute about 15% of the nuclear bolometric luminosity. We constrain the minimum age of the cluster by noting that the cluster would have to be $< 10^7$ years old in order to contribute more than 50% of the nuclear bolometric luminosity.

References

Antonucci, R. R. J., and Miller, J. S. 1985, ApJ, 297, 621
Cid Fernandes, R., and Terlevich, R. 1991, in Relationships between Active Galactic Nuclei and Starburst Galaxies, Filippenko, A. ed., A.S.P Conf. series, 31, 241
Cid Fernandes, R., and Terlevich, R. 1995, MNRAS, 272, 423
Dallier, R., Boisson, C., and Joly, M. 1996, A&AS, 116, 239
De Marchi, G., and Paresce, F. 1996, in Science with the Hubble Space Telescope - II, Benvenuti, P., Macchetto, F. D., and Schreier, E. J. eds., p. 310
Dressler, A. 1984, ApJ, 286, 97
Faber, S. et al. 1997, AJ, preprint
Genzel, R., Hollenbach, D., and Townes, C. 1994, Rep. Prog. Phys., 57, 417
Genzel, R., Weitzel, L., Tacconi-Garman, L. E., Blietz, M., Cameron, M., Krabbe, A., Lutz, D., and Sternberg, A. 1995, ApJ, 444, 129
King, I. R., Piotto, G., Cool, A. M., Anderson, J., and Sosin, C. 1996, in Science with the Hubble Space Telescope - II, Benvenuti, P., Macchetto, F. D., and Schreier, E. J. eds., p. 297
Kormendy, J., and Richstone, D. 1995, ARA&A, 33, 581
Koski, A. T. 1978, ApJ, 223, 56
Leitherer, C., and Heckman, T. 1995, ApJS, 96, 9
Maiolino, R. et al. 1997, in press
Malkan, M., and Filippenko, A. 1983, ApJ, 275, 477

Meynet, G., Maeder, A., Schaller, G., Schaerer, D., and Charbonnel, C. 1994, A&AS, 103, 97
Miller, G. E. and Scalo, J. M. 1979, ApJS, 41, 513
Neugebauer, G. et al. 1980, ApJ, 238, 502
Oliva, E., and Moorwood, A. F. M. 1990, ApJ, 348, L5
Oliva, E., Origlia, L., Kotilainen, J. K., and Moorwood, A. F. M. 1995, A&A, 301, 55
Origlia, L., Moorwood, A. F. M., and Oliva, E. 1993, A&A, 280, 536
Quirrenbach, A. et al. 1997, in Proceedings of the NGC 1068 workshop, Schloß Ringberg, Gallimore, J. and Tacconi, L. eds., in press
Rees, M. 1984, ARA&A, 22, 471
Sternberg, A., and Kovo, O. 1996, in preparation
Telesco, C. M., Becklin, E. E., Wynn-Williams, C. G., and Harper, D. A. 1984, ApJ, 282, 427
Terlevich, E., Diaz, A. I., and Terlevich, R. 1990, MNRAS, 242, 271
Terlevich, R. 1991, in Relationships between Active Galactic Nuclei and Starburst Galaxies, Filippenko, A. ed, A.S.P Conf. series, 31, 133
Thatte, N., Kroker, H., Weitzel, L., Tacconi-Garman, L. E., Tecza, M., Krabbe, A., and Genzel, R. 1995, Proceedings of the SPIE, 2475, 228
van der Hulst, J. M., Terlouw, J. P., Begeman, K., Zwitser W., and Roelfsema, P. R. 1992, in Astronomical Data Analysis Software and Systems I, eds. D. M. Worall, C. Biemesderfer and J. Barnes, A.S.P. Conference Series, 25, p. 131
Weitzel, L., Krabbe, A., Kroker, H., Thatte, N., Tacconi-Garman, L.E., Cameron, M., and Genzel, R. 1996, A&AS, 119, 531

ISO Observations of Quasars and Quasar Hosts

Belinda J. Wilkes

Smithsonian Astrophysical Observatory, 60 Garden St., Cambridge, MA 02138, USA

Abstract. The Infrared Space Observatory (ISO), launched in November 1995, allows us to measure the far-infrared (far-IR) emission of quasars in greater detail and over a wider energy range than previously possible. In this paper, preliminary results in a study of the 5–200 μm continuum of quasars and active galaxies are presented. Comparison of the spectral energy distributions show that, if the far-IR emission from quasars is thermal emission from galaxian dust, the host galaxies of quasars must contain dust in quantities comparable to IR luminous galaxies rather than normal spiral galaxies. In the near-IR, the ISO data confirm an excess due to a warm 'AGN-related' dust component, possibly from the putative molecular torus. We report detection of the high-redshift quasar, 1202-0727, in the near-IR indicating that it is unusually IR-bright compared with low-redshift quasars.

1 Introduction

Quasars are multi-wavelength emitters, emitting roughly equal amounts of radiation throughout the whole electromagnetic spectrum from far-IR through to γ-ray energies. 10% are also strong radio emitters. To understand the energy generation mechanisms at work, it is first essential to obtain multi-λ data covering the full spectral range of the emission. We now have a good understanding of the spectral energy distributions (SEDs) of low-redshift quasars and active galaxies. However, in the far-IR this has been limited by the short lifetime and wide beam of the ground-breaking IRAS satellite. Now, more than 10 years later, ISO is providing us with the chance for a second, more detailed look at the far-IR sky, allowing us to extend our knowledge to IR-fainter and higher redshift sources and to longer and shorter wavelengths (5–200μm).

To this end we are observing a sample of quasars and active galaxies with the photometer on ISO (ISOPHOT). The sample was originally designed to include $\sim$ 130 quasars and active galaxies covering the full range of redshift and of known SED properties. With the reduced in-flight sensitivities of ISOPHOT, our program has been reduced significantly and will likely include $\sim$ 50 objects, not all with full wavelength coverage. The sample will include full wavelength coverage for a well-defined subset of optically-selected, PG quasars (Laor et al. 1996), along with a few high-redshift quasars, X-ray selected Seyfert 1 galaxies, and red quasars.

One question that the ISO data will address is particularly relevant to this conference, namely the contribution of the host galaxy in the far-IR. Figure 1 shows SEDs of spiral and elliptical galaxies superposed on the SED of a median low-redshift quasar (from Elvis et al. 1994). The plot clearly shows the near-IR

($\sim 1-2\mu m$ "window" on the host galaxy which has been used to great advantage (McLeod & Rieke 1994, Dunlop et al. 1993). Although the strength of the far-IR peak, due to cool dust, is as yet unknown, Figure 1 demonstrates that this is the most likely wavelength range for a second "window" on the host galaxy.

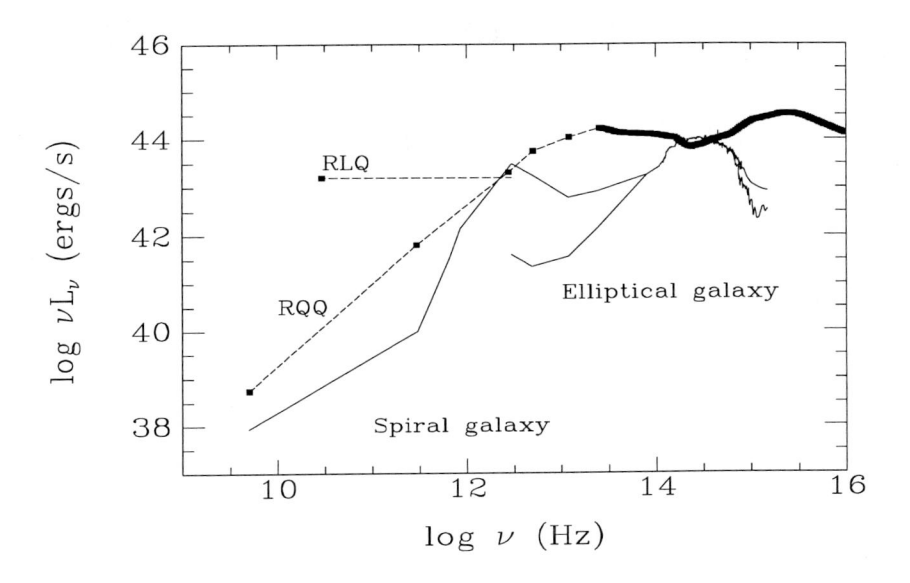

Fig. 1. The Median SED of a low-redshift quasar superposed on the SED of a spiral galaxy showing the well-explored, near-IR "window" ($\sim 1\mu m$) and the potential far-IR "window" on a quasar's host galaxy (courtesy Kim McLeod).

2 ISO Observing Program

ISOPHOT observations are being made in eight broad bands covering the full energy range of the instrument, 5–200 μm. The detector/filter combinations are: P1:5,7,12 μm; P2: 25 μm; C100: 60,100 μm and C200: 135,200 μm. Until October 1996, AOTs P03 and P22 were used in rectangular chopping mode with a chopper throw of 180" in all cases. The apertures, chosen to match the instrument field of view or the point spread function as applicable, are 52" for $\lambda \leq 12\mu m$, 120" for $\lambda = 25\mu m$ and the array size for the long wavelength points. For the largest and/or brightest sources, we use staring mode with a separate sky observation. Following the recommendations of the ISOPHOT team, we have re-specified our remaining time to observe a smaller subset of objects, re-observing where necessary, using small rasters whose dimensions depend on the detector in use. These observations are scheduled to begin in early 1997 and should provide the reliable long-wavelength data which is currently lacking.

3 Analysis of ISOPHOT data

ISOPHOT suffers from several well-known problems which complicate the data analysis and limit (currently) the accuracy with which fluxes can be determined (for details see ISOPHOT Observers Manual and associated updates):

- The responsivity of all detectors drifts significantly following a change in the incident signal, for example when pointing to a new source or changing filters. This drift is difficult to calibrate and the analysis software does not yet support fitting for chopped observations such as ours. We have thus concentrated our efforts on analysing objects with observations sufficiently long that the detector reaches a stable portion of the drift curve, generally $\gtrsim 128$ secs total time.
- The internal (FCS) calibrators needed to be re-calibrated in-orbit following a change of state of FCS1 in February 1996. The combination of the shortness of the FCS observations (16/32 secs) taken during an individual observation and the lack of a definitive re-calibration led us to use the default responses for all detectors in this analysis. There is a drift $\sim 30\%$ in the detector responsivity as a function of time during an orbit so our flux normalisation errors are expected to be of this order.
- At long wavelengths (C200 detector in particular) the two adjacent beams are large enough that part of the detector lies within a part of the telescope beam which is significantly vignetted. This leads to an asymmetry in the derived fluxes across the detector and a $\sim 20\%$ flux correction. The correction for this affect has not yet been released and has not been applied to our data.

Our analysis was performed using the PHOT Interactive Analysis Package (PIA), an IDL-based system provided by the PHOT team. We carried out the following steps: non-linearity correction, read-out de-glitching, 1st order fit to ramps, dark current subtraction, de-glitching to delete highly discrepant points, deletion of data during strong detector drifts and of remaining highly-discrepant points, background subtraction, and calibration using the default responsivities. Background subtraction is done using the average of the background in the chopper plateaux before and after each source plateau. On a non-linear drifting curve, this is not accurate and adds noise to the signal which could be reduced by fitting the background and source curves separately and then subtracting them. Currently points on the drift curve are deleted, reducing the potential S/N of the observations once more sophisticated analysis is carried out.

From the subset of our sources with relatively long exposure times, we chose four objects covering a range of luminosity and redshift: PG1244+026, a radio-quiet Seyfert 1 galaxy; PG1543+489, a radio-quiet quasar; 3C249.1 (PG1100+772), a radio-loud quasar; and 1202-0727, a high-redshift, radio-quiet quasar. The observational details are provided in Table 1. For these sources we found reliable detections in most/all the short wavelength bands (5-25 μm). In the long wavelength bands, however, only two of the sources are detected, the other two have negative "detections" in all four long wavelength bands.

Subsequent analysis of additional sources not reported here has shown a large number of negative signals at long wavelengths, particularly 135,200 μm. We are currently investigating the cause and, since the negative signals are often at levels similar to the positive ones, are treating all our long wavelength data with scepticism. The most likely cause is cirrus confusion and would imply significant structure on the scale of $\sim 3'$ (our chop distance). However further investigation is required to confirm this.

Table 1. Details of the ISOPHOT observations.

Name	z	ISO date	AOT[1]	Filter μm	Time sec[2]	Filter μm	Time sec[2]	Filter μm	Time sec[2]	Filter μm	Time sec[2]
PG1244+026	0.048	14/07/96	PHT03	4.85	16	7.3	256	12	256	25	256
			PHT22	60	64	100	64	135	64	200	64
PG1543+099	0.400	30/05/96	PHT03	4.85	256	7.3	128	12	128	25	256
			PHT22	60	64	100	64	135	64	200	128
3C249.1	0.313	17/06/96	PHT03	4.85	256	7.3	256	12	256	25	256
(PG1100+772)			PHT22	60	128	100	128	135	256	200	256
1202−0727	4.69	19/07/96	PHT03	4.85	512	7.3	512	12	512	25	512
			PHT33	60	512	100	128	135	512	200	512

1: AOT: Astronomical Observation Templates
2: on-source time

4 Spectral Energy Distributions

We have combined our ISO results with data from other wavelengths collected by ourselves and from the literature to generate SEDs of the four objects (Figure 2). To investigate possible contributions from the quasar host galaxy, particularly in the far-IR, SEDs for several kinds of galaxies were generated using data from the literature (McLeod, private communication). The IRG and ULIRG templates correspond roughly to $L_{IR} \sim 10^{10-11}$ $L_\odot$ and $10^{11.5}$ $L_\odot$ respectively. Superposed on each quasar SED, we have shown various galaxy SEDs as labelled and a median SED for low-redshift quasars (Elvis et al. 1994) for direct comparison (Figure 2).

PG1244+026 is a low luminosity active galaxy ($L \sim 10^{44}$ erg s^{-1}), officially classified as a Seyfert 1 galaxy and with low redshift (z=0.048). Figure 2 shows an L* spiral galaxy which is consistent with the AGN SED $\sim 1\mu m$ and with the cool dust contribution in the far-IR. Between 5 and 100 μm the quasar SED is dominated by an "AGN" component believed to originate in warm dust within the putative molecular torus. This component peaks $\sim 25 - 60\mu m$ in PG1244+026.

PG1543+489, also an optically selected PG quasar, has a luminosity $\sim 10^{45.5}$ erg s^{-1} (a bona fide quasar) and a redshift of 0.400. In this case an L* galaxy

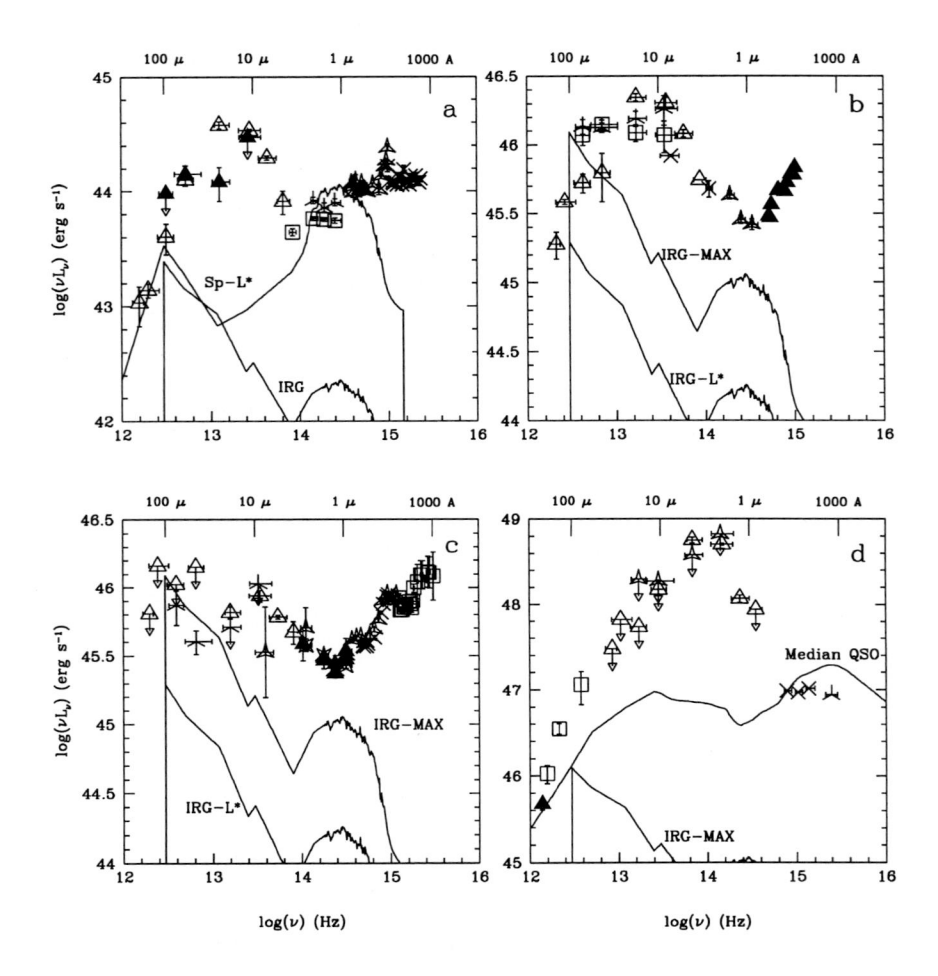

Fig. 2. The far-IR – ultra-violet SEDs of a: PG1244+026, b: PG1543+489, c: 3C249.1, d: 1202-0727, with various galaxy and quasar SEDs (as labelled) superposed. Each different dataset uses a different symbol, the ISO points are always indicated by open triangles.

would make no significant constribution in the near-IR. A galaxy with the maximum host galaxy luminosity seen to date (McLeod & Rieke 1994) is a factor ~ 4 too low at $1\mu m$. An amount of dust comparable to an IR-bright galaxy is necessary to explain the far-IR emission ($L_{IR} \sim 10^{10-11}$ $L_\odot$). Once again a mid-IR bump due to warm dust is apparent, peaking $\sim 100\mu m$.

3C249.1 is a lobe-dominated, radio-loud quasar at a reshift 0.389. The ISO short wavelength detections once again show a typical mid-IR bump with a peak

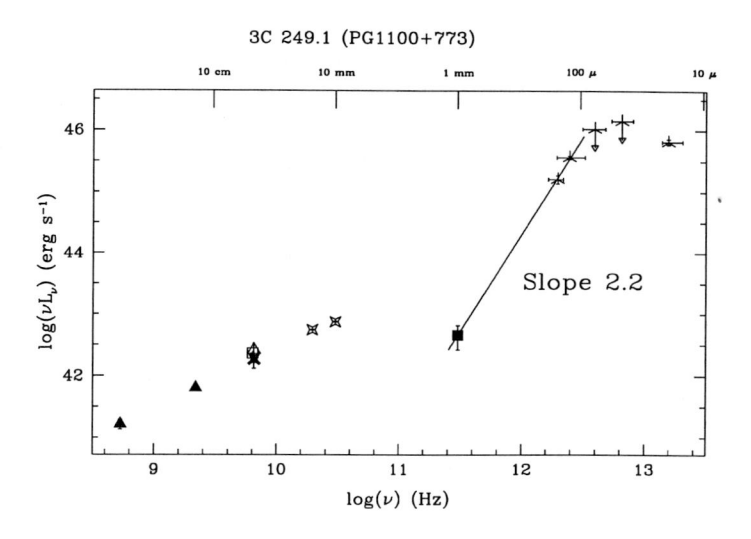

Fig. 3. The radio–far-IR SED of 3C249.1 showing the current limits on the slope of the far-IR turnover.

< 100μm. The IRAS upper limits (Elvis et al. 1994) are very low (~ 20 mJy at 100 μm) and inconsistent with the ISO detections. Re-analysis of the IRAS survey data using the current software (XSCANPI) yields upper limits which are largely consistent, as shown in Figure 2c. Unfortunately we currently have only weak upper limits on the far-IR emission of this source from ISO indicating that the data could be consistent with an L*, IR-bright galaxy but not one with maximal luminosity in the near-IR. However since we have no estimates of the host galaxy from the near-IR, this provides a very weak constraint on the amount of dust. Figure 3 shows the far-IR SED with a marginal detection at 1mm (Antonucci et al. 1990). Assuming that the 1mm flux represents the long wavelength tail of the far-IR emission, our current data give an upper limit to the slope of the far-IR turnover of $\alpha < 2.2$ ($f_\nu \propto \nu^\alpha$). The discontinuity between the far-IR and radio emission in the SEDs of lobe-dominated, radio-loud quasars is generally interpreted as evidence for differing emission mechanisms and thus thermal IR emission (Antonucci et al. 1990). However, the flat lower limit to the far-IR turnover in this source prevents us from ruling out non-thermal synchrotron emission.

1202-0727 is one of the highest redshift quasars known (z=4.690). It is an extremely interesting source, mentioned a number of times in these proceedings (Barvainis, Yamada). It is a double source with 4" separation at mm wavelengths, has strong CO emission (Ohta et al. 1996, Omont et al. 1996), a sub-mm spectrum which suggests emission from 50–100 K dust (Isaak et al. 1994) and a Lyα emission companion 2" away with the same redshift (Hu et al. 1996). The source has a very high luminosity, $\sim 10^{47}$ erg s^{-1}, such that the contribution from its host galaxy in the rest-frame optical and near-IR is several orders of magnitude below that seen by ISO (12, 25 μm observed frame, Figure 2d). The mid-IR bump, with a broad peak from 4–80 μm in the rest frame, is two orders of magnitude stronger than that of the low-redshift median (Fig. 2d). The rest-frame far-IR emission determined by the sub-mm observed frame data (Isaak et al. 1994) would require a host galaxy similar to an ultra-luminous IR galaxy (ULIRG) in a pure dust scenario. The crude far-IR upper limits from ISO suggest that our planned re-observation could strongly constrain the mid-IR SED.

5 Conclusions

Although the current status of the ISO far-IR data limits the usefulness of ISO to study the host galaxy dust contribution, we can already demonstrate that, if the far-IR emission of bona fide quasars (L> 10^{44} ergs^{-1}) is from the host galaxy, these galaxies are unusually far-IR bright, comparable to IR-bright galaxies or ULIRGs ($L_{IR} \sim 10^{10-11.5}$ $L_{\odot}$). The ISO data also allow us to investigate the mid-IR "AGN" bump in quasars covering a range of redshift and luminosity. We plan to use these data to test and constrain current models of emission from a molecular torus (Pier & Krolik 1992, Efstathiou & Rowan-Robinson 1995).

We have detected the z=4.69 quasar, 1202-0727, in the rest-frame near-IR at a level far above that seen in typical low-redshift quasars. Observations of more high-redshift quasars are necessary to determine whether the near-IR emission is unusual, as are many other aspects of this source. For pure host galaxy far-IR emission, the host must be comparable to an ULIRG to explain the mm data (Isaak et al. 1994).

Acknowledgements

I would like to thank all my collaborators on this project, in particular Drs. Kim McLeod, Jonathan McDowell and Martin Elvis at CfA and our other ISO co-Is. Thanks are also due to the ISOPHOT team in Heidelberg and the ISO centers at VILSPA and IPAC for their prompt and invaluable help in response to my frequent email messages. The financial support of NASA grant NAGW-3134 is gratefully acknowledged.

References

Antonucci, R., Barvainis, R. & Alloin, D. 1990, ApJ, 353, 416
Dunlop, J. S., Taylor, G. L., Hughes, D. H. & Robson, E. I. 1993, MNRAS 264, 455
Elvis, M., Wilkes, B. J., McDowell, J. C., Green, R. F., Bechtold, J., Willner, S. P., Cutri, R., Oey, M, S., and Polomski, E.
Hu, E. M., McMahon, R. G. & Egami, E. 1996, ApJ, 459, L53
Isaak, K. G., McMahon, R. G., Hills, R. E. & Withington, S. 1994 MNRAS, 269, L28
"ISOPHOT Observers Manual" and associated updates found on WWW: http://isowww.estec.esa.nl:80/ISO/iso_manuals.html

Laor, A., Fiore, F., Elvis, M., Wilkes, B.J., & McDowell, J.C. (1996) ApJ. *in press*
McLeod, K.K. & Rieke, G. 1995, ApJ, 441, 96
McLeod, K.K. private communcation
Ohta, K, Yamada, T., Nakanishi, K., Kohna, K., Akiyama, M. & Kawabe, R, 1996, Nature, 382, 4260
Omont, A., Petitjean, P., Guilloteau, S., McMahon, R. G., Solomon, P. M. & Pecontal, E. 1996, Nature, 382, 4280
Pier, E. A. & Krolik, J. H. 1992, ApJ 401, 99
Efstathiou, A. & Rowan-Robinson, M. & 1995, MNRAS 273, 649
XSCANPI, Interactive software for the analysis of IRAS data provided by IPAC via "telnet xscanpi.ipac.caltech.edu"

ISO Observations of Seyfert Galaxies

Jose Miguel Rodríguez Espinosa and Ana María Pérez García

Instituto de Astrofísica de Canarias, 38200-La Laguna, Tenerife, Spain

Abstract. We present ISOPHOT observations of several objects from the CfA sample of Seyfert galaxies. Data for over 50% of the objects in the sample have been acquired so far. In all instances the far IR emission can be accounted for by assuming two different emission components.

1 Introduction

The Infrared Space Observatory, launched in November 1995, has been performing scientifc observations since about February-March 1996, and is expected to last till about December 1997, some six months beyond the nominal lifetime of the satellite.

Data for the objects in the CfA sample of Seyfert galaxies are being obtained as part of the ISOPHOT Guaranteed time granted to the authors because of their involvemet in the construction of ISOPHOT-S, a spectrophotometer working in the range from 2.5 to 12 μm which is a subsystem of ISOPHOT. ISOPHOT-S was built at the Instituto de Astrofísica de Canarias.

2 Data

We are working on a project consisting of a study of the morphologies, physical properties and energetics of the CfA sample of Seyfert galaxies. This sample contains 48 galaxies from the CfA redshift survey, and have been selected for having emission lines in their spectra (Huchra & Burg, 1992). The CfA sample is complete to m_w =14.5, and has received a lot of attention in the recent past. The sample has been observed in the radio (Kukula et al, 1995; Kukula et al, 1994a,b), mid-IR (Edelson et al, 1987), near-IR (Peletier et al, 1997, McLeod & Rieke, 1995), optical (Pérez García et al, in preparation) and UV (Edelson et al, 1990). The observations reported here are the first to be available in the far IR regime for most of the objects.

Far IR observations in a number of filters (table 1) ranging from 12 to 205 μm have been obtained with ISO, in order to accurately define the far IR spectral energy distribution (SED) of these objects, and in combination with the optical data, determine the origin of the far IR emission, which is still somewhat debated.

Most objects in the CfA sample are bright enough to be seen even in the raw data products (e.g. fig 1). The data have been reduced with the PHT Interactive Analysis (PIA) tool, kindly provided to us by the MPIA and ESA. Starting from Edited Raw Data (ERD) we have corrected for a number of effects in

the data, like non-linearity of ramps, cosmic particle hits on the detectors, and
detector drifts. Dark current is substracted when we correct the object files of its
background, since both the object and background frames were acquired with
the same integration time.The photometric calibration was achieved using an
updated responsivity value kindly provided to us by J. Acosta (a member of the
ISOPHOT team at Vilspa). The far IR calibration is still being refined, and we
have thus adopted 30% as the error for all the calibrated data.

Filter	λ_c (μm)	$\Delta\lambda$ (μm)
16	15.14	2.86
25	23.81	9.18
60	60.8	23.9
90	95.1	51.4
120	119.	47.3
135	161	82.5
180	185.5	71.7
200	204.6	67.3

Table 1. List of filters used in the observations

All objects observed so far show good detections at all the observed wavebands. The ISO data agree relatively well with preexisting IRAS data in those objects for which IRAS data were available. In many objects he values at 16μm tend to be somewhat higher than the interpolated IRAS data, although that is not always the case (e.g. Mrk 817, NGC7469, NGC4388). We are, however, carefully investigating this effect.

3 Results

Two interesting features are readily seen in the data when represented in a F_ν versus λ plot, namely a far IR turnover in the SED, and a mid IR bump (Rodríguez Espinosa et al 1996). It is important to note that this behaviour is present in all objects observed so far. Figure 2 shows some examples, including NGC3227 and NGC4151, which are well known objects, and 1058+45 , which is a fainter and less known object. A question arises as to whether these behaviour is real or is rather an artifact produced by the high 16μm point together with the lack of any intermediate data points between 25 and 60 μm. We claim that that is not the case as the 25μm point agrees in most cases quite well with the IRAS 25 μm measurement (e.g. NGC 3227 and NGC7469, fig 2). Another example of this behaviour, with a far IR turnover and a mid IR bump is seen in the KAO data of NGC 1068 obtained by Telesco et al (1984) We therefore conclude that the presence of two humps, possibly corresponding to two distinct emitting regimes, is an universal feature in Seyfert galaxies.

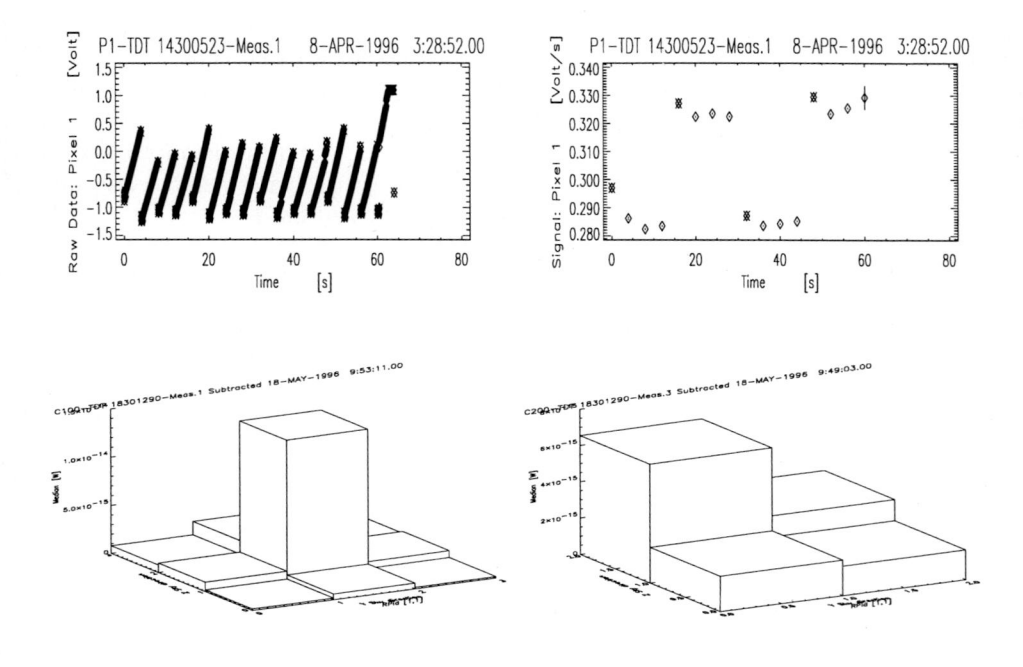

Fig. 1. Edited Raw Data (ERD) of NGC3079 at 16μm (a) and Signal per Ramp Data (SRD) for this same object (b), after correction linearity and deglitching of the ERD data. These measurements were done in chopping mode. Figures (c) and (d) shows the signal (in watts) of NGC7469 with the C100 (60μm) and C200 (180μm) arrays. These measurements were done in staring mode, and the backgrounds have been substracted.

4 Discussion

In order to explain the two broad features seen in the SED of Seyfert Galaxies, we have fitted two emissivity weighted blackbody functions with very good results (e.g. figure 3). These two balckbodies can be characterized by temperatures of around 150 K for the warmer BB and around 30K for the colder one.

The most likely explanation for the cold component is emission by dust heated in starforming regions. This idea is supported by the temperature range obtained from our blackbody fit to the long wavelength data. Indeed the range 25 to 35 K is typical of the dust generally found in HII region/molecular clouds complexes (Telesco et al, 1980).

The nuclear origin of the warm dust emission is supported by the relatively high temperature of the dust (150~180 K), far warmer that typical dust grains temperatures (20-50 K) in conventional star-forming regions and galactic molecular clouds (Telesco et al, 1980). A recent work of Giuricin et al (1995) support this idea. These autors have found that small aperture 10μm luminosities of a

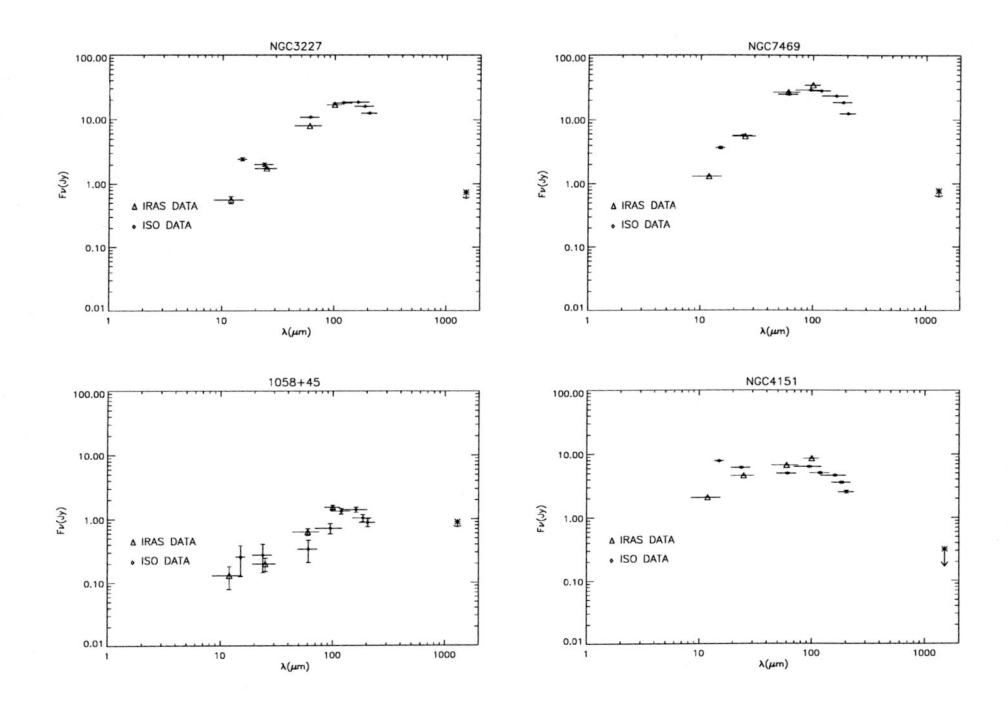

Fig. 2. Mid and far IR spectral energy distribution (SEDs) of some objects of the CfA sample. ISO data are represented by diamond while IRAS data are shown with triangles. The 1.3mm upper limit, from Edelson *et al* (1987), is shown with a star. Horizontal error bars indicate the FWHM middling of the different filters.

sample of over 100 active galaxies correlate very well with their IRAS luminosities at 12 and 25μm, while the correlation is poorer with the 60 and 100μm luminosities. Other explanation for this warm dust is the possible presence of a thick torus that would be responsible for the mid-IR emission in AGNs. Indeed a combination of both mechanisms is also a plausible scenario.

5 Conclusions

The far IR emission from Seyfert galaxies can be explained as originating in two different regimes of thermal emission, a cold one very likely produce by the emission of cold dust heated in star forming regions; and a warmer region which most likely is due to emission by dust heated directly by the active nucleus.

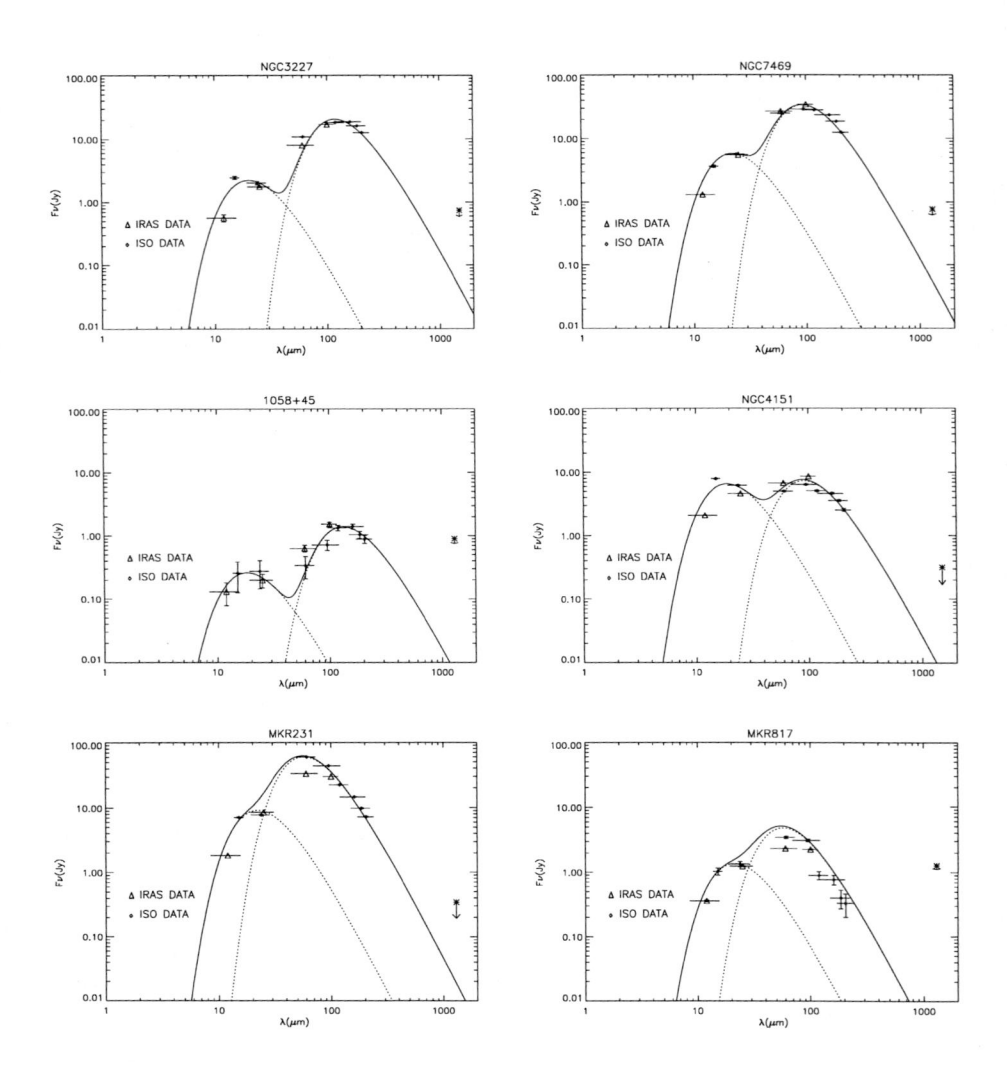

Fig. 3. IR SEDs of of several galaxies from the CfA sample, fitted with two emissivity weighted blackbodies (dotted lines), one representing warm dust and a second one representing cold dust (see text).

References

Edelson, R.A., Pike, G.F., Krolik, J.H. (1990): ApJ, **359**, 86.
Edelson, R.A., Malkan, M.A., Rieke, G.H. (1987): ApJ, **321**, 233.
Giuricin, G., Mardirossian, F, Mezzetti, M. (1995): ApJ,**321**, 233.
Huchra, J., Burg, R. (1992): ApJ,**393** 90
Kukula, M.J., Pedlar, A., Baum, S.A., O'Dea, C.P (1995): MNRAS, **276**, 1276.
Kukula, M.J, Pedlar, A., Baum, S., O'Dea, C., Unger, S. (1994): IAUS, **159**, 514.
Kukula, M.J, Pedlar, A.,Unger, S., Baum, S., O'Dea, C. (1994): ApJSS, **216**, 371.
McLeod, K.K., Rieke, G.H. (1995): ApJ, **441**, 96.
Pérez García, A.M, Rodríguez Espinosa, J.M., *in preparation*
Peletier, R.F., Knapen, J.H., Shlosman, I., Nadeau, D., Doyon, R., Rodríguez Espinosa, J.M., Pérez García, A.M. (1997): ApJL, *submitted*
Telesco, C.M., Becklin, E.E., Wynn-Williams, C.G. (1984): ApJ, **282**, 427.
Telesco, C.M., Becklin, E.E., Wynn-Williams, C.G. (1980): ApJ, **241**, L69.

Origin of Spread in the $B - K$ Color of Quasars

R. Srianand

IUCAA, Post Bag 4, Ganesh Khind, Pune 411 007 (INDIA)

Abstract. Recently Webster et al. (1995) have shown that there is excess reddening in radio selected quasars relative to the optically selected population. If the reddening is universal to all quasars, then optical surveys which depend on the UV excess, say, would be seriously incomplete. Based on our model calculations we suggest that the required amount of reddening can not be produced by the dust in the intervening damped Ly α absorbers. The optical depth of dust intrinsic to a quasar, which is required to produce the observed spread in optical-to-near-IR colors, is estimated for different extinction curves. Results of photoionization models suggest that, for a wide range of ionization parameter and metallicity, the gas associated with dust will produce Lyman limit as well as saturated heavy element absorption at the redshift of the quasar. The distribution of colors in quasars with associated absorption suggests that the reddening is not dominated by dust intrinsic to the quasars. We present evidence for the aspect dependence of optical-to-near-IR colors in the radio-loud quasars. We propose the relativistic beaming in flat spectrum radio sources to be a source of the large $B - K$ colors found by Webster et al.

1 Intrinsic Spectral Energy Distribution

Since quasars have strong emission lines and different components of the continuum emission (like the big blue bump, Balmer continuum etc) we use a more realistic spectrum for a model quasar as different features will contribute to the observed colors at different redshifts. We use the composite spectrum compiled by Francis et al. (1991) for the rest wavelength range between 800Å and 6000Å. We have assumed that SED in the near-IR to be a power law with spectral index between 0.5 and 2.0, allowing for the uncertainties in the IR spectral shape. This range is necessary to cover the observed IR colors (i.e. $J - K$ and $H - K$) of low redshift quasars in our sample. We have estimated various standard colors by passing the composite spectrum through the filter response function of corresponding pass bands.

2 The Effect of Dust in the Damped Ly α Absorbers

It is widely believed that the damped Ly α absorbers (DLAs) at high redshifts are the progenitors of the present day galaxies. It is also confirmed that these absorbing clouds have small amount of dust in them. In what follows we construct a realistic model for dust in these absorbers and estimate its contribution to the spread in $B - K$ color.

The average optical depth of dust present in the DLAs along the line of sight to a quasar at any redshift can be written as

$$\bar{\tau}(z) = 104h \int_0^z dz' \frac{k(z')\Omega_{HI}(z')(1+z')}{\sqrt{1+2q_o z'}} \xi\left(\frac{\lambda_B}{1+z'}\right), \quad (1)$$

where h is Hubble's constant in units of $100 \, \text{km sec}^{-1} \, \text{Mpc}^{-1}$, $\xi(\lambda)$ is the ratio of the extinction at any wavelength λ to that in the B-band, k is the dimensionless dust to gas ratio and $\Omega_{HI}(z)$ is the mean comoving density of H I in DLAs, in units of the present critical density.

The observed flux $f_o(\lambda_o)$ and the emitted flux $f_e(\lambda_e)$ are related by,

$$f_o(\lambda_o) = \frac{f_e(\lambda_e)}{1+z_e} \exp\left[-\int_0^{z_{em}} dz_a \, \bar{\tau}_a\left(\frac{\lambda_o}{1+z_a}\right)\right], \quad (2)$$

Thus for a given extinction curve the effect of dust in the DLAs can be modeled with two parameters (k and Ω_{HI}).

Wolfe et al. (1995) have shown that the Ω_{HI} evolves with redshift as

$$\Omega_{HI}(z) = \Omega_{HI}(0) \exp(\alpha z) \quad (3)$$

where $\Omega_{HI}(0)$ is the density of neutral hydrogen gas at $z=0$ and α is a constant characterizing the rate of evolution. They found, for $q_o = 0.5$, $\Omega_{HI}(0) = 0.23 \pm 0.08 \times 10^{-3} h^{-1}$ and $\alpha = 0.70 \pm 0.15$. Pei, Fall and Bechtold (1991) have obtained the values of k in the DLAs to be $0.35^{+0.24}_{-0.09}$ for Galactic extinction curves. We use this value of k with $\Omega_{HI}(0)$ and α obtained by Wolfe et al. (1995) in what we call "low reddening models".

Note $\Omega_{HI}(z)$ is calculated from DLAs in the foreground of optically selected quasars and it is probably biased by the very obscuration and the observed values of Ω_{HI} and k may be lower limits. We get an upper limit on Ω_{HI} from big bang nucleosynthesis calculations. Boesgaard and Steigman (1985) estimated the baryon density, Ω_b, based on nucleosynthesis as $0.034-0.048$. At $z=3.0$ Steidel (1990) estimated the Ω_{HI} in Lyman limit systems (LLS), Ω_{LLS}, to be 0.004. The contribution of luminous matter is estimated to be $\Omega_{lb} \simeq 0.002$ (Persic and Salucci (1992)). The baryonic density in the IGM, Ω_{IGM}, estimated using the Gunn-Peterson effect is less than 0.004 (Giallongo et al. (1994)). One can write

$$\Omega_b = \Omega_{IGM} + \Omega_{LLS} + \Omega_{HI}^{DLy\,\alpha} + \Omega_{lb} \quad (4)$$

and thus the maximum value of Ω_{HI} that can be in the DLAs, $\Omega_{HI}^{DLy\,\alpha}$, at $z \simeq 3$ is 0.032.

If we assume the form of equation (3) is valid in the biased distribution of DLAs too, then the maximum inferred Ω_{HI} at $z=3$ will give $\alpha \simeq 1$ for the $\Omega_{HI}(0)$ obtained from radio observations of nearby galaxies (Rao and Briggs (1993)). In the "maximum reddening models" we use $\alpha = 1$, $\Omega_{HI}(0)$ given by Rao and Briggs (1993) and dust to gas ratio $k = 0.8$ as observed in our Galaxy.

It is clear from the figure 1 that for the range of near-IR spectral indices, the models with no reddening give a lower envelop of the $B-K$ color at all redshifts.

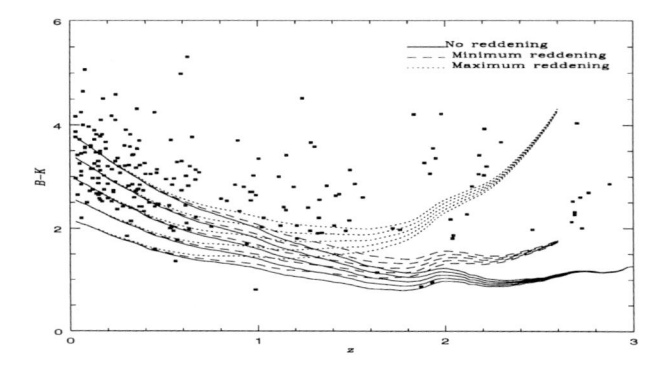

Fig. 1. Effect of dust in the Damped Ly α clouds.

Low as well as maximum reddening models fail to produce sufficient reddening at low redshift as seen in the observed distribution. Note that our maximum extinction model assumes that all the unobserved baryons are in DLAs with maximum dust to gas ratio and still fails to produce the observed spread in the $B - K$ color at low redshift. Thus our analysis clearly shows the observed spread at low z is not due to the reddening of quasars by the intervening objects.

3 Effect of Dust Intrinsic to Quasars

Dust can be intrinsic to the quasar either associated with central region or with the host galaxy. We estimate the amount of dust, closed to the quasar, needed to explain the spread in the $B - K$ colors. Since we do not know the exact extinction curve for the associated dust we used mean extinction curves of our galaxy, LMC and SMC (Savage and Mathis (1979), Nandy et al. (1981) & Pervot et al. (1984)). We can write the optical depth of dust at different wavelengths in terms of optical depth in the B band, τ_B, as,

$$\tau_\lambda = \tau_B \, \xi(\lambda). \qquad (5)$$

The estimated maximum B band optical depth required to produce the upper envelop of the observed color distribution, τ_B^{max}, and the total column density, N_H^{total}, for k similar to (A) and one tenth (B) of the dust-to-gas ratio in the interstellar medium of Galaxy, LMC and SMC are given in table 1.

3.1 Dust in the BLR

It is believed that surface covering factor of clouds in the BLR is ~ 0.10. Thus only a few percent of the quasars will have a broad line emitting clouds along our line of sight. AGN observations do not show any large reduction in the strengths of resonance lines like Ly α, C IV etc., compared with calculated intensity of

Table 1: Parameters of intrinsic dust

Dust model	τ_B^{max}	log (N_H^{total}) for Model A	Model B
Galactic	0.35	20.60	21.60
SMC	0.25	21.70	22.70
LMC	0.30	21.00	22.00

the intercombination lines like C III]λ1909. Therefore the amount of dust in the BLR clouds is unlikely to be significant. Reverberation mapping studies of BLR in nearby AGN shows the radius of the BLR is $\sim 0.06 L_{46}^{0.5}$ pc (Netzer and Laor (1993)), where L_{46} is the bolometric luminosity in units of 10^{46} erg s^{-1}. Laor and Draine (1993) have shown that the sublimation radius from the central engine at which even large graphite grains will sublimate is $\sim 0.2 L_{46}^{0.5}$ pc. Thus dust cannot survive inside the BLR and the required reddening is not caused by the BLR clouds.

3.2 Dust and Associated Gas Outside BLR

The obscuring dust, if associated with quasar, should lie outside BLR. In this section we study the ionization structure of the gas associated with the dust, using photoionization models (using cloudy). The absorber is assumed to be a plane parallel slab with total hydrogen column density obtained using Galaxy extinction curve (given in Table 1). We use a range of ionization parameters ($-1.5 < log(\Gamma) < -3.5$). The metal abundances and dust-to-gas ratio are assumed to be similar to (Model A) and one tenth (Model B) of the average ISM values of our Galaxy. The incident ionizing radiation was assumed to be like a typical AGN spectrum.

The results of photoionization calculations suggest that, for the range of parameters considered here, the neutral hydrogen optical depth at the Lyman limit will be always greater than one and most of the observable heavy element transitions like C IV and Mg II will be in the saturation portion of the curve of growth. This will mean an associated absorption at the redshift of the quasar. In figure 2a and 2b we have plotted $B - K$ color as a function of redshift of the quasars in our sample for which Lyman limit and heavy element absorption line information are available respectively. We denote the quasars with associated absorption with crosses. It is clear from the figures that there is no tendency for red quasars to show associated absorption.

In order to produce $B - K \geq 3$ one needs $\tau_B > 0.25$ for our model composite spectrum. For dust-to-gas ratio similar to the Galactic ISM this will correspond to a total hydrogen column density $> 3 \times 10^{20}$ cm^{-2}. Photoionization models predict column density of C IV and Mg II to be greater than 10^{14} cm^{-2} for a wide range of parameters. This will mean all quasars with $B - K > 3.0$ should show associated intrinsic absorption. There are 18 quasars with $B - K > 3$ in our sample, of which only 4 show associated absorption. Using 5σ upper limits

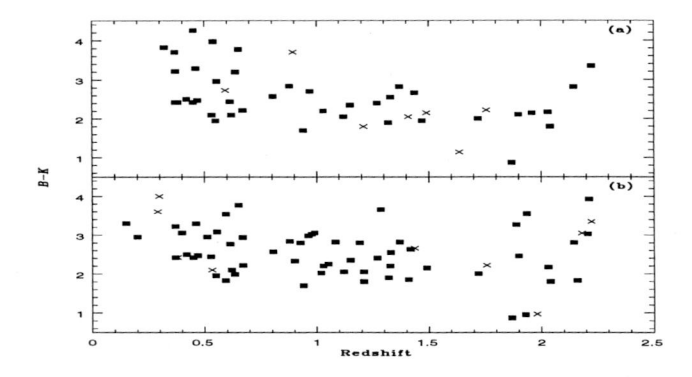

Fig. 2. Intrinsic absorption

on the equivalent widths of Mg II (0.3 Å) and C IV (0.15 Å) absorption lines for these quasars, we estimate an upper limit to dust optical depth of $\tau_B < 10^{-3}$. Thus our results suggest that the intrinsic spread in the $B - K$ color of 2 mag seen in our sample is produced by effects other than reddening. Note that our sample does not have very red quasars (i.e. $B - K > 5.0$) as in the case of the Webster et al. sample. The presence of associated absorption in these quasars will favor the dust models and can give a tight bound on the dust optical depth.

4 Discussion

Various results discussed above suggest that the spread in $B - K$ color of quasars up to ~ 2 mag is caused by effects other than reddening due to line of sight dust. One possible alternative is that most of the observed spread in $B - K$ in flat spectrum quasars is due to some property peculiar to those objects. The large values of $B - K$ in quasars selected in the flat-spectrum surveys could also be due to a beamed component at the near-IR wavelengths. We have obtained the value of R (defined as the ratio of radio core flux density to the extended radio lobe flux density) for most of the quasars in our sample from the literature. In the relativistic beaming model for radio sources R is related to the angle between the radio axis and the line of sight. In Figure 3 we have plotted $B - K$ colors of quasars in the sample against R. It can be seen from the figure that, on an average, the core dominated quasars (CDQs) tend to be redder than the lobe dominated quasars (LDQs). The Spearman rank correlation test also suggest a weak correlation, in spite of the large spread shown in Figure 3.

Using composite spectra of the Molonglo sample of quasars, Baker and Hunstead (1995) have shown the extinction to be aspect dependent, with the reddening decreasing with R. Their sample was originally selected at 408 MHz, where steep spectrum, extended emission dominates. The typical value of $\log R$ in the LDQs is between -2 and -1. When we consider only quasars, in our sample, with

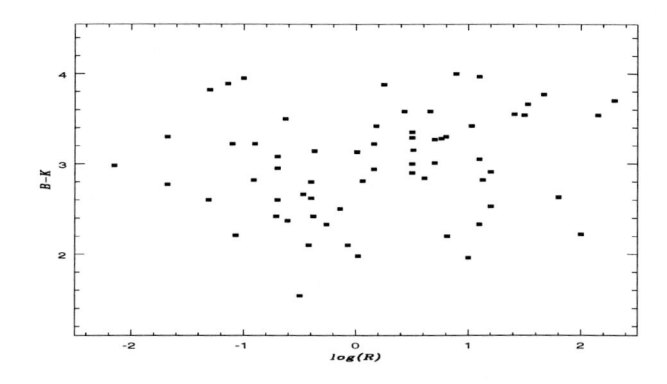

Fig. 3. Orientation dependence of $B - K$ color

$\log R > -1.0$ (i. e., consider only marginally lobe-dominated and CDQs) the correlation between $\log R$ and $B - K$ color increases and becomes more significant. When we consider only LDQs (i.e. quasars with $\log R < 0$) we find significant anti-correlation between $\log R$ and $B - K$ colors. Thus our results suggest that both extremely lobe-dominated as well as extremely core dominated quasars tend to be red. While obscuration is the cause in the case of LDQs, in CDQs non-thermal beamed emission may be the source of the reddening. Note that our sample is drawn from different sources in the literature and is affected by unknown selection biases. Also our data set does not have the very red quasars which are present in the Webster et al. (1995) sample. Observing very large values of R for quasars with high $B - K$ and with no associated absorption lines in a complete sample will confirm the beaming model.

References

Baker, J., & Hunstead, R. W. 1995, ApJ, 452, L95.
Boesgaard, A. M., & Steigman, G. 1985, ARA&A., 23, 319.
Francis, P. J. et al., 1991, ApJ, 373, 465.
Giallongo et al., 1994, ApJ, 425, L1.
Laor, A., & Draine, B. T. 1993, ApJ, 402, 441.
Nandy, K., et al., 1981, MNRAS, 196, 955.
Netzer, H., & Laor, A. 1993, ApJ, 404, L51.
Pei, Y. C., Fall, S. M., & Bechtold, J. 1991, ApJ, 378, 6.
Persic, M., & Salucci, P. 1992, MNRAS, 258, 14p.
Pervot, et al., 1984, A&A, 132, 389.
Rao, S., & Briggs, F. H. 1993, ApJ, 419, 515.
Savage, B. D., & Mathis, J. S. 1979, ARA&A, 17, 73.
Steidel, C. C. 1990, ApJS,74,37
Webster, et al., 1995, Nature, 375, 469
Wolfe et al., 1995, ApJ, 454, 698.

Using HI to Probe AGN Hosts and Their Nuclei

Carole G. Mundell

University of Manchester, NRAL, Jodrell Bank, Macclesfield, Cheshire, SK11 9DL

Abstract. The λ21-cm neutral hydrogen (HI) spectral line is a valuable probe of the distribution and kinematics of gas in AGN hosts. I present results which cover a wide range of size-scales and physical processes. These include interactions on scales of 100 kpc, inflow along a galactic bar on scales of 10 kpc and evidence of a circumnuclear torus on scales of 10–100 pc. These studies of the triggering/fuelling chain in nearby AGN hosts indicate the potential for similar studies of more distant AGN such as quasars, given instruments such as the proposed square kilometre array.

1 Introduction

The host galaxy of an Active Galactic Nucleus (AGN) represents a vast reservoir of potential fuel, but one of the key problems in AGN physics is how to transport that fuel from the outermost regions of the galaxy, down 6 orders of magnitude in distance scales, to the centre to fuel the AGN. In particular, the fuelling of an AGN requires that the gas is delivered to the centre with essentially zero angular momentum, in order to form (and re-fuel) an accretion disc around a central black hole (Shlosman et al., 1990). It has been suggested that tidal interactions between galaxies may play a role in this process, either directly, when gas from the companion, or outer regions of the host galaxy, is tidally removed and deposited onto the nucleus, or by triggering instabilities (such as bars) which provide non-circular motions and inflows of galactic gas (Shlosman et al., 1989).

Seyfert galaxies are the closest and most common type of AGN and, although their active nuclei are relatively weak, they exhibit many of the properties of their more luminous counterparts, particularly quasars. As such they represent excellent laboratories for the study of the AGN phenomenon and its relationship to the host galaxy environment, which is difficult to observe in more distant and powerful AGN. Statistical studies have shown that many Seyferts are distorted spirals, are barred, have nearby 'companion' galaxies or are in interacting systems, and are often gas-rich (e.g., Heckman, 1978; Simkin, et al., 1980; Keel et al., 1985; MacKenty, 1990). In the case of Seyfert galaxies, significant circumstantial evidence exists to suggest that interactions are closely linked to the phenomenon of nuclear activity, but attempts to establish a conclusive relationship between the two are highly controversial (Fuentes-Williams & Stocke, 1988; Laurikainen et al., 1994; Rafanelli, et al., 1995). Similarly, bars are thought to be strongly related to starburst activity, but their role in Seyfert activity is less well established (Ho, 1996).

λ21-cm neutral hydrogen (HI) spectral line is a very good tracer of galactic structure and dynamics over many size scales and HI emission and absorption studies together can span over 6 orders of magnitude in distance scales. Since HI is often the most spatially extended, observable component of a galaxy's disk, it is particularly sensitive to interactions (Gallagher et al., 1981), and galaxies which are optically classified as isolated may in fact show signatures of interaction in their HI distribution (e.g. tidal tails extending ~100's kpc). HI synthesis observations also provide 2-dimensional kinematic information which is vital in constraining models of the interaction.

Neutral gas may also react in a highly non-linear way to even small deviations in axial symmetry making it a good tracer of barred potentials (Teuben et al., 1986). Gas streaming in galactic bars may play an important role in the early stages of the fuelling process (Simkin et al., 1980), with significant angular momentum loss occurring when two different families of gas orbits meet and form dissipative shocks, allowing gas to move inward (Prendergast, 1983; Athanassoula, 1992).

On the smallest scales, HI absorption studies, with MERLIN and the VLBA, provide a direct probe of the obscuring torus invoked in AGN unification schemes (Mundell et al., 1995a; Conway & Blanco, 1995). Column densities and kinematics in the torus can be measured on scales <10 pc in nearby objects. Absorption studies also distinguish between infall and outflow from the nucleus.

To date, sensitivity limitations have resulted in few detailed synthesis studies of HI in Seyferts (Brinks & Mundell, 1997, and references therein). I present some recent observations of HI in some well-known Seyfert galaxies which form part of an ongoing project to study the distribution and kinematics of HI in AGN.

2 NGC3227

NGC3227 is an interacting Seyfert galaxy and, as part of a detailed optical and radio study (Mundell et al, 1995b), the HI distribution and kinematics have been studied with angular resolution ranging from 12" to 60" (Mundell et al., 1995c). The low-resolution images (Fig. 1) show plumes of HI extending to ~70 kpc north and ~31 kpc south of the galaxy which may be a consequence of interaction. At higher resolution, the galactic disc and bar are resolved, and we see approximate solid-body rotation and an anomalous-velocity cloud (~150 km/s above the systemic velocity of NGC3227) situated north-west of the disc at the base of the northern plume (Fig. 1). The cloud, which shows some evidence for rotation, might be a gas-rich dwarf galaxy and may either be partly responsible for the interaction or has been formed as a consequence of it.

In the disk of NGC3227, a Z-shaped HI bar crosses the nucleus from NW to SE and seems to be a continuation of a CO bar reported by Meixner et al. (1990). Clearly, the interaction could have played a major role in the formation of the bar and it is interesting to note that the bar is approximately perpendicular to the nuclear optical outflow cone and subarcsecond radio jet (Mundell et al., 1995b), suggesting that in some way the structures may be related.

158   C.G. Mundell

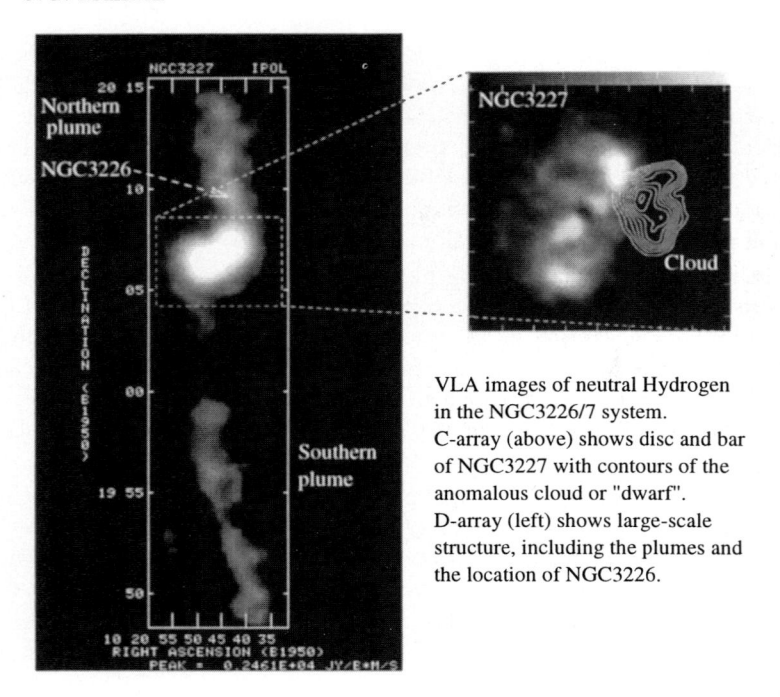

Fig. 1. Integrated HI distribution of NGC3227

VLA images of neutral Hydrogen in the NGC3226/7 system. C-array (above) shows disc and bar of NGC3227 with contours of the anomalous cloud or "dwarf". D-array (left) shows large-scale structure, including the plumes and the location of NGC3226.

3 NGC4151

NGC4151 is a well-studied nearby gas-rich spiral galaxy with a prominent central bar and a Seyfert 1.5 nucleus. The bar is a 'fat' oval structure with dimensions of $2' \times 3'$ and is elongated along PA $\sim 130°$. Fig. 2 shows our new high-resolution HI image of the the bar (resolution $6'' \times 5''$), clearly resolving the two small regions of enhanced emission which lie in the NW and SE corners of the bar, offset from the bar major axis. The gas distribution in the bar closely resembles simulations of gas moving in periodic orbits in a *weak* bar potential (Athanassoula, 1992), in which shocks form along the leading edges of the bar when two of the main families of gas orbits, x_1 and x_2, meet.

In order to determine whether shocks are present in the gas in NGC4151 we examined HI emission along the equivalent of a $6''$-wide slit, at several locations and orientations, across the length of each arc, to look for the predicted velocity jumps that are characteristic of shocks in galactic bar gas. When the slit is positioned perpendicular to each arc and across its brightest peak the maximum velocity jump is seen, and an example of this for the NW arc is shown in Fig. 2. Similarly oriented slits centred at different positions along both arcs showed a similar velocity structure. The observations bear a striking resemblance to the theoretical predictions of gas behaviour in a weak barred potential (Fig. 2(iv)), in which the maximum velocity jump in the simulated data is also found when the

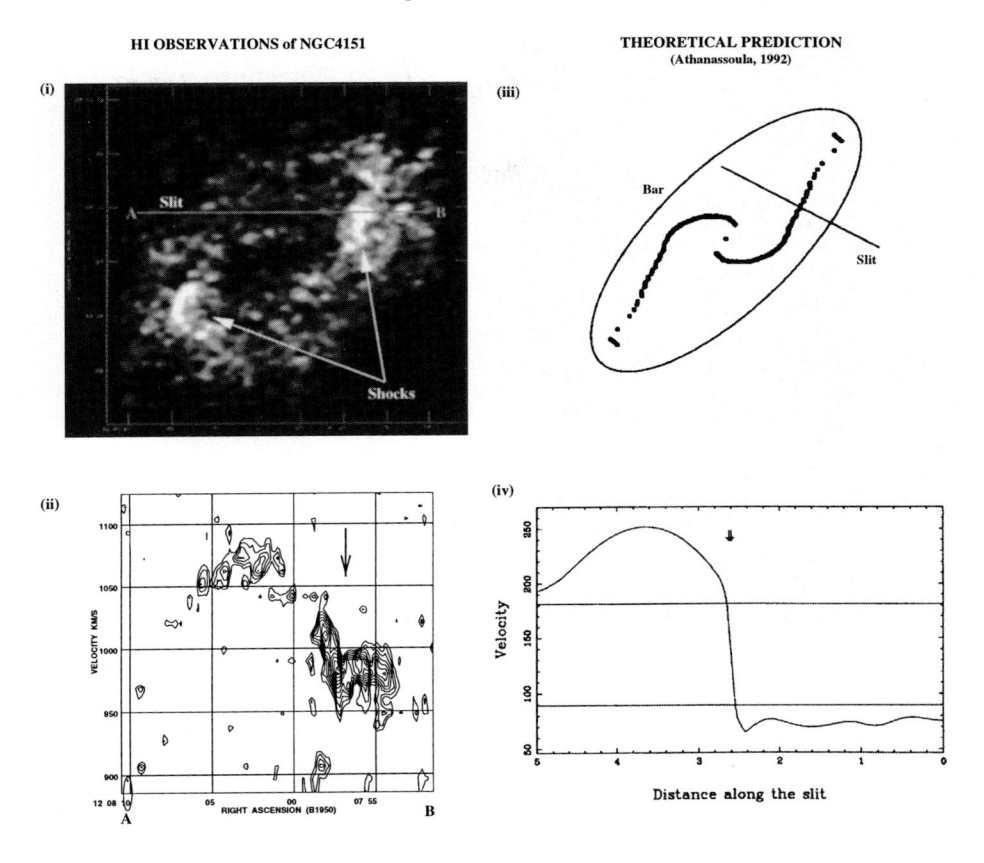

Fig. 2. Comparison of observations with numerical simulations of gas flows in a weak barred potential. (i) HI in the Bar of NGC4151 with shocks and slit position indicated; (ii) Velocity profile across NW shocks taken along the slit shown in (i); (iii) Numerical simulations of gas in a general weak bar potential from Athanssoula (1992); (iv) predicted velocity profile across a typical leading-edge shock in a weak bar (Athanassoula, 1992).

slit is perpendicular to the shock. A residual velocity field, produced by removing a circular rotational component from the observed velocity field, reveals strong deviations from the local bar field, in the shock regions. Assuming trailing spiral arms these deviations are consistent with gas inflow along the shocks towards the nucleus (Mundell & Shone in preparation).

Neutral gas within the central $0.1''$ of NGC4151 has been detected in recent subarcsecond MERLIN observations of HI absorption against the nuclear radio jet (Mundell et al., 1995a). The results are interpreted as evidence for a circumnuclear torus, 50pc in size, which may form part of the nuclear fuelling chain and be responsible for collimating the nuclear UV radiation.

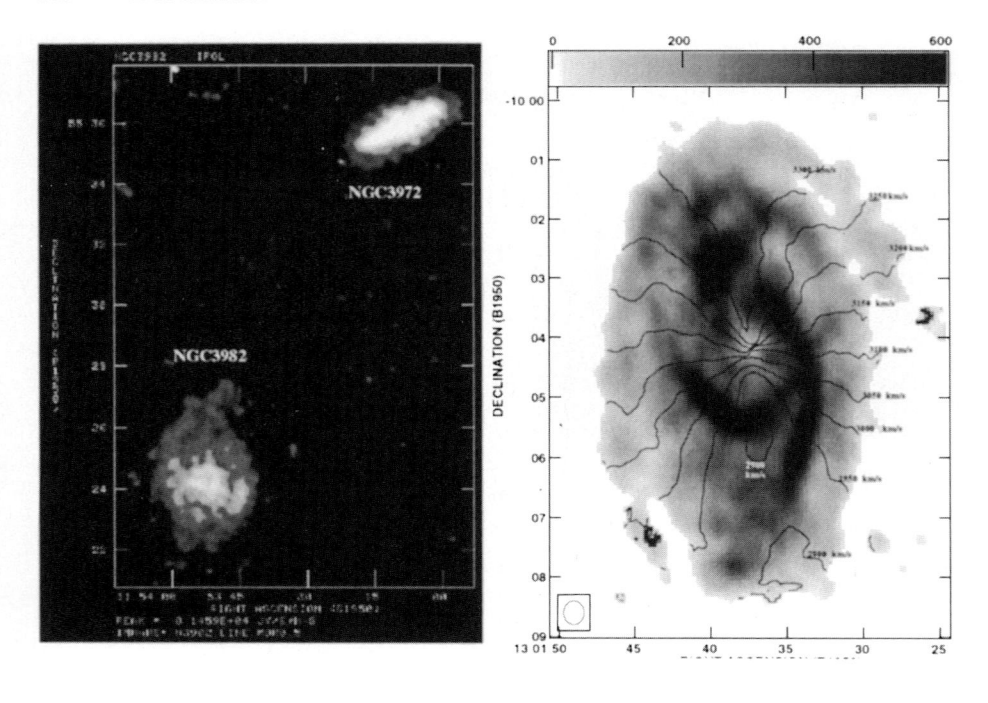

Fig. 3. HI in NGC3982; optically classified as isolated, this Seyfert may be interacting with gas-rich NGC3972. NGC4939 HI distribution with the velocity contours superimposed.

4 NGC3982, NGC4939 and NGC5506

Preliminary results are presented for NGC3982, NGC4939 and NGC5506 (Figs. 3, 4). Both NGC3982 and NGC5506 show evidence of tidal disturbance. In NGC3982, a ring of strong HI emission is coincident with the circumnuclear starburst ring (Pogge, 1989) and the velocity field of the disk is relatively undisturbed. NGC4939 appears to be isolated but twisting of the iso-velocity contours in the central ~20″ suggests the possible presence of a bar. In addition to the disk emission in NGC5506, strong absorption is seen against the radio continuum nucleus and the central kinematics are complex.

In summary:
Perhaps the most important conclusion to draw from these new HI synthesis studies of Seyferts is that nothing catastrophic appears to be happening in these systems and, kinematically, their galactic disks seem remarkably undisturbed. The high sensitivity and high resolution now achievable in HI studies permit detailed, spatially resolved examination of individual gas features which produced uninterpretable complex kinematics seen in previous low resolution HI studies.

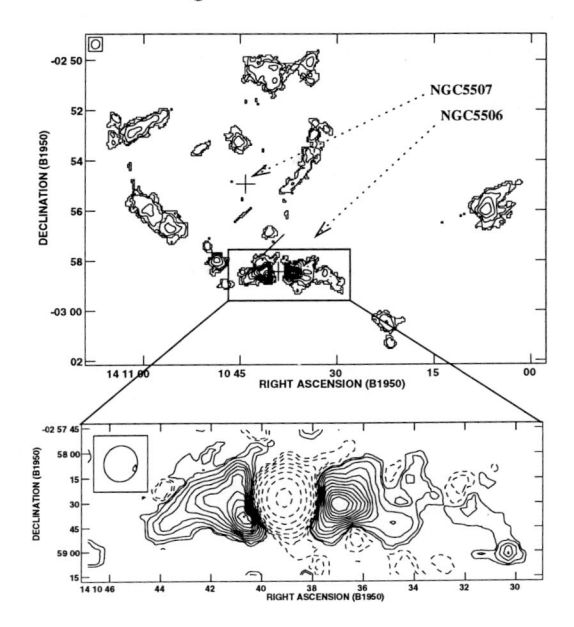

Fig. 4. HI emission and absorption in the edge-on Seyfert NGC5506. HI tidal tails provide clear evidence for interaction but no HI is detected in NGC5507.

References

Athanassoula, E., 1992, MNRAS, 259, 345
Brinks, E. & Mundell, C.G., 1997, in "Minnesota Lecture Series on Extragalactic HI", A.S.P. Conference Series, Vol. 106, p268
Conway, J.E. & Blanco, P.R. 1995, ApJ, 449, L131
Fuentes-Williams, T. & Stocke, J.T., 1988, AJ, 96, 1235
Gallagher, J.S., Knapp, G.R. & Faber, S.M., 1981, AJ, 86, 1781
Heckman, T.M., 1978, PASP, 90, 241
Ho, L.C., 1996, *PASP*, 108, 637
Keel, W.C., Kennicutt, R.C., Hummel, E. & van der Hulst, J.M., 1985, AJ, 90, 708
Laurikainen, E., Salo, H., Teerikorpi, P., Petrov, G., 1994, A&AS, 108, 491
MacKenty, J.W., 1989, ApJ, 343, 125
Meixner, M., Puchalsky, R., Blitz, L., Wright, M. & Heckman, T., 1990, ApJ, 354, 158
Mundell et al., 1995a, MNRAS, 272, 355
Mundell et al., 1995b, MNRAS, 275, 67
Mundell et al., 1995c, MNRAS, 277, 641
Pogge R.W., 1989, ApJ, 345, 730
Prendergast, K.H., 1983, IAU Symp. 100, p215
Rafanelli, P., Violato, M. & Baruffolo, A., 1995, AJ, 109, 1546
Shlosman I., Frank J. & Begelman M.C., 1989, Nature, 338, 45
Shlosman, I., Begelman, M.C. & Frank, J., 1990, Nature, 345, 679
Simkin, S.M., Su, H.J. & Schwarz, M.P., 1980, ApJ, 237, 404
Teuben, P.J. et al., 1986, MNRAS, 221, 1

The Large-Scale Environments of Low Redshift Radio-Quiet Quasars

P. Goldschmidt[1] and L. Miller[2]

[1] Astrophysics Group, I.C.S.T.M., Prince Consort Rd., London SW7 2BZ
[2] Dept. of Physics, Keble Rd., Oxford, OX1 3RH

Abstract. We have imaged areas of sky around ~ 35 radio-quiet quasars with redshifts $0.3 \leq z \leq 0.4$ and spanning a wide range in optical luminosity. We find that the relative surface density of objects as a function of radial position from the quasar is marginally higher than that predicted by the galaxy-galaxy correlation function. We estimate the overdensities around individual quasars by using a maximum likelihood method, in order to look for any link between the richness of environments and quasar luminosity. There is marginal evidence for a non-random distribution on the luminosity-environment plane, but the link between environment and luminosity is clearly not straightforward. We then test the hypothesis that quasars have close nearest neighbours by comparing measured distances to nearest neighbours to predicted distances. We find that, whilst a subset of our data do have close nearest neighbours, there is no evidence for the population as a whole that quasar activity is linked to interactions.

1 Introduction

The space density of optically selected quasars has been observed to evolve strongly with redshift, in the sense that quasars were either more common or more luminous in the past compared to the current epoch (e.g. Schmidt 1968, Boyle et al. 1988). However the physical causes of this evolution are not at all well understood. One possibility is that quasar activity is dependent upon environment. Ellingson et al. (1991, EYG) quantified the richness of radio-loud quasars' environments and found that the properties of these quasars appeared linked to those of the surroundings.

Less detailed studies have been carried out on the environments of radio-quiet quasars even though they make up the bulk of the population. This is because initial work by EYG, Boyle & Couch (1993) and Smith et al. (1995) estimating the strength of the cross-correlation amplitude suggests that the environments are poor, and comparable to those of field galaxies. Therefore any trend with luminosity and/or redshift will be harder to detect.

In this paper we use deep optical images of the environments of ~ 35 low redshift radio-quiet quasars to compare the measured surface density of objects as a function of angular distance from the quasar with that predicted from the galaxy-galaxy correlation function, for the population as a whole. Then we estimate the richness of the environments of individual quasars using a maximum likelihood method to test the hypothesis that quasar luminosities and environments are linked. Finally we compare the measured distances to the nearest neighbours with estimated distances to test the hypothesis that quasars are more likely to have close near neighbours than their larger scale environments suggest.

2 Comparison of Surface Density with Galaxy Angular Correlation Function

We estimated whether there was any evidence for an excess of galaxies around quasars for the sample as a whole. This was done by calculating the surface density of objects as a function of radial distance from the quasar and averaging over the whole sample of quasars. The results were compared to the expected value of the surface density which was calculated from the galaxy-galaxy correlation function.

The surface density of objects as a function of radial distance from the quasar is assumed to have the form

$$n(\theta)d\Omega = \rho(1 + A\theta^{1-\gamma})d\Omega \qquad (1)$$

where $n(\theta)d\Omega$ is the number of objects at angular distance θ in angular area $d\Omega$ from the quasar, ρ is the background density of objects, A is the amplitude and $1-\gamma$ is the index of the angular correlation function. Because we have not carried out star-galaxy separation, our estimate of ρ will be biassed high. Therefore we normalise this estimate by dividing by the total number of objects found at large distances from the quasar, and simply calculate the *relative* overdensity at a given angular separation.

Efstathiou (1995) presents a summary of different values of A from estimates of the angular correlation function for faint galaxies. Because of the uncertainty in A we compared our data to 2 different values, representing the two extremes which are shown in fig. 1 as the solid and dotted lines. Figure 1 also shows the measured surface density of objects as a function of angular distance for the whole combined dataset, where the variance on each point has been calculated using 4 subsets of the data. Each subset has been normalised to the data in the furthest bin so that the variance in this bin is zero. In order to compare the data directly with the model we integrated the model to estimate the mean value per bin. Taking the highest value of A from Efstathiou (1995) the expectation value for the first bin is 1.05, 1.9σ from the measured value.

We conclude that overall there is a marginal excess of galaxies near quasars in comparison to that predicted by the galaxy correlation function. The probability that the environments of quasars are the same as those of field galaxies as measured by the galaxy-galaxy correlation function is 6%. The empirical variance of the relative surface density of objects near the quasars is $\sim 50\%$ higher than that estimated using Poisson statistics. This might indicate that quasars are found in environments of different richnesses.

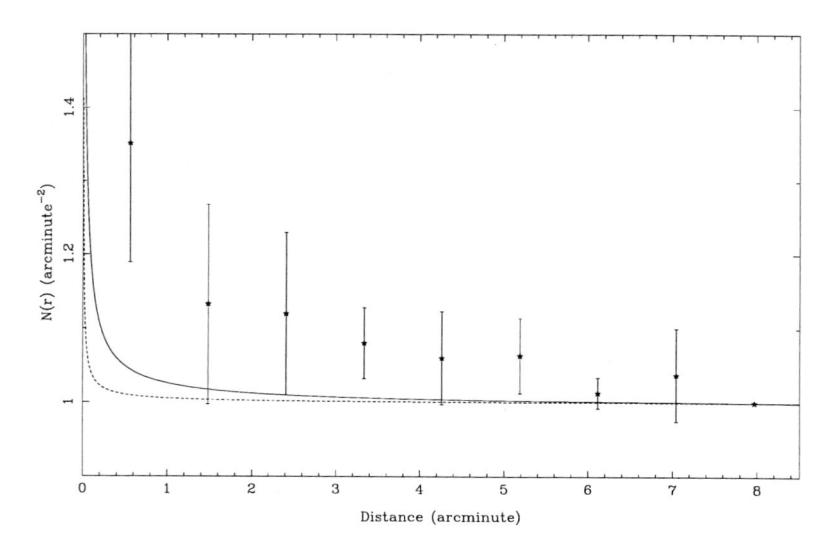

Fig. 1. Surface density of detected objects, normalised to that found in the furthest bin. The lines are the surface densities calculated using the galaxy-galaxy angular covariance amplitudes from Efstathiou (1995)

3 Estimate of Quasars' Environments from Modelling Number Counts

We used maximum likelihood to estimate the excess number of galaxies in the quasar environment, given the observed fluxes of the galaxies and a model for the galaxy luminosity function at the quasars' redshifts. This was done by considering that an observed object near a quasar is either a foreground object or at the redshift of the quasar. Therefore the probability of observing that object can be predicted from the probability (flux) distributions for the two different populations and we can use maximum likelihood to to estimate the most likely model forms for these distributions.

We divided the data into two subsets, one including objects greater than 1 arcminute from the quasar (corresponding to ~ 0.5 Mpc at the median redshift of the quasar sample) and the other including objects within that radius. The differential number counts for the two samples are shown in fig. 2 together with the models described below. The number counts for the objects within 1 arcminute of the quasar show a systematic increase above the background counts at $20 \leq R \leq 23$.

We assumed a model form for the background number counts $N(m)dm$;

$$N(m)dm = 10^{(-0.4\alpha(m-M_b^*)^{0.5})}dm. \qquad (2)$$

For the "excess" galaxies we assumed a Schechter function form;

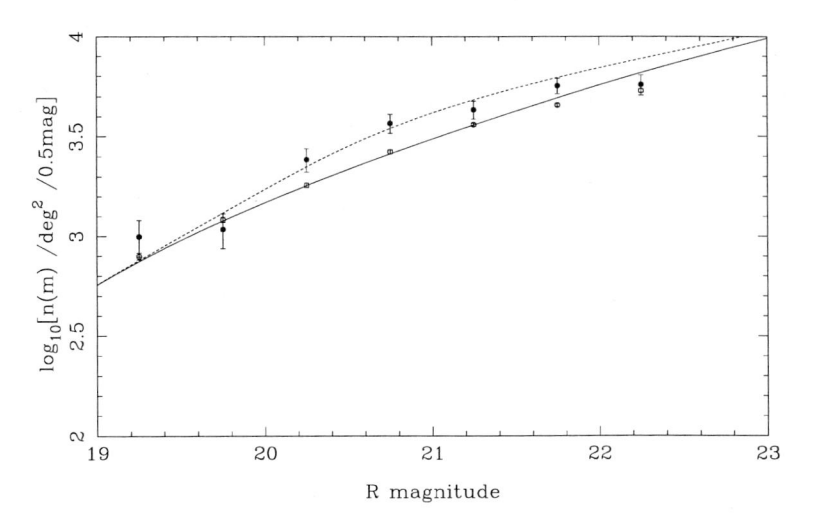

Fig. 2. Number counts for background objects (empty dots) and objects within 1 arcminute of the quasar (filled dots), together with models fitted.

$$G(m)dm = \Phi_q 10^{(-0.4(\beta+1)(m-M_q^*))} \exp[-10^{-0.4(m-M_q^*)}]dm \qquad (3)$$

where Φ_q is the normalisation.

We assumed that the shapes of the models for both the background and excess number counts are universal and that only the normalisation varies from field to field. We therefore used the combined datasets for all the quasar fields to estimate the parameters of the models. The best-fit solution is $\alpha = -2.5 \pm 0.01$, $M_b^* = 17.96 \pm 0.02$, $\beta = -0.83 \pm 0.5$, and $M_q^* = 21.42 \pm 0.5$. The values for β and M_q^* are in agreement with the Schechter function fit in Ellis et al. (1996) to galaxies with $0.15 \le z \le 0.35$. Note that each quasar has its own background field, i.e. we estimated Φ_q with respect to the *local* background surface density. The advantages of this method are that it minimises the effect of contamination due to foreground clusters, since these will contaminate the background counts as well, also there is no need to distinguish between stars and galaxies in our datasets since the local stellar density will be the same for both quasar and background field.

We calculated Φ_q out to $r = 1$ Mpc. These are plotted for individual objects against optical luminosities in fig. 3. From this figure we can see that there is no obvious trend of richness of environment with luminosity, although it is possible that fainter quasars sample a wider range of environments than more luminous ones. The most general way of testing for a link between Φ_q and quasar luminosity was to see if the Φ_q distribution in fig. 3 was consistent with being random or not. For each quasar luminosity we generated random values of Φ_q

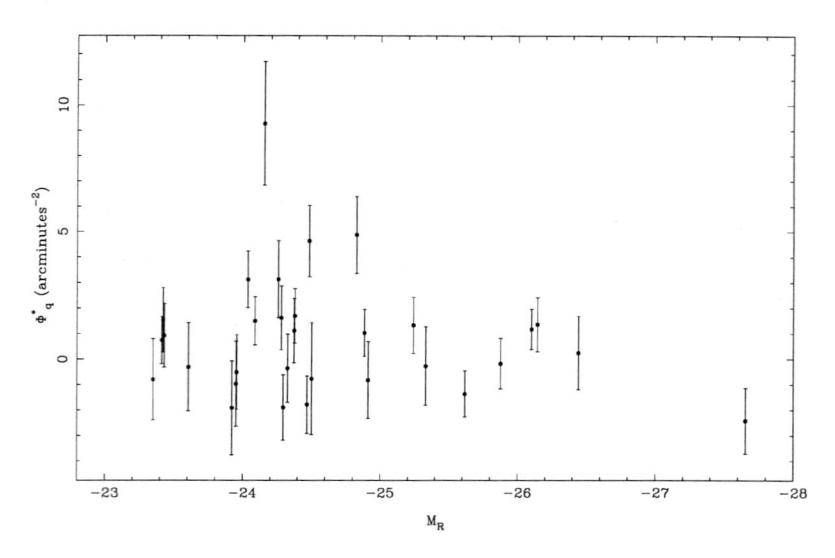

Fig. 3. Overdensities calculated within a radius of 1 Mpc, plotted against quasars' luminosities.

and compared this 2D distribution in Φ_q and M_r against the real distribution using a KS test. The test showed that the two distributions were marginally inconsistent; there was a probability of 5% that the two distributions agreed.

4 Analysis of Nearest Neighbour Distribution

It has been suggested that quasar activity is triggered by interactions/mergers with neighbouring galaxies. One way of testing this using a sample of quasars is to compare measured distances to nearest neighbours with those predicted from the more extensive distribution of objects. If quasar activity is triggered by interactions/mergers *and* if the two phenomena are contemporaneous then the then the measured nearest neighbour distances should be systematically smaller than those predicted.

In general, the angular nearest neighbour distance θ can be estimated by maximising the probability $p(\theta)d\theta$ that an object is found at θ given the surface density of objects ρ;

$$p(\theta)d\theta = 2\pi\theta\rho \exp\left[-\pi\theta^2\rho\right]d\theta. \qquad (4)$$

We used (4) to estimate θ for each quasar, given ρ, modified to take into account of the strength of clustering found around each quasar.

The data were not ideal in that the light from some of the quasars spilled over into a large area and thus for these objects the expected nearest neighbour distance was smaller than the minimum radius which we could search. This affects about 25% of our data and thus these objects are not included in the analysis. If the interaction/merging hypothesis were true, then the mean value of the ratio of expected distance to measured distance should be significantly greater than 1. There are clearly some quasars in our sample, such as the IRAS-detected quasar 1402+4341, which do have significantly close near neighbours and these objects merit more detailed study. However, we found for the sample as a whole, that the mean value is 1.05±0.56, and thus there is no direct evidence that quasar activity in radio-quiet quasars is linked with interactions/mergers.

Acknowledgements

We acknowledge allocations of telescope time at the INT. The INT is operated by the Royal Greenwich Observatory in the Spanish Observatorio del Roque de los Muchachos of the Instituto de Astrofísica de Canarias. Data reduction was carried out on STARLINK.

References

Boyle, B.J., Shanks, T. & Peterson, B., (1988), MNRAS, **235**, 935.
Boyle, B.J. & Couch, W.J., (1993), MNRAS, **264**, 604.
Efstathiou, G., (1995), *in Wide Field Spectroscopy and the Distant Universe,* eds. Maddox & Aragon-Salamanca (World Scientific)
Ellingson, E., Yee, H.K.C. & Green, R.F., (1991), ApJ, **371**, 49 (EYG).
Ellis, R.S., Colless, M., Broadhurst, T., Heyl, J. & Glazebrook, K., (1996), MNRAS, **280**, 235.
Schmidt, M., (1968), ApJ, **151**, 353.
Smith, R., Boyle, B.J. & Maddox, S.J., (1995), MNRAS, **277**, 270.

High Velocity Resolution Observations of the ISM in NGC 4151

M.W. Asif[1,2], S.W. Unger[1], A. Pedlar[2], C.G. Mundell[2], A. Robinson[3] and N.A. Walton[1]

[1] Isaac Newton Group, Apartado 321, 38780 Santa Cruz de La Palma, Tenerife, Canary Islands, Spain
[2] Nuffield Radio Astronomy Laboratories, University of Manchester, Jodrell Bank, Macclesfield, Cheshire SK11 9DL, U.K.
[3] Division of Physical Sciences, University of Hertfordshire, College Lane, Hatfield, Herts AL109AB, U.K.

Abstract. We have used the Utrecht Echelle Spectrograph to obtain high velocity resolution (6 kms^{-1}) observations of the extended narrow line region in the Seyfert nucleus of NGC4151. We compare the velocity structure of the ionised gas with neutral hydrogen observations and confirm that the photoionisation of quiescent ambient gas is the prefered mechanism to account for the extended region of ionised gas associated with this source. We observe line-splitting in the ENLR of NGC4151 which is consistent with gas associated with an ionisation front expanding outwards from the ENLR.

1 Introduction

In this poster we present the results of a more stringent test of the photoionisation hypothesis, using the Utrecht Echelle Spectrograph (UES) on the William Herschel Telescope (WHT) to provide a velocity resolution of 6 kms^{-1} in conditions of 1.5 arcsec seeing. The UES observations are compared with new VLA neutral hydrogen observations with an angular resolution of ∼6 arcsec. With these data we can measure small differences in velocity between the neutral and ionised gas as well as differences in linewidth, and we can do so for distinct spatial components within the ENLR.

2 Observations and Reduction

The spectra were taken in service time on the night of the 21st March 1994. The instrument was the Utrecht Echelle Spectrograph (UES), used at the Nasmyth focus of the 4.2 metre William Herschel Telescope at the Observatorio del Roque de los Muchachos, La Palma. The slit was aligned at a position angle of 48^0 through the centre of NGC 4151. Two exposures of 1200 secs were obtained. Calibration frames; bias frames, tungsten lamp flat field frames and a Thorium-Argon arc frame to wavelength calibrate the data, were also taken. The thorium-argon lamp frame was reduced in an identical manner to the object frames and the spectral resolution determined from the profile of the arc lines was in good

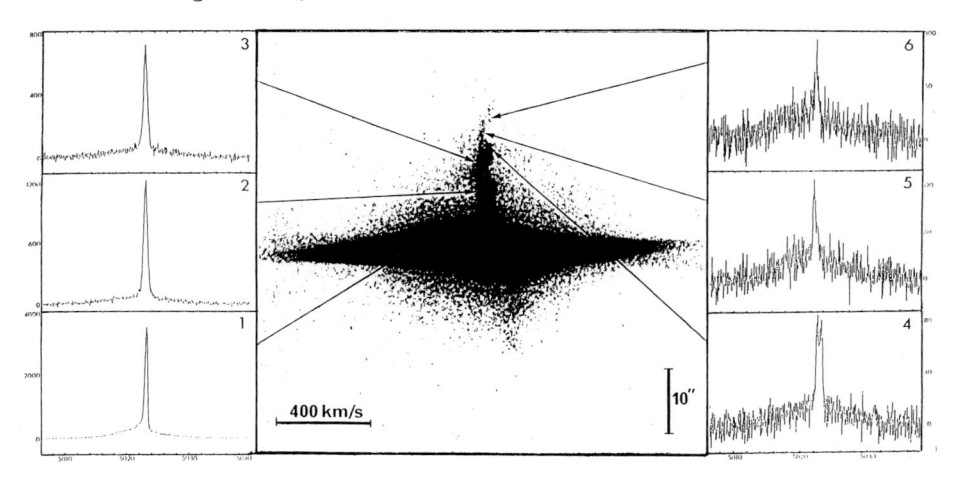

Fig. 1. A greyscale image of the UES longslit spectrum, together with spectra of distinct regions within the extended emission line region.

agreement with the nominal resolving power of the UES. Figure 1 shows a greyscale image of the longslit spectrum, together with spectra of the distinct regions within the extended emission line region after spatial binning to increase signal-to-noise.

NGC4151 was observed for a total of 18 hours in April 1993 using the VLA in B configuration. A velocity cube was produced with an angular resolution of 6.1 x 5.1 arcsec (PA=-78°), a velocity resolution of 12 kms^{-1} and a sensitivity of 0.3 mJy/beam. Full details of the image processing etc of these data are given in Mundell PhD Thesis 1995. In Figure 2 we show a position velocity plot of HI brightness along the same PA as the Echelle slit.

3 Results and Discussion

In order to quantify the velocity information we have fitted gaussians to the Echelle long slit data and the results are shown in Table 1. The properties of the [OIII] knots in Table 1 have been plotted on the HI position-velocity plot in Figure 2. It can be seen that there is generally good agreement between the [OIII] and HI velocities, the [OIII] components lying close to peaks in the HI position velocity plot.

By analogy with HII regions, we expect the ionised gas in the ENLR to be surrounded by an ionisation front. If we assume that the ionisation front is D-type (associated with dense clumps of gas) then it is expected to expand into the ambient interstellar medium at a velocity equal to the isothermal sound speed in the ionised gas. Robinson *et al.*, 1994 determined the average temperature and density of the ionised gas in the ENLR to be 17000K and 250 cm^{-3} respectively, implying an isothermal sound speed of 17 kms^{-1}, in extremely close agreement with the observed line splitting. Thus the observed line-splitting is consistent

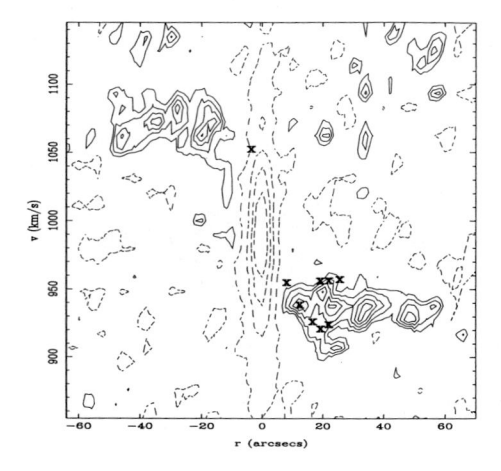

Fig. 2. A position velocity plot of HI along PA 48° passing through the radio nucleus. The crosses represent the positions and velocities of the distinct ENLR components from the UES. Increasing r towards South-west.

Component	Offset From Nucleus (arcs)	Peak Counts	Velocity (km/s)	FWHM (km/s)	Sum of Spatial Extent (arcs)
1	8.1	3109	954.54	31.36	4.10
2	12.4	1096	938.24	44.6	2.92
3	16.7	60.04	925.84	45.5	5.26
4a	19.3	6.48	920.54	28.67	3.51
4b	19.3	6.35	955.57	29.57	3.51
5a	22.0	9.38	923.70	27.18	5.26
5b	22.0	4.10	955.75	21.01	5.26
6	25.6	4.56	956.65	44.38	3.80

Table 1. Properties of the ENLR components.

with gas associated with an ionisation front expanding outwards from the ENLR. The ionisation front must be moving close to the line of sight, since otherwise projection effects would decrease the observed line splitting. This suggests that the ENLR itself lies close to the plane of the sky and is also consistent with photoionisation of material in the disk.

References

Mundell, C.G., PhD Thesis, Univ. of Manchester (1995)
Robinson, A., et al., 1994, A&A, 291, 351

Mg_2 Index Map of the Centre of NGC 7331

C. del Burgo, E. Mediavilla, S. Arribas, and B. García-Lorenzo

Instituto de Astrofísica de Canarias, E-38200 La Laguna, Tenerife, Spain

Abstract. Simultaneous bidimensional spectroscopy of the circumnuclear region ($12'' \times 9''$) of the active galaxy NGC 7331 has been obtained using an optical fibre system (2D-FIS). We have used these data to obtain the Mg_2 index map of the centre of this galaxy. A radial variation from 0.32 (centre) to 0.25 ($r = 6''$) was found, which supports the hypothesis that there has been large-scale star formation at the centre of NGC 7331.

1 Introduction

NGC 7331 is a nearly edge-on Sb galaxy, which could host a 'dead' QSO. Models that include a dark central point mass show that the presence of a $\sim 10^9$ $M_\odot$ black hole can be excluded in the case of NGC 7331 (Bower et al. 1993). However, more recently, the VLA observations by Cowan, Romanishin & Branch (1994) resulted in the detection of an unresolved radio source in the nucleus of NGC 7331, suggesting that this galaxy may indeed harbour a massive black hole.

Young & Scoville (1982) found that NGC 7331 has a CO 'ring' at $r \sim 45''$, and suggested that it could be produced by exhaustion of the gas to form stars in the central region. Telesco, Gatley & Stewart (1982) have pointed out that colour effects due to spatial variation in the stellar population may be present in the circumnuclear region of NGC 7331. In order to investigate this, we have obtained a map of Mg_2 index, which should not depend strongly on dust reddening.

2 Observations and Results

The present data were obtained using 2D-FIS (2-Dimensional Fiber ISIS System, García et al. 1994), which links the Cassegrain focus of the 4.2-m WHT with the ISIS double spectrograph. The core of the system consists of a bundle formed by 125 optical fibres, 95 of which form an array of $9.4'' \times 12.2''$ on the sky. In the blue arm of ISIS, a 600 groove mm^{-1} grating allowed high-resolution work (~ 1.5 Å) in the spectral range 4590–5403 Å. With the above configuration, three consecutive exposures, of 1800 s each, were taken for the central region of NGC 7331. The mean seeing value during observing was about $1.25''$.

The Mg_2 index for each individual spectrum was computed following the definition of Worthey et al. (1994). The Mg_2 map (Figure 1, left) indicates a radial decrease from the centre to the outer regions. To discuss the data quantitatively we have averaged the Mg_2 map in circular rings around its maximum obtaining the radial profile, $Mg_2(r)$, plotted in Figure 1 (right). As we can see, the Mg_2

index falls from 0.32 (in the centre) to 0.25 (at $r > 4''$). According to the recent calibration obtained by Casuso et al. (1996), this behaviour may be interpreted as: i) a change in metallicity (e.g. from about $Z_\odot$ to 0.6 $Z_\odot$, if an age ≥ 8 Gyr is assumed), ii) a change in age (e.g., from ~ 15 Gyr to ~ 5 Gyr, if 1.25 $Z_\odot$ is assumed), or iii) a combined change in age and metallicity.

Therefore the present observations suggest that a large-scale process of stellar formation has taken place in the circumnuclear region of NGC 7331 in accordance with the hypothesis of Young & Scoville (1982), and the stellar population in the outer parts could be younger than in the centre.

We thank A. García for his help with 2D-FIS and all the staff of the ING/ORM for their kind support. This work has been partially supported by the DGYCIT (PB93-0658).

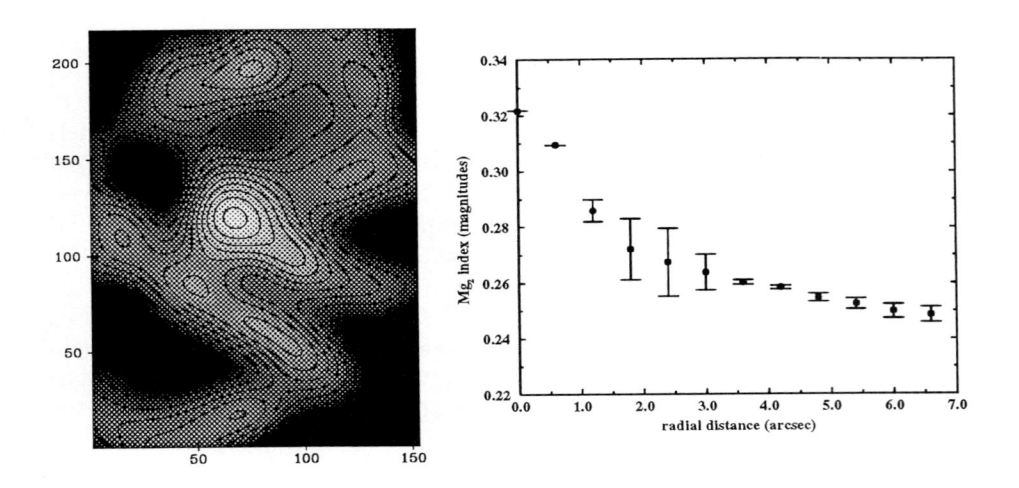

Fig. 1. Left: Mg_2 map: north (top); east (left); 20 pixels = 1 arcsec. Right: Radial profile of Mg_2 index.

References

Bower, G. A., Richstone, D. O., Bothun, G. D., and Heckman, T. M. (1993): ApJ, **402**, 76
Casuso, E., Vazdekis, A., Peletier, R. F., and Beckman, J. E. (1996): ApJ, **458**, 533
Cowan, J. J., Romanishin, W., and Branch, D. (1994): ApJ, **436**, L139
García, A., Rasilla, J. L., Arribas, S., and Mediavilla, E. (1994): Instrumentation in Astronomy, Vol 2198, ed. D. L. Crawford and C. R. Craine (Hawaii: SPIE), 75
Telesco, C. M., Gatley, I., and Stewart, J. M. (1982): ApJ, **263**, L13
Worthey, G., Faber, S. M., Gonzalez, J. J., and Burstein, D. (1994): ApJS, **94**, 687
Young, J. S., and Scoville, N. (1982): ApJ, **260**, L41

Stellar Dynamics of Two AGNs

Francisco Prada[1], and Carlos M. Gutiérrez[1]

Instituto de Astrofísica de Canarias, E-38200 La Laguna, Tenerife, Spain

We obtained long-slit spectroscopy of the LINER NGC 7331 and the Seyfert II NGC 5728 at the 4.2 m William Herschel Telescope with the ISIS spectrograph. We have determined the stellar Line of Sight Velocity Distribution (LOSVD), using the near-IR CaII triplet ($\sim$8500 Å), along the major axis of these two galaxies. The analysis of NGC 7331 shows (see Figure 1) that, in the radial range between 5″ and 20″, the LOSVD of the absorption lines has two components. This LOSVD can be decomposed (see Figure 2) into a fast-rotating component with $v/\sigma > 3$, and a slower rotating, retrograde component with $v/\sigma \approx$ 1-1.5. A two-dimensional bulge-disk decomposition of the near-infrared K-band image shows that the radial surface brightness profile of the counter-rotating component follows that of the bulge, while the fast-rotating component follows the disk. At the radius at which the disk starts to dominate, the isophotes change from being considerably boxy to being very disky. This makes NGC 7331 the first spiral galaxy known to have a boxy, fairly warm, counter-rotating component which is dominating the central regions.

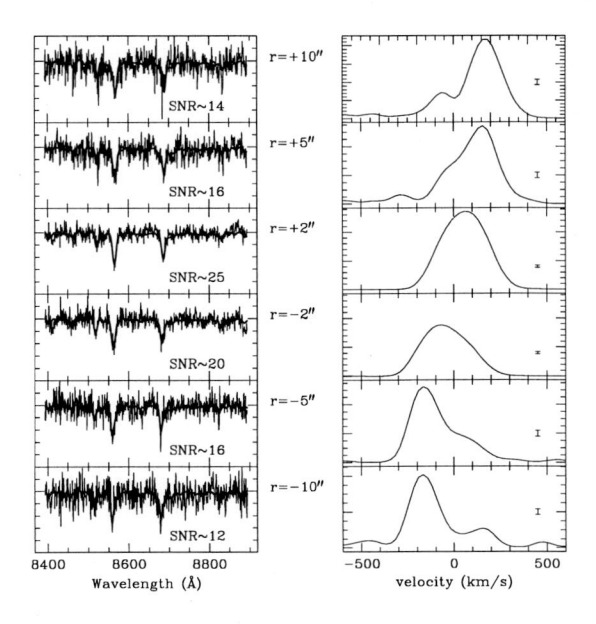

Fig. 1. (*Left:*) The original and recovered spectra in the region of the CaII IR triplet along the major axis of NGC 7331. The SNR is indicated in each of the panels. (*Right:*) The corresponding LOSVD.

Preliminary kinematic analysis of the galaxy Seyfert II NGC 5728 is also presented. This galaxy shows a nuclear bar twisted ∼ 60 degrees with respect to the main bar. The LOSVD again shows the presence of two components in the nuclear regions. In the inner 3 arcsecs the main component seems to be associated with the nuclear bar; beyond this radius this component follows the main bar. A counter-rotating component is present in the inner ∼ 10 arcsecs which could be associated with a hidden structure.

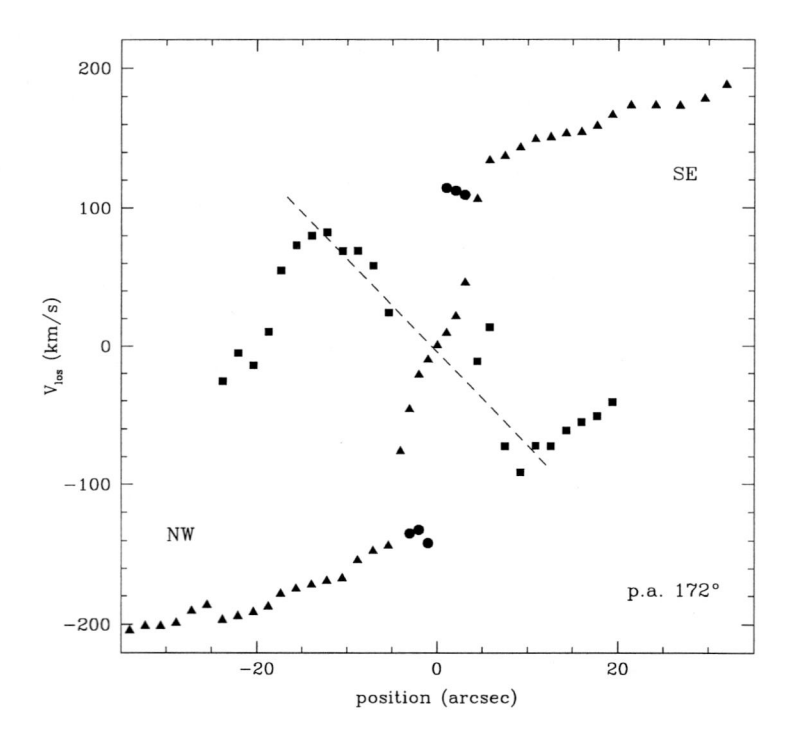

Fig. 2. Decomposition of the LOSVD along the major axis of NGC 7331. Triangles and squares refer to the brighter and fainter components respectively. The dashed line is a model for the bulge component. The circles correspond to regions where no unambiguous decomposition was possible.

References

F. Prada, C.M. Gutiérrez, R. F. Peletier, & C. D. McKeith, 1996, ApJ, 463, L9

Part 4

THE RADIO LOUD/QUIET DICHOTOMY

Infrared Imaging and Off-Nuclear Spectroscopy of Quasar Hosts

Marek J. Kukula[1], James S. Dunlop[1], David H. Hughes[1], Geoff Taylor[2] and Todd Boroson[3]

[1] Institute for Astronomy, University of Edinburgh, Royal Observatory, Edinburgh EH9 3HJ, UK
[2] Astrophysics Group, Liverpool John Moores University, Byrom St, Liverpool L3 3AF, UK
[3] NOAO, PO Box 26732, Tucson, AZ 85726-6732, USA

Abstract. We present the results of two complementary ground-based programmes to determine the host galaxy properties of radio-quiet and radio-loud quasars and to compare them with those of radio galaxies. Both infrared images and optical off-nuclear spectra were obtained and we discuss the various strategies used to separate the quasar-related emission from that of the underlying galaxy. However, the key feature of this project is the use of carefully matched samples, which ensure that the data for different types of object are directly comparable.

1 Introduction

The paper briefly describes the results of a continuing long-term project to study the host galaxies of powerful AGN. The aim of the project is two-fold: to test the scheme for radio-loud quasars (RLQs) and radio galaxies (RGs) which attemps to unify the two types of object via orientation effects, and to investigate the extent to which the host galaxy influences the radio properties of the AGN by comparing the hosts of radio-loud and radio-quiet quasars (RQQs).

Using ground-based observations we have approached the question of host galaxy properties from two independent directions: near-infrared (K-band) imaging, to determine the host morphologies and luminosities, and off-nuclear optical spectroscopy to investigate their star-formation histories. A more detailed description of the near-infrared imaging can be found in Dunlop *et al.* (1993) and Taylor *et al.* (1996).

2 Sample Selection

In the past, comparative studies of quasars and radio galaxies have often been hampered by the use of poorly matched samples, sometimes using wildly differing selection criteria. A key feature of the current project was the selection of three *carefully matched samples* of RQQs, RLQs and powerful (FRII) radio galaxies.

In order to ensure that the samples were directly comparable with each other the RQQs and RLQs were selected to have identical distributions in the $V - z$

plane. Meanwhile the RG sample was chosen to match the radio luminosity − redshift ($L_{5GHz} - z$) and spectral index - redshift ($\alpha - z$) distributions of the RLQs. The objects are all of relatively low redshift and cover a narrow range in z ($0.1 < z < 0.3$). Both of the quasar samples were drawn largely from the (optically-selected) Bright Quasar Survey (Schmidt & Green 1983) and consist of objects at the fainter end of the quasar luminosity function ($-26 < M_B < -23$).

3 Near-IR Imaging with UKIRT

There are several advantages to working at near-infrared wavelengths. Quasars are, by definition, heavily nuclear-dominated objects in the optical, making galaxy magnitude and morphology determination extremely sensitive to the estimated strength of the core component and the reliability of the adopted form of the point-spread function. Quasars are relatively blue objects ($f_\nu \simeq$ const) whereas the luminosity of the host galaxy is expected to peak at near-infrared wavelengths (Sanders et al. 1989), making the near-infrared the waveband of choice for minimising the nuclear:host ratio. Working in the near-infrared also helps to avoid contamination of the images by strong emission lines and/or light from regions of enhanced starformation, both of which could mask the true nature of the underlying galaxy. Finally, the high sky background in the infrared - a major drawback in many ways - does at least mean that the signal:noise ratio of an image is not compromised by subdividing the integration into sufficiently small sub-integrations to avoid saturation of the quasar nucleus, a point of some importance when attempting to determine the correct form of the PSF.

3.1 Observations and Modelling

The observations were made in K-band (2.2μm) using the 62×58 array IRCAM on the United Kingdom Infrared Telescope (UKIRT). A library of ~ 100 standard star images was compiled from which the PSF most suitable for a particular quasar image could subsequently be selected. The procedure adopted for the data reduction is described in detail by Taylor et al. (1996).

In our first attempt to analyse the quasar images a simple PSF subtraction was used (Dunlop et al. 1993) but this gave rise to a number of problems; most significantly there was an inevitable oversubtraction of galaxy light in the centre of the image. The adopted solution was to use two-dimensional modelling of the surface brightness distribution in order to properly decouple the shape of the host galaxy from that of the PSF (Taylor et al. 1996). This allows one to extrapolate smoothly into the central regions of the galaxy and thus estimate the luminosity of each host in a self-consistent manner.

The modelling algorithm fits five parameters to the data: nuclear luminosity, host galaxy luminosity, galaxy scale-length, galaxy position angle and axial ratio. In order to determine the morphological type of the host galaxy one further parameter is required: the index β, which describes the form of the galaxy luminosity profile ($\mu(r)/\mu_0 = exp((-r/r_0)^\beta)$). For the present study we decided to consider only the alternatives of $\beta = 1$ (an exponential disc) and $\beta = 1/4$ (a de Vaucouleurs law), and thus to confine our morphological investigation to determining whether a given host galaxy is dominated by a disc or a spheroidal component. For each quasar the fitting procedure was carried out twice, once with $\beta = 1$ and again with $\beta = 1/4$, and the two model fits were then compared with the original image to determine which was the most successful.

In view of the large number of free parameters involved in the model fitting procedure, as well as the notorious sensitivity of such methods to uncertainties in the adopted form of the PSF, the process was subjected to rigorous testing in order to establish the degree of reliability of the fits under a wide range of starting conditions. To this end a series of synthetic quasar+host combinations were constructed and convolved with a range of PSFs. The accuracy with which the modelling algorithm was able to recover the 'true' parameters of the artificial galaxies could then be measured in a self-consistent fashion. These tests show that the host galaxy parameters derived from the model fits are typically accurate to within 10%.

However, as expected, the uncertainty increases with the nuclear:host ratio of the quasar and this also has a very strong effect on the degree of confidence with which an exponential disc profile can be distinguished from a de Vaucouleurs law. We decided to adopt a pessimistic approach and to reject as unreliable the morphological classification of objects for which the nuclear:host ratio exceeded a very conservative limit. Effectively, this means that for $z \simeq 0.1$ objects with $L_{nuc} : L_{host} > 10$ are excluded; for $z \simeq 0.3$ the limit becomes an even more stringent $L_{nuc} : L_{host} > 5$. Fortunately, the low nuclear:host ratio of quasars at near infrared wavelengths means that the majority (31/40) of the objects in our samples survive this selection procedure, and in these cases we are confident that the morphological preference displayed by the fitting algorithm is both valid and meaningful. This would not have been the case in the optical, where typical values of $L_{nuc} : L_{host}$ exceed 10 (in B-band; see Taylor & Dunlop 1997).

3.2 Results

The principal results to emerge from this study can be summarized as follows:

(i) RGs, and the hosts of both RLQs and RQQs are all **luminous galaxies** with $L \geq L^*$ at K ($< M_K > \simeq -26$).

(ii) RGs and the hosts of RLQs and RQQs are all **large galaxies** with a half-light radius (*ie* the radius containing half of the total galaxy luminosity) $r_{1/2} \geq 10$ kpc.

(iii) The basic parameters of the host galaxies are no different from those of other comparably large and luminous galaxies. In particular the hosts of all three types of AGN display a $\mu_{1/2} - r_{1/2}$ relation which is identical in both slope and normalisation to that displayed by brightest cluster galaxies - objects which are thought to be the product of successive merger events. This suggests that, regardless of their current interaction status, the host galaxies of powerful AGN have all experienced merger events in the past.

(iv) Essentially all of the RGs and RLQ hosts are best described by a de Vaucouleurs law, consistent with unification of powerful radio-loud AGN via orientation. Thus it appears that an elliptical host galaxy is necessary for an active galaxy to produce a radio luminosity in excess of $L_{5GHz} \simeq 10^{24} \mathrm{WHz}^{-1} \mathrm{sr}^{-1}$.

(v) Slightly more than half of the RQQs appear to lie in galaxies which are dominated by an exponential disc. Those RQQs which have elliptical hosts are in general more luminous than those which reside in discs. A significant fraction of the RQQ population may at least be capable of producing powerful radio emission.

(vi) The majority of the radio galaxies in our sample contain additional nuclear flux at K in excess of that expected from the best fitting $r^{1/4}$-law model. These unresolved nuclear components may simply be indicative of central cusps in their starlight, but their colours and magnitudes are consistent with dust-reddened quasars (Taylor & Dunlop 1997).

3.3 Comparison with HST Results

In general the findings of our ground-based imaging programme are in good agreement with those of recent optical HST studies (*eg* Hutchings *et al.* 1994, Hutchings & Morris 1995, Disney *et al.* 1995). The red colours and large scale-lengths of the hosts as determined from our ground-based data almost certainly explain the failure of earlier HST programmes to detect some of these galaxies in the optical (*eg* Bahcall, Kirhakos & Schneider 1995). Subsequent re-analysis of these HST images has revealed that large, luminous host galaxies are indeed present (McLeod & Rieke 1995, Bahcall, Kirhakos, Saxe & Schneider 1997).

4 Off-Nuclear Optical Spectroscopy

A completely independent way to characterise the host galaxies of AGN is via analysis and classification of their stellar populations. The aim of the observations described in this section is to obtain high signal-to-noise spectra of the quasar hosts which could then be used to determine the composition, age and evolutionary history of their stellar components.

Previous attempts to take optical spectra of quasar 'fuzz' were severely hampered by scattered light from the quasar itself which effectively swamped the starlight from the surrounding galaxy and prevented any meaningful analysis from being carried out. However, the deep near-infrared images of our quasar samples presented us with a unique opportunity to circumvent this problem: armed with knowledge of the extent and orientation of the host galaxy on the sky we were able to choose a slit position which was far enough from the nucleus to avoid the worst excesses of scattered quasar light, but which simultaneously maximised the amount of galaxy light falling onto the slit.

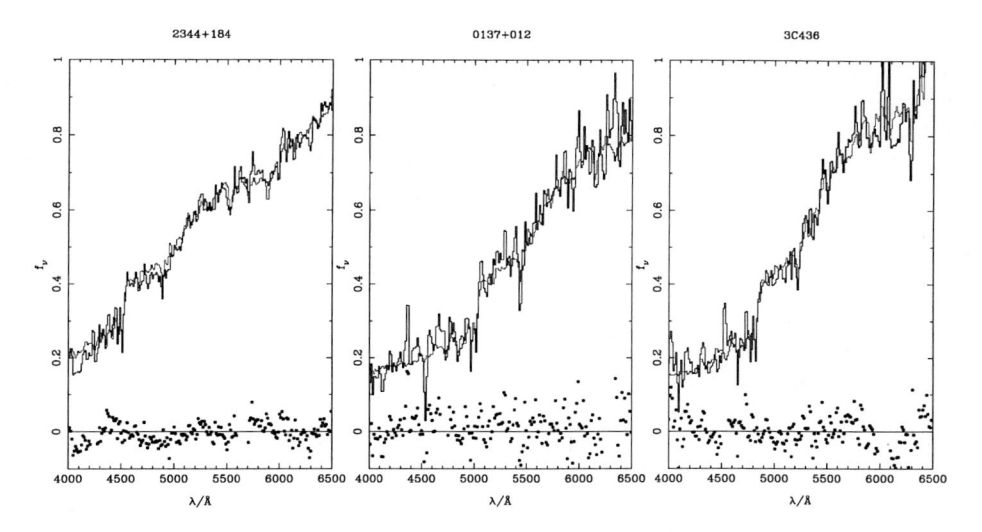

Fig. 1. Off-nuclear spectra of three active galaxies: the RQQ 2344+184 ($z = 0.137$), the RLQ 0137+012 ($z = 0.258$) and the radio galaxy 3C436 ($z = 0.215$). Each spectrum is best described by the combination of an old burst model and a very blue component (probably scattered quasar light). The observed spectrum is shown as a heavy line, the model by a thin line and the residuals by dots.

4.1 Observations

Initial observations were carried out on 10 objects using the Mayall 4-m telescope at Kitt Peak and covering a wavelength range of 3500-7500Å. With the slit positioned 5″ from the nucleus, starlight was easily detected in all 10 objects and the spectra were of sufficient quality to allow us to fit spectrophotometric models to the stellar populations.

Subsequent observations were carried out on the 4.2-m William Herschel Telescope (WHT) on La Palma. The availability of the ISIS double-beam spectrograph on the WHT enabled us to extend our wavelength range into the red down to ~ 9000Å - the extra wavelength coverage being particularly useful for constraining models of galaxy spectrophotometric evolution. To date twenty five of our objects have been observed.

4.2 Initial Analysis

For the purposes of this early analysis we have attempted to fit a simple 'burst' model (Guiderdoni & Rocca-Volmerange 1987) to the off-nuclear spectrum, varying the age of the model to obtain the best fit. The burst model assumes that all starformation occurs in a single burst of activity lasting ~ 1 Gyr, and that the stellar population evolves passively thereafter. We are currently working towards applying more sophisticated and realistic models, but we note that the fits obtained using this very simple scenario are surprisingly good (Figure 1).

Whilst longwards of the 4000Å break the spectra are clearly dominated by starlight, at shorter wavelengths an additional blue component becomes prominent in some of the off-nuclear quasar spectra. This leads to poor fits from the burst model and to extremely young ages for the stellar population.

Since the presence of this blue component is often accompanied by the appearance of broad emission-lines, particularly Hβ, we conjectured that it was probably the result of residual scattered quasar light which, although relatively low-level, was still sufficient to dominate the combined spectrum at $\lambda_{rest} < 4000$Å. In order to test this theory we took a nuclear spectrum of the RQQ 0054+144 and scaled it to match the height of the broad Hβ feature in the off-nuclear spectrum of the same object. Subtraction of the scaled nuclear spectrum caused a marked improvement in the quality of the resulting fit and, as expected, a substantial increase in the age of the best fitting model (from 5 to 13 Gyrs).

We therefore decided to carry out a two-component fit to the off-nuclear spectra, using a combination of a burst model and a flat-f_ν component to simulate scattered quasar light, and allowing the age of the burst and the amplitude of the flat component to vary freely. For the sake of consistency this procedure was used on all the objects in our sample, including the radio galaxies and those quasars which appeared to lack a strong blue component. This approach appears to have been vindicated by the fact that in cases where a blue component was not obviously present in the spectrum the modelling algorithm invariably achieved a good fit without resorting to the addition of a strong 'scattered quasar' component. As a final check, we applied the same procedure to spectra of M32, a dwarf elliptical companion of the Andromeda galaxy, and M33, a nearby late-type spiral. In both cases a good fit was obtained without recourse to a flat-f_ν component, and sensible ages were obtained from the models (old for the dwarf elliptical and relatively young for the spiral).

4.3 Preliminary Results

The galaxy ages obtained from the best fitting models are shown in Figure 2 along with the derived ages for M32 and M33. The histograms for the RLQs and RGs are statistically indistinguishable and the host galaxies are generally rather old, red systems, consistent with unification of the two types of AGN. The histogram for the RQQs shows a prominent tail of younger, bluer galaxies - three galaxies have ages < 6 Gyr even after the removal of any scattered quasar light. A comparison with our near-infrared images shows that all of these 'young', blue galaxies appear to have close companions and/or display distorted morphologies, implying that they are currently (or have recently been) involved in interactions or mergers.

However, the general trend is that the hosts of all three types of AGN are dominated by an old stellar population. In the RG and RLQ samples 80% and 75% respectively of the host galaxies have ages ≥ 11 Gyr, whilst for the RQQ sample the proportion with ages ≥ 11 Gyr is still 50% (with the current, rather crude level of analysis, distinguishing between ages greater than 11 Gyr is a

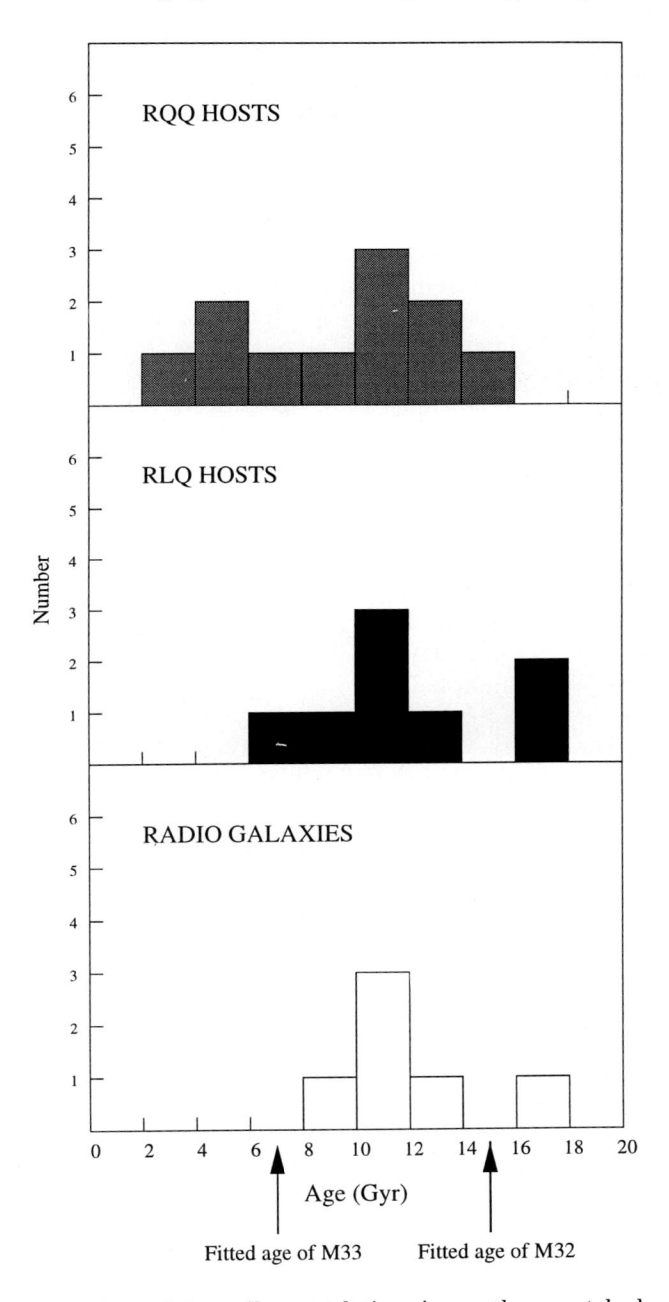

Fig. 2. Age distributions of the stellar populations in our three matched samples. Note that the RQQ sample contains a significant proportion of young, blue hosts whereas the RLQ and RG samples have distributions which are indistinguishable from each other and consist of older, redder galaxies. Also indicated in this figure are our fits to M33 (a late-type spiral) and M32 (a dwarf elliptical).

highly model-dependent affair). Other than the fact that the youngest RQQ hosts appear to be interacting galaxies there is no obvious correlation between the age of the hosts and their morphological type.

5 Summary

This has proved to be a very fruitful project. Many interesting (and some unexpected) results have emerged from the near-infrared imaging study and, although the off-nuclear spectroscopy is still very much a work in progress, the fact that we have been able to isolate starlight in all of the spectra taken so far is a very encouraging result.

A consistent picture is emerging from the data. It appears that RLQ hosts and RGs are indeed the same type of galaxy - large luminous spheroidal systems with old, red stellar populations - consistent with the unified scheme. The hosts of RQQs are also large and luminous, and can be either disc-dominated or spheroidal systems. There seems to be a tendency for the most luminous RQQs to occur in elliptical hosts. Ages of the RQQ hosts cover a wide range and the bluest galaxies all seem to be undergoing interactions.

In the immediate future we have been awarded 34 orbits on the HST to observe our three AGN samples in R-band. Not only will these images enable us to determine the optical morpholgies of the hosts with a level of detail which is impossible from the ground, but by providing us with reliable optical luminosities for the host galaxies they will enable us to bridge the gap between our two ground-based datasets, allowing us to calculate $R - K$ colours for the galaxies and thus to test whether a particular spectrophotometric model can explain the shape of a galaxy spectrum from optical through to near-infrared wavelengths.

References

Bahcall, J. N., Kirhakos, S., Schneider, D. P. (1995): ApJ, **450**, 486
Bahcall, J. N., Kirhakos, S., Saxe, D. H., Schneider, D. P. (1997): preprint
Disney, M. J., et al. (1995): Nature, **376**, 150
Dunlop, J. S. et al. (1993): MNRAS, **264**, 455
Guiderdoni, B., Rocca-Volmerange, B. (1987): A&A, **186**, 1
Hutchings, J. B. et al. (1994): ApJ, **429**, L1
Hutchings, J. B., Morris, S. C. (1995): AJ, **109**, 1541
McLeod, K. K., Rieke, G. H. (1995): ApJ, **454**, L77
Sanders, D. B., et al. (1989): ApJ, **347**, 29
Schmidt, M., Green, R. F. (1983): ApJ, **269**, 352
Taylor, G. L., Dunlop, J. S., Hughes, D. H., Robson, E. I. (1996): MNRAS, **283**, 930
Taylor, G. L., Dunlop, J. S. (1997): in preparation

The Radio Properties of Radio-Quiet Quasars

Marek J. Kukula[1], James S. Dunlop[1], David H. Hughes[1] and Steve Rawlings[2]

[1] Institute for Astronomy, University of Edinburgh, Royal Observatory, Edinburgh EH9 3HJ, UK
[2] Department of Astrophysics, Nuclear and Astrophysics Laboratory, University of Oxford, Keble Rd, Oxford OX1 3RH, UK

Abstract. Although radio-quiet quasars (RQQs), which constitute the majority of optically-identified quasar samples, are by no means radio silent the properties of their radio emission are only poorly understood. We present the results of a multi-frequency VLA study of 27 low-redshift RQQs. In general, we find that the properties of the radio sources in RQQs are consistent with them being weak, small-scale ($\sim$ 1 kpc) jets similar to those observed in nearby Seyfert galaxies. We conclude that a significant fraction of the radio emission in RQQs is directly associated with the central engine and is not a result of stellar processes in the surrounding galaxy. There appears to be no difference between the radio properties of RQQs in elliptical and disc galaxies, implying that the relationship between the host galaxy and the 'radio loudness' of the active nucleus is not straightforward.

1 Introduction

Studies of nearby active galaxies have shown that the radio-loud objects (*ie* radio galaxies) are invariably elliptical systems whereas the radio-quiets (Seyferts) tend to be spirals, suggesting that it is some property of gas-rich disc galaxies which inhibits the formation of large, powerful radio sources. Two factors have encouraged the extension of this result to the higher redshifts and larger nuclear luminosities typical of quasars: the success of Unified Schemes which link radio galaxies and radio-loud quasars (RLQs) via beaming effects and viewing angle and the fact that Seyfert 1 nuclei and radio-quiet quasars (RQQs) form a continuous sequence in terms of their optical luminosities and have identical emission line characteristics. Thus we have a picture in which RLQs occur in elliptical galaxies whilst RQQs are found in discs.

However, recent studies have shown that quasars occur in a wide variety of environments (*eg* Bahcall *et al.* 1997). In particular it seems that *not all* RQQs lie in disc systems, with as many as 50% occurring in elliptical galaxies (*eg* Véron-Cetty & Woltjer 1990). Indeed, there is some evidence that elliptical galaxies might account for all of the most optically luminous RQQs (Taylor *et al.* 1996). Clearly the simple 'radio-loud $\equiv$ elliptical, radio-quiet $\equiv$ disc' picture can no longer be supported, and it has become more important than ever to determine in what respects radio-quiet quasars are different from their radio-loud counterparts. Unfortunately, the most obvious wavelength regime in which they differ - the radio - is also the regime in which least is known about the properties of RQQs. Most radio surveys of optically-selected quasar samples have lacked the sensitivity to detect the majority of radio-quiet objects, and have provided only limited information on their radio structure and spectral index.

To remedy this situation we embarked on a multi-frequency, high-resolution radio survey of 27 low-redshift ($0.1 \leq z \leq 0.3$), low-luminosity ($M_V > -26$) RQQs using the Very Large Array (VLA). Observations were made in 'A' configuration at 1.4, 4.8 and 8.4 GHz, with angular resolutions of 1.4, 0.4 and 0.24" respectively, and radio emission was detected in 75% of the quasars. The survey will be discussed in more detail by Kukula et al. (1997).

2 Results of the Survey

A principal goal of the survey was to investigate the origin of the radio emission in RQQs and to attempt to distinguish between radio emission produced by stellar processes (*eg* circumnuclear starburst regions) in the body of the host galaxy and emission which is directly associated with the quasar (*eg* radio jets). Our findings can be summarized as follows:

(1) In the majority of objects we detect an unresolved (≤ 1 kpc) radio component which is coincident with the optical position of the quasar.

(2) The radio spectral indices of the RQQs are generally steep ($\alpha \sim 0.7$, where $S \propto \nu^{-\alpha}$), although two objects, which also exhibit unusually large radio luminosities and optical variability, have flat radio spectra. Following Falcke, Sherwood & Patnaik (1996) we interpret these flat-spectrum objects as RQQs in which a relativistic jet is closely aligned to the line of sight, leading to doppler boosting of the radio emission.

(3) Lower limits on the brightness temperatures of the radio sources place them at the upper end of the range expected for emission related to stellar processes ($T_B \sim 10^5$ K). In many objects the brightness temperature is almost certainly many orders of magnitude greater than this, in which case the radio emission *cannot* be produced by a starburst. Observations with greater angular resolution will be required in order to confirm these results.

(4) In five objects we are able to resolve radio structure, which takes the form of double, triple and linear sources on scales of a few kiloparsecs. These images resemble early, low-resolution maps of nearby Seyfert nuclei - objects which have since been shown to contain small-scale (≤ 1 kpc), highly-collimated radio jets. The current maps could therefore be taken as evidence that collimated ejection of radio plasma from the central engine is also occurring in RQQs but high-resolution VLBI observations will be necessary in order to confirm this interpretation.

(5) The distribution of radio luminosities in the RQQ sample forms a natural extension to that of Seyfert 1 nuclei. There appears to be a correlation between radio luminosity and the optical absolute magnitude of the quasar, implying a close relationship between the central engine and the mechanism responsible for the bulk of the radio emission.

We therefore conclude that in the majority of RQQs a significant fraction of the overall radio emission comes from a compact nuclear source which is directly associated with the quasar's central engine. By analogy with Seyfert galaxies this nuclear source probably takes the form of a small-scale radio jet, qualitatively similar to the more powerful jets observed in RLQs and radio galaxies.

3 Radio Emission and the Hosts of RQQs

Seventeen of the RQQs in the present sample were also included in our near-infrared imaging study of quasar hosts (Taylor et al. 1996; described elsewhere in this volume by Kukula et al.). These observations showed that slightly less than half of the RQQs occur in galaxies in which the dominant stellar component has a spheroidal rather than an exponential (disc) distribution.

We were therefore able to carry out a study of the relationship between the host morphology and the radio properties of the AGN and to investigate the extent to which the radio sources in RQQs with spheroidal hosts could be distinguished from those in discs. The AGN traditionally associated with elliptical hosts (ie radio galaxies and RLQs) produce large, powerful radio sources and this suggests that the RQQs in elliptical galaxies, whilst technically 'radio quiet', might still harbour radio sources which differ in size and luminosity from those in disc galaxies.

However, in our sample we were able to find *no* clear separation between the radio properties of the two types of radio-quiet quasar. The RQQs with elliptical hosts *do not* contain larger or more luminous radio sources than those in disc galaxies. Both types of RQQ are equally likely to contain extended radio structure and to show evidence for collimated radio jets. The radio sources in elliptical galaxies show no tendency to have flatter spectra than those in discs (as might have been expected if the jets in elliptical galaxies are more likely to be relativistic).

Although the current sample is small, and is limited to RQQs of relatively low optical luminosity, our radio survey clearly demonstrates that having an elliptical host does not automatically confer a large radio luminosity on the active nucleus. A significant number of ellipticals with active nuclei are *not* producing large, powerful radio sources but contain small, weak sources which appear to be identical to those in disc systems. Further, more detailed studies of the host galaxies will be required in order to determine if and how these ellipticals differ from those containing radio loud AGN.

References

Bahcall, J. N., Kirhakos, S., Saxe, D. H., Schneider, D. P. (1997): preprint
Falcke, H., Sherwood, W., Patnaik, A. R. (1996): ApJ, **471**, 106
Kukula, M. J., et al.: this volume
Kukula, M. J., Dunlop, J. S., Hughes, D. H., Rawlings, S. (1997): in preparation
Taylor, G. L., Dunlop, J. S., Hughes, D. H., Robson, E. I. (1996): MNRAS, **283**, 930
Véron-Cetty, M.-P., Woltjer, L. (1990): A&A, **236**, 69

HST Imaging of Redshift $z > 0.5$ 7C and 3C Quasars

Stephen Serjeant[1], Steve Rawlings[2], Mark Lacy[2]

[1] Astrophysics Dept., Imperial College London, Blackett Laboratory,
  Prince Consort Road, London SW7 2BZ, England
[2] Astrophysics Dept., Oxford University, Nuclear and Astrophysics Laboratory,
  1 Keble Road, Oxford, OX1 3RH, England

Abstract. We present preliminary results from HST imaging of radio-loud quasar hosts, covering a $\sim \times 100$ range in radio luminosity but in a narrow redshift range ($0.5 < z < 0.65$). The sample was selected from our new, spectroscopically complete 7C survey and the 3CRR catalogue. Despite the very large radio luminosity range, the host luminosities are only weakly correlated (if at all) with radio power, perhaps reflecting a predominance of purely central engine processes in the formation of radio jets, and hence perhaps also in the radio-loud/-quiet dichotomy at these redshifts. The results also contradict naive expectations from several quasar formation theories, but the host magnitudes support radio-loud Unified Schemes.

1 Introduction

The strong evolution of quasars from redshifts $z = 2$ to $z = 0$ is very well documented but poorly understood. Their evolution may reflect changing merger rates (*e.g.* Carlberg 1990) or may reflect processes in galaxy formation *via* the changing formation efficiency of nuclear black holes (*e.g.* Haehnelt & Rees 1993). Alternatively, Small & Blandford (1992) suggest that the largest galaxies, hosting the brightest quasars, formed first (contrary to "bottom-up" hierarcichal structure formation); quasar evolution is then ascribed to a luminosity-dependent transition from continuous to intermittent activity. Physically, this could be interpreted an initial quasar phase in newly-forming galaxies followed by merger-driven events.

Each of these models has a wide parameter space in which to accommodate strong positive quasar evolution, which reflects the essential lack of a physical understanding. Nevertheless, the various classes of models make widely different predictions for the immediate environments of the active nuclei. For example, in merger-driven quasar evolution one expects tidal features or disturbed morphologies in the host galaxies, possibly more disturbed at higher z; in models where a quasar stage is common in bottom-up galaxy formation, one expects that quasars of a given luminosity have increasingly smaller host galaxies, with increasing redshift, but in the Small & Blandford (1992) model one predicts an opposite trend.

Host galaxies also provide a useful test of radio-loud unified schemes (*e.g.* Antonucci 1993). If dust shrouded quasars are universal in radiogalaxies, then there must be identical host galaxy properties in radioquasars and radiogalaxies. However, in making such comparisons one must ensure the quasar and radiogalaxy samples are well-matched in some isotropic quantity, such as hard X-ray luminosity or radio lobe luminosity. Unfortunately, the only large, complete (published) sample of lobe-dominated quasars is from the 3CR catalogue, in which radio luminosity is tightly correlated with redshift. This makes it difficult to address the evolution of any parameter, such as host galaxy luminosity, because of the possibility of radio luminosity dependence. In this paper we present a complete sample which breaks this degeneracy by comparing 3CR quasars with a new coeval, spectroscopically complete sample $\sim$ 100 times fainter in radio luminosity.

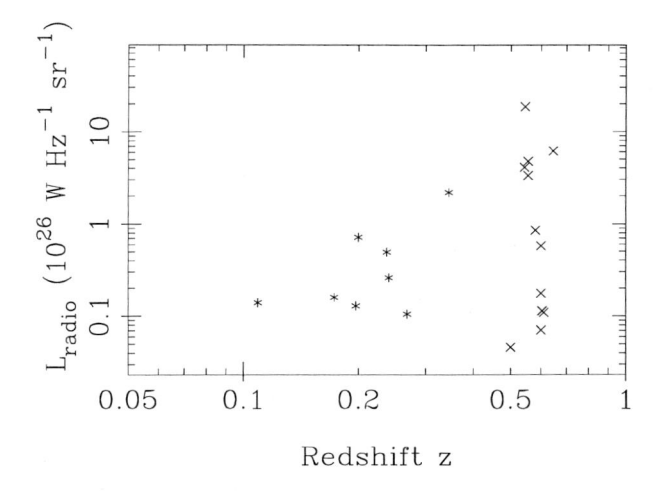

Fig. 1. Radio luminosity-redshift plane for our sample (crosses) and steep-spectrum quasars from Dunlop *et al.* (1993)

2 Data Acquisition and Analysis

2.1 Sample Definition

We have recently completed a spectroscopic campaign on the INT and NOT of steep-spectrum quasars (SSQs) from the 151MHz 7C catalogue (McGilchrist *et al.* 1990), which has a limiting flux density of $S_{151} = 0.1$Jy. This sample is ideal for comparisons with coeval quasars from the 178MHz 3CR catalogue (Laing, Riley & Longair 1983), since the high flux limit of $S_{178} = 10$Jy in the latter gives a wide dispersion in radio luminosity at any epoch. VLA snapshots of 7C obtained by us confirmed that the 3C and 7C SSQs are from the same parent population; both samples consist of FRII sources with cores. As we will show below, this allows us to decouple luminosity dependence from evolution.

We were awarded twelve orbits to image a subsample of 3C and 7C SSQs with the HST WFPC2. Our sample selection was driven by several competing requirements:

- spectroscopic completeness;
- freedom (as far as possible) from beaming and gravitational lensing biases;
- a narrow redshift interval, to counter any potential differential evolution;
- a choice of filter and redshift range avoiding strong emission lines;
- as red a filter as possible to maximise the contrast between the quasar and its host;
- the highest accessible redshifts for reliable detection of host galaxies.

Our choice of SSQs (as opposed to flat-spectrum quasars) ensured our samples were largely free from lensing and beaming since the steep-spectrum radio fluxes are dominated by extended, optically thin synchrotron emission.

Based on crude models of PSF-subtracted frames, we estimated that SSQ hosts should be detectable at $z \lesssim 0.6$ if Unified Scheme predictions hold, *i.e.* that the host galaxies are giant ellipticals. Our targets were therefore confined to the interval $0.5 < z < 0.65$, to be imaged in the F675W filter which successfully avoids the [OII] 3727Å, [OIII] 5007Å and Hβ emission lines. The complete 3CR and 7C samples contain 12 steep-spectrum quasars in this redshift interval. The very wide range in radio luminosity of our sample is demonstrated by the vertical dispersion in figure 1.

Also plotted on figure 1 are steep-spectrum quasars from the sample of Dunlop *et al.* (1993). Using samples such as these we can therefore make comparisons between well-matched quasars between epochs (*i.e.*, in a horizontal slice of figure 1), which for the first time are free of the luminosity dependence which marrs studies of 3C alone.

Moreover, even at the median redshift of our sample ($z \sim 0.6$) the quasar number density is already around an order of magnitude larger than the present (*e.g.* Dunlop & Peacock 1990). Our sample is therefore (in principle at least) capable of probing any environmental causes of quasar evolution.

Fortuitously, two of the 3CR targets (3C334 and 3C275.1) also have ground based detections of giant host galaxies, which was not realised when defining the survey selection criteria. These detections provide important corroborations of our photometry and quasar subtraction technique. It is perhaps significant that these large host galaxies, among the largest for any quasars, are at the highest possible redshifts for optical ground-based detections.

2.2 Data Acquisition

For the majority of the targets we broke our observations of each quasar (one orbit per quasar) into four separate exposures. Two of the four exposures were offset by 5.5 pixels in both directions, to assist with the positioning of the quasar in the central pixel (recall that the Planetary Camera slightly undersamples). Although this strategy might be expected to present problems with the cosmic

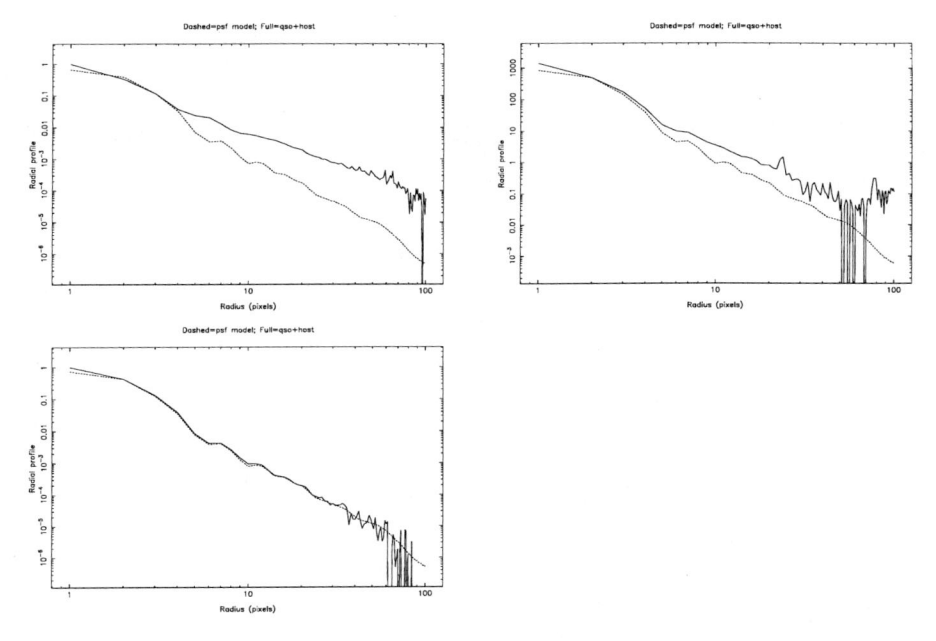

Fig. 2. Radial profiles for 3c275.1 (top left), the brightest host in our sample, and for a more typical target (7C2886, top right). The uprise at large radii in the latter is due to a companion galaxy. In both cases the Tiny Tim PSF is also plotted, normalised to the first unsaturated pixel. Also shown is the empirical PSF compared to the Tiny Tim model (bottom).

ray subtraction, we found that the cosmic rays could be identified and removed adequately using the CRREJ algorithm in the IRAF IMCOMBINE task, despite having only two frames per position.

Sky subtraction was achieved by selecting regions of the planetary camera frame free of bright objects, and constructing a histogram of pixel values (choosing bin sizes and widths to minimise the digitisation effects); the sky levels were then estimated by fitting Gaussians to these histograms.

2.3 Point Spread Functions

Since the Planetary Camera undersamples the point spread function, the pixel to pixel variations may depend strongly on the position of the quasar within the central pixel. We therefore expected that our QSO subtraction would be improved by determinining this position *via* a cross correlation of the quasar frame with a model oversampled point spread function from Tiny Tim version 4.0 (Kirst 1995). The oversampled model would need to be rebinned to the cruder Planetary Camera pixel scale at each proposed centre, and a least squares solution found for the central position. One small complication in this cross correlation is the existence of a non-negligable pixel to pixel scattering function. The resampled point spread functions must therefore be convolved with the appropriate kernel from Kirst (1995) before comparisons are made with the data frame.

To complement the Tiny Tim models, we also constructed an empirical point spread function from our standard star observations, masking out saturated pixels before coaddition. An undersampled empirical point spread function for this filter was unfortunately not available.

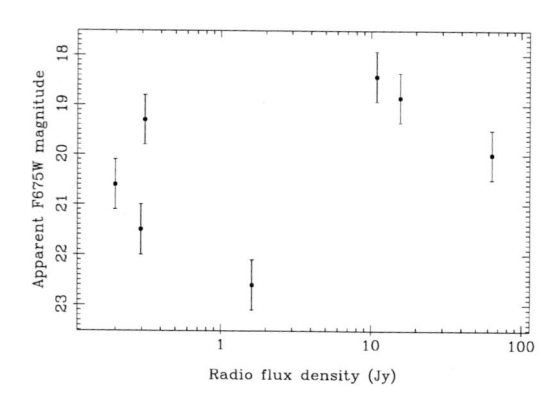

Fig. 3. Apparent F675W magnitudes of the hosts measured to date. Note the lack of, or only weak presence of, a correlation with radio flux. The F675W magnitude system is about half a magnitude fainter than Cousins R.

3 Results

3.1 Robustness of Host Magnitudes

We attempted QSO subtractions using PSFs obtained by three methods: an arbitarily centred model PSF, the empirical PSF discussed above, and a PSF centred on the QSO by cross correlation above with saturated pixels masked out. For the first two, we determined the PSF normalisation from the radial profiles, using the first unsaturated pixel. In the case of the cross-correlation method, these normalisations were in excellent agreement with those obtained independently from the cross-correlation itself. Although specific features such as diffraction spikes are better reproduced, the total magnitudes were (reassuringly) found to be insensitive to the PSF centering. We also attempted normalising to the diffraction spikes, but the results were much less satisfactory. Nevertheless, the host magnitudes were not found to be sensitive to the masking -or not- of the diffraction spikes. The results for the quasars examined to date are shown in figures 2 and 3. The previous ground-based detections of host galaxies in 3C275.1 and 3C334 are well reproduced by our algorithms. From the dispersion between the different methods, we estimate an error of $\sim \pm 0.2$ in our host galaxy magnitudes.

We used azimuthal averaging to reach fainter flux levels, as used by several other authors (figure 2). We might hope to obtain a morphological classification from fits to the radial profiles (*e.g.* exponential disk, de Vaucoleurs profile); this involves fitting to the curvature of the PSF-subtracted profile, which is not as well constrained as the total host magnitude. Furthermore, the reasonable inclusion of a bulge component to a disk model was found to introduce too many free parameters. This will be discussed in more detail in Serjeant *et al.* (1997), which also compares one- and two-dimensional fits to our data.

4 Discussion and Conclusions

Remarkably, the host magnitudes are only weakly dependent -if at all- on radio luminosity, despite the very wide dispersion in radio luminosity ($\sim \times 100$; see figure 3). This may reflect a predominance of purely central engine processes in the formation of radio jets, and hence perhaps also in the radio-loud/-quiet dichotomy at these redshifts. If so, we predict that the radio-quiet and radio-loud hosts should be similar by redshifts $z \simeq 0.6$.

We can compare figure 3 with the R magnitudes of radiogalaxies in our redshift range from the samples of Lacy *et al.* 1993 and Eales 1985. Both the radiogalaxies from these studies and our SSQ samples were selected in an orientation-independent manner; these radiogalaxies have $18.5 < R < 21.0$ which is clearly well reproduced by our data, supporting radio-loud Unified Schemes.

The results also contradict naive expectations from several quasar formation theories. In both the Haehnelt & Rees (1994) model (if applicable at this z) and the Small & Blandford (1992) model strong trends of host properties are expected with redshift (see introduction), whereas giant ellipticals appear to host most radio-loud AGN at $0.2 < z < 0.7$ (*e.g.* Taylor *et al.* 1996).

References

Antonucci, R., 1993, Ann. Rev. Astron, Astrop. 31, 473
Carlberg 1990 ApJ 350, 505
Dunlop, J.S., & Peacock, J.A., 1990, MNRAS 247, 19
Dunlop, J.S., *et al.*, 1993, MNRAS 264, 45
Eales, 1985, MNRAS 217, 167
Haehnelt, M., Rees, M., 1993, MNRAS 263, 168
Kirst, J., 1995, *The Tiny Tim User's Manual Version 4.0*
Lacy *et al.* 1993, MNRAS 263, 707
Laing, Riley & Longair 1983, MNRAS 204, 151
McCleod & Reike 1995, ApJL, 454, 77
McGilchrist *et al.*, 1990, MNRAS, 246, 110
Small, T., Blandford, R., 1992, MNRAS 259, 725
Serjeant, S., Rawlings, S., Lacy, M., 1997, in preparation
Taylor *et al.* 1996, MNRAS 283, 930

HST Imaging of BL Lac Objects

R. Falomo[1], C.M. Urry[2], J. Pesce[2], R. Scarpa[1,2], A. Treves[3] and M. Giavalisco[4]

[1] Osservatorio Astronomico di Padova, Italy
[2] Space Telescope Science Institute, USA
[3] University of Milan, Italy
[4] Carnegie Observatories, USA

Abstract. We present preliminary results on the HST images of six BL Lac objects (0814+49, 0828+49, 1308+32, 1538+14, 1823+56, and 2254+08) extracted from the complete sample of radio selected BL Lacs. Surrounding nebulosity is observed around 4 sources, while two (0814+49 and 1308+32) remain unresolved. The detected host galaxies are luminous ($M_I <= -24.5$) and rather smooth ellipticals that are well centered around the bright BL Lac nucleus. In four cases faint companion galaxies at a projected distance of 50-100 Kpc are found.

1 Introduction

BL Lac Objects are AGN supposedly dominated by a relativistic jet. This explains the high contrast between the stellar like central source and the host galaxy, and the weakness of lines, so that the redshift may be difficult to establish. The study of host galaxies are of basic importance for understanding the nature of the BL Lac phenomenon. It is relevant to questions like whether relativistic jet formation is associated to a given type of galactic structure, and whether interaction with companion galaxies plays a role in the triggering of the BL Lac activity.

The unprecedented resolution of the refurbished HST allows one to trace the host galaxies much closer to the nucleus than in the case of ground based observations. This permits the study of the hosts at larger z ($\lesssim 1$), and for near-by objects to constrain galaxy models effectively.

For these reasons systematic programs of HST imaging of BL Lacs have been started. Similar programs aimed at studying the host of quasars have been undertaken by other groups (see e.g. Bahcall et al. 1997 and references therein, Disney et al. 1995, these proceedings). We have focussed on objects selected through radio observations, while Jannuzi et al (private communication) have considered mainly X-ray selected BL Lacs. Our targets consist of six objects taken from the complete 1 Jy sample of Stickel et al (1991), spanning the redshift range z= 0.2 to 1. Three objects have already been discussed in detail (0814+425, 1823+568, 2254+074; Falomo et al 1997). Here we report on a preliminary analysis of the other three, 0828+493, 1308+32, 1538+14, and briefly discuss some properties of the entire sample.

2 Observations and Image Analysis

All observations were obtained with the BL Lac target centered in the Planetary Camera (PC) chip and through the F814W filter ($\sim$ 8140 Å). In order to obtain a final image well exposed both in the inner, bright nucleus and in the outer regions where the host galaxy emission should be more relevant, we set up for each object a series of exposures typically from a few tens of seconds to exposures of 500-1000 s duration.

Each BL Lac object was observed for a total of two orbits, with each orbit consisting of the same sequence of exposures. Between orbits, we rolled the telescope by $\sim$ 20°, so that the diffraction spikes rotated relative to the sky. This technique allows us to check the reliability of faint features close to the diffraction spikes and to detect easily any asymmetric emission (e.g., jets, distortions, companions) by comparing images with different roll angle. Table 1 summarizes the objects observed and their main properties. The final average images were calibrated using the prescription of Holtzman et al. (1995), converted to the I band in the Cousins system. Additional information on data analysis are given in Falomo et al. 1997.

The combined image (single roll angle) for each object is reported in figure 1. After centering the PSF model with the BL Lac center we have subtracted a scaled PSF so that, after subtraction, the residual image (galaxy) continued to exhibit fluxes increasing monotonically towards the center. The resulting PSF subtracted images are reported in figure 1b. From this procedure it is seen that 4 sources are clearly embedded in a nebulosity while in two cases (0814+42 and 1308+32) no evidence of extended emission is found.

To derive the properties of the host galaxies we have then performed surface photometry of the galaxy directly on the unsubtracted average images, masking out the diffraction spikes. We used an interactive numerical mapping package to derive isophotes, fitted with ellipses down to $\mu_I = 25$ mag arcsec^{-2}. In addition we have also computed azimuthally averaged radial brightness profiles that allows us to improve the surface photometry analysis in the very inner region (where ellipse fitting may not work) and at low surface brightness levels (where isophotes are rather noisy).

The luminosity radial profiles are shown in Figure 2 together with a scaled profile of the PSF. This comparison of the profiles confirms that both 0814+42 and 1308+32 are consistent with a point source. For the 4 resolved objects, we modeled the radial profile with either a de Vaucouleurs ($r^{1/4}$) law or an exponential disk, convolved with the PSF, plus the contribution of the point source in the nucleus. In all cases the elliptical model resulted a better representation than the disk model. In Table 1 we give the derived properties of the host galaxy. In the cases of unresolved objects we determined an upper limit (90%) of the luminosity of the host assuming an elliptical galaxy with effective radius 10 kpc.

3 Results of Individual Sources

0814+425 In our *HST* images we find no evidence of a host galaxy, despite the

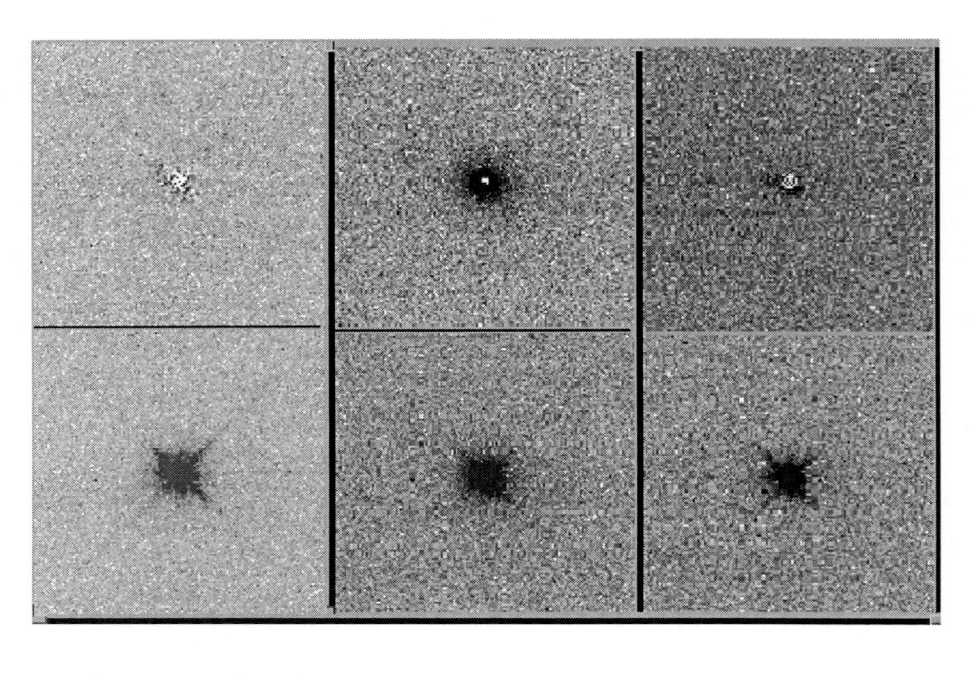

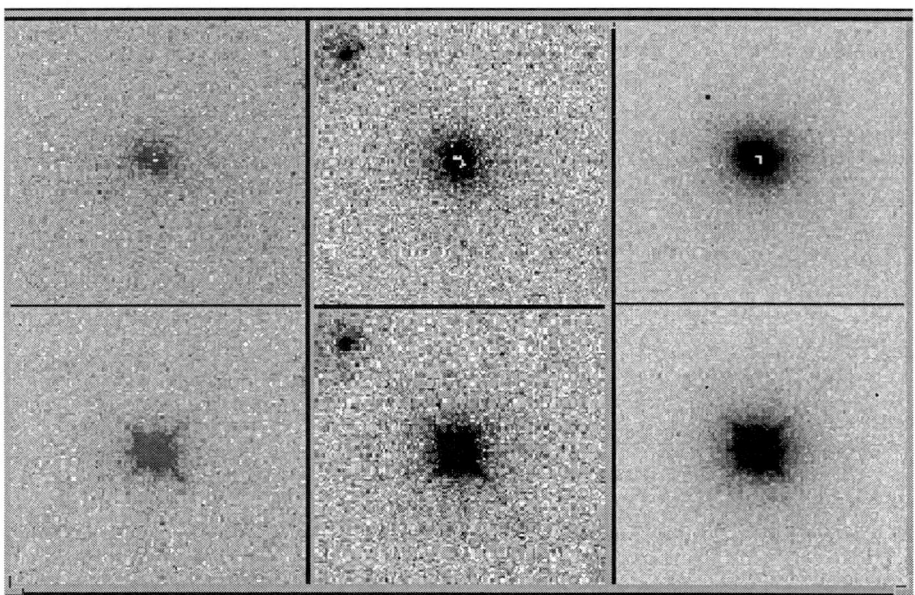

Fig. 1. Central part of the images of BL Lacs observed with *HST* WFPC2 (F814W filter). Upper panels show images after subtraction of a scaled PSF. Lower panels give the unsubtracted images. From left to right are: (upper figure) 0814+42, 0828+49, 1308+32, (lower figure) 1538+14, 1823+56 and 2254+07.

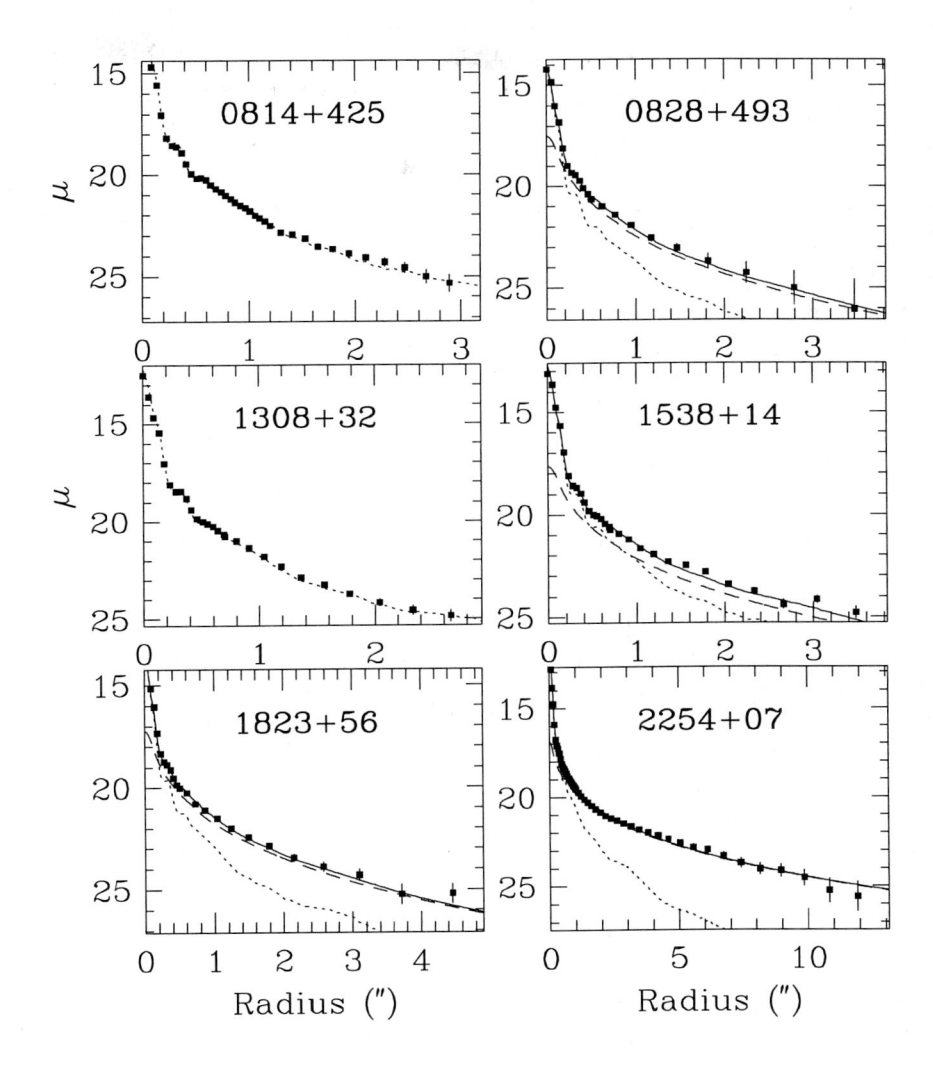

Fig. 2. Radial surface brightness profile (filled squares) of the BL Lac objects together with the best fit model (solid line). The point source is modeled by a scaled PSF (dotted line) while the host galaxy (dashed line) is given by an $r^{1/4}$ law convolved with the PSF. Where the dashed lines are not visible, they lie beneath the solid line.

relatively low redshift ($z = 0.258$) proposed by Wills & Wills (1976). This result is supported by the PSF subtracted image (see Figure 1) and by the good match between the radial profile and a scaled PSF (Figure 2). Assuming the BL Lac is really at $z = 0.258$ we can estimate an upper limit for the host galaxy (elliptical galaxy with effective radius $r_e = 10$ kpc) magnitude . We find that a galaxy with fairly modest luminosity, $M_V \lesssim -20.8$ mag (including a K-correction of 0.24), would have been detected at the 90% confidence level. Alternatively the reported redshift could be wrong and the host galaxy could be much more luminous. In this case assuming a typical host galaxy of magnitude $M_V = -23.3$ mag, the BL Lac would have to be at redshift $z \gtrsim 0.6$ in order for the galaxy to remain undetected in our images.

0828+49 This source is clearly resolved and its luminosity profile is well described by an elliptical model with parameters: $m_{tot} = 19 \pm 0.4$ mag and $r_e = 4$ Kpc . This yields a luminous smooth host galaxy with $M_I = -24.6$ (k-corr = 0.5). Ground based observations (Wurtz, Stocke and Yee, 1996) in the r-gunn filter show the object only marginally resolved. From the latter data an absolute magnitude $M_r = -23.4$ was derived. A value that is consistent with our measurement if a (r-I) ~ 1.3 color is assumed.

1308+32 This is a high redshift (z=0.997) object. Stickel et al. 1993 on the basis of deep ground based images of 1308+32 have claimed the presence of a foreground galaxy superposed onto the BL Lac source and proposed the source as a microlensed candidate. In our HST image we do not see convincing evidence of surrounding nebulosity down to $\mu_I \approx 25$. Our data are still consistent with the BL Lac being hosted by a luminous ($M_I > -25$) galaxy. These results are in agreement with deep R band subarcsec images of this source obtained at Nordic Optical Telescope (see Falomo et al. 1996). In the immediate environment of the BL Lac object there are several faint galaxies; the closest companion is a $m_I \sim 21.5$ galaxy at 5.5 arcsec (projected distance 60 Kpc).

1538+14 The object ($z = 0.605$) is spatially resolved. Fit with elliptical model yields $m_{tot} = 18.9$ and $r_e = 6$ Kpc. Little deviations from the fit are present between 1 and 2 arcsec and need further investigation. Also in this case a luminous host ($M_I = -25.2$) is found. Four faint (22-24 mag) galaxies are detected at projected distance < 60 Kpc.

1823+56 Despite the relatively high redshift ($z = 0.664$; Lawrence et al. 1986) this BL Lac object is clearly resolved in our *HST* image (see Figure 1). Surface photometry indicates that the host galaxy is slightly elongated (ellipticity $\epsilon \sim 0.1$) at position angle $\sim 120°$. The radial profile is well fitted by a model consisting of an elliptical galaxy plus point source (Figure 2). The best fit galaxy has total apparent magnitude $m_I^{tot} = 18.3 \pm 0.3$ mag. At the reported redshift, this would correspond to an absolute visual magnitude in the rest frame of $M_V = -24.7 \pm 0.3$ mag.

2254+07 This is a relatively nearby BL Lac object with a well determined redshift ($z = 0.19$). As is apparent from Figure 2, the galaxy is detected out to ~ 12 arcsec in our combined *HST* image. The radial profile is well fitted by an elliptical model plus a point source, yielding an apparent galaxy magnitude

$m_I = 15.9 \pm 0.3$ mag, corresponding to $M_V = -23.5 \pm 0.3$ mag. A disk model gives a significantly worse fit than the de Vaucouleurs law (see Table 2) especially in the region close to the nucleus. The object has been well resolved from ground based imaging but models could not be well constrained. (Stickel et al. 1988; Falomo et al. 1995b)

Properties of BL Lac Host Galaxies

Object	z	Date	Tot. Exp.	M_I(host)	Type	Note
0814+425	0.258?	13 Nov 95	2x1960	>-22.1	U	
0828+49	0.548	19 Feb 96	2x2060	-24.6	Ell.	$\epsilon \sim 0.1$
1308+32	0.997	04 Mar 96	2x1800	> -25.5	U	
1538+14	0.605	21 Mar 96	2x1670	-25.2	Ell.	$\epsilon \sim 0.1$
1823+568	0.664	02 Oct 95	2x2120	-26.0	Ell.	$\epsilon \sim 0.1$
2254+074	0.190	12 Oct 95	2x1520	-24.8	Ell.	$\epsilon < 0.02$; twist

U = Unresolved; Ell. = Elliptical Host Galaxy

4 Conclusions

High resolution images of a six BL Lac objects extracted from the 1Jy sample (Stickel et al. 1991) have shown that 4 are hosted by a luminous ($M_V \lesssim -23.5$) elliptical galaxy. The host galaxy is always well centered (within 0.05″) around the BL Lac nucleus and appears rather smooth and unperturbed. In several cases close (projected distance < 100 Kpc) companion galaxies are detected and could represent the 'remnant' of past interactions with the main galaxy. Distortions in some of the companion galaxies (as e.g. in the case of 1823+56) give support to this idea.

References

Disney, M. J., et al. 1995, Nature, 376, 150
Falomo, R., et al. 1996 Science with the HST - II, Paris, Ed. Benvenuti et al. p. 226.
Falomo, R. 1997, MNRAS, in press.
Falomo, R., et al. 1997, Ap.J., in press.
Holtzman, J. A. et al. 1995, PASP, 107, 1065
Lawrence, C. R., et al. 1986, AJ, 91, 494
Stickel, M., Fried, J. W., & Kühr, H. 1988, A&A, 191, L16
Stickel, M. et al. 1991, ApJ, 374, 431
Stickel, M., Fried, J. W., & Kühr, H. 1993, A&AS,
Wurtz, R., Stocke, J. T., & Yee, H. K. C. 1996, ApJS, 103, 109

Near-IR Imaging of BL Lac Host Galaxies

Jochen Heidt[1]

Landessternwarte, Königstuhl, D-69117 Heidelberg, Germany

Abstract. The first results of a study of the host galaxies of BL Lac objects in the near-IR from the complete 1 Jy sample are presented. In particular, the observations of BL Lac, S4 1823+568 and PKS 0139-097 are shown and compared with optical observations. Whereas BL Lac itself is hosted by a galaxy with a luminosity typically found for BL Lac hosts, S4 1823+568 is extremely bright. No host galaxy for PKS 0139-097 could be detected. However, previously unknown galaxies very close (≤ 3") to the center were found. These galaxies may be responsible for the Mg II absorption feature detected in the spectrum of PKS 0139-097. PKS 0139-097 is a further BL Lac, which might be affected by gravitational microlensing.

1 Introduction

Studies of the host galaxies and close surroundings of active galactic nuclei (AGN) may provide important insights into the mechanism, which provides material for fuelling the very center of the AGN, finally giving rise to the extreme properties of this kind of extragalactic objects. They also allow to probe Unified Schemes, which relates the different types of AGN basically through different viewing angles together, may find objects, whose properties are affected by lensing effects and finally allow to study galaxy evolution over a wide redshift range.

This is especially true for the BL Lac objects, the most extreme class of AGN. They are characterized by strong and rapid variability across the whole electromagnetic spectrum, high and variable polarization as well as (nearly) featureless optical spectra. In many cases superluminal motion has been observed. Due to these observed properties it is nowadays believed that BL Lac objects are AGN, which are seen very close along the line-of-sight of a relativistic jet. As a result the radiation of this jet overwhelms in most cases the light from the underlying host galaxy being a challenge for each observer to separate the nuclear contribution from the stellar light.

Nevertheless, several studies in recent years using large telescopes with excellent seeing conditions have shown the possibility to resolve the host galaxies of BL Lac objects up to redshifts of ~ 0.5 (e.g. Wurtz et al., 1996, Falomo, 1996, Abraham et al., 1991). In general, they were hosted by giant ellipticals with typical luminosities around $M_V = -23$. In a few cases they may also be hosted by disc-type systems (Wurtz et al., 1996, McHardy et al., 1994).

Near-IR observations of BL Lac host galaxies offer several advantages over those carried out in the optical regime. First, the spectral energy distribution of elliptical galaxies peaks somewhere in the NIR ($\sim 1.6 \mu$m, McLeod & Rieke, 1994), hence the ratio of the stellar to the nuclear light is higher than in the optical. Secondly, the red stellar population and therefore the mass tracing component of the galaxies is observed. Thirdly, the K correction is much better than in the optical. Finally, due to the brightness of the sky in the NIR, which limits typical integration times to a few seconds and the characteristics of the NIR cameras it is easily feasible to achieve very deep images without saturating the core. This is essential in order to determine the shape of the point spread function (PSF) properly. With the rapid evolution of NIR cameras in the last years larger cameras become avaiable, which allow to place the BL Lac object and useful PSF stars on the same frame.

Within this context an extensive survey of the host galaxies of the complete sample of BL Lac objects from the 1 Jy catalogue (Stickel et al., 1991) by means of deep near-IR imaging using telescopes on Calar Alto, Spain and ESO, Chile has been started. Here the observations of BL Lac, S4 1823+568 and PKS 0139-097 carried out in September 1995 using the Calar Alto 2.2m telescope are presented. During the observations the NIR camera MAGIC equipped with a 256 × 256 NICMOS3 array (0$''$636/pixel) and a K' filter was used. Details of the observations and data reduction can be found in Heidt (1996). In this paper $H_0 = 50 \mathrm{kms}^{-1} \mathrm{Mpc}^{-1}$ and $q_0 = 0$ is assumed.

2 BL Lac

BL Lac is the prototype of the BL Lac objects and one of the BL Lacs with the lowest redshift of the 1 Jy sample ($z = 0.069$). The total integration time of this field was 3 hours, the limiting magnitude of the final frame is m(K') ~ 20.5. It is shown in figure 1 (left). As can be seen from that image, BL Lac is embedded in a well resolved galaxy. In order to derive the properties of the galaxy, the azimuthally averaged surface brightness profile (SBP) following the procedure outlined by Bender & Möllenhoff (1988) was extracted from the image. Prior to the fitting procedure, three foreground stars as well as a diffraction spike from the bright star southeast from BL Lac were masked out. Next, a de Vaucouleurs profile between 3" and 10" from the center (in order to minimize seeing effects) was fitted to the SBP. An effective radius of r_e ~ 6.8 kpc and an apparent magnitude of $m_{K'}$ ~ 12.1 for the host galaxy have been calculated from the fitting procedure resulting in M_K ~ −26.2±0.2 (including K correction adopted from Neugebauer et al., 1985). Neugebauer et al. (1985) determined M_K ~ −26.1±0.3 from photometer measurements using diaphragrms, in excellent agreement with the results presented here. From the derived parameters, a model for the host galaxy of BL Lac was constructed and subtracted from the image. The result is shown in figure 1 (right). The model galaxy fits the observations quite well. Note that even the diffraction spike is continuous along the region where the host galaxy was subtracted. Stickel et al. (1991) presented data for the host galaxy of BL Lac derived from R band measurements. They found r_e ~ 10.5 kpc and an apparent magnitude of m_R ~ 15.1 resulting in an absolute magnitude of M_R ~ −23.9±0.2. This gives an R−K colour index of ~ 2.3, which indicates that the host galaxy is slightly bluer than a "normal" elliptical galaxy, but is not an extreme case.

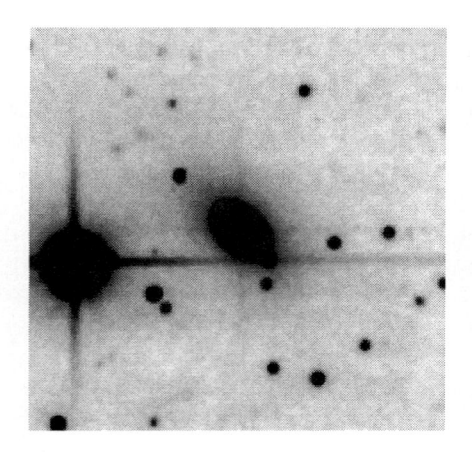

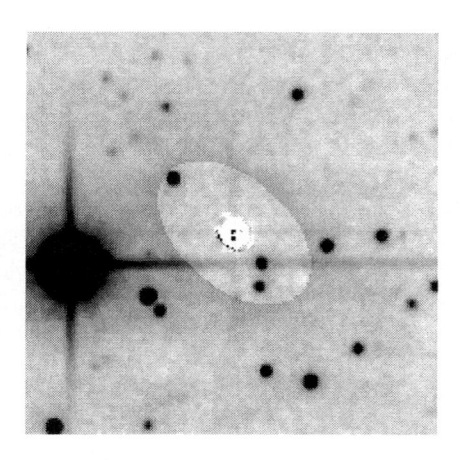

Fig. 1. Left) The field of BL Lac in K'. North is up and east to the left, the image covers 1' on each side. Right) The same field after the subtraction of a model galaxy using the parameters derived from the fitting procedure.

3 S4 1823+568 = 4C 56.27

The BL Lac S4 1823+568 has a redshift of z = 0.664 (Stickel et al., 1993). It was observed for a total of 3 hours. The final frame has a limiting magnitude of m(K') $\sim$ 20. The host galaxy is clearly resolved as can be seen from figure 2 (top, left). The are also numerous faint galaxies in the field. Very interesting is a nonstellar feature 5" east of S4 1823+568 (projected distance $\sim$ 46 kpc). This object is asymmetric with a faint elongation to the south indicating possibly tidal interaction with the BL Lac. This is shown in figure 2 (top, right), where the close vicinity of S4 1823+568 is presented. The asymmetry was also noticed by Falomo et al. (1997) based on HST imaging. The integrated magnitude of this feature is m(K') $\sim$ 17 thus it should be possible to measure the redshift of this source with 3m class telescopes in order to check a physical connection with the BL Lac.

The image of S4 1823+568 was analyzed using a numerical fitting package developed by J. Fried (c.f. his contribution in the proceedings for details). After extracting the SBP, the profile was modelled by a scaled PSF plus a de Vaucouleurs profile convolved with the PSF. The results of the fitting procedure are shown in figure 2 (bottom). The derived parameters for the host galaxy are r_e = 14.5±2.8 kpc and m(K') = 15.1. Using the K corrections of Neugebauer et al. (1985) the absolute magnitude is M(K')= −29.1±0.3. This indicates that S4 1823+568 is hosted by a very luminous galaxy. Falomo et al. (1996) gave M(V) = −24.7±0.3 from their HST data resulting in a V-K colour index of $\sim$ 4.4. Thus the host galaxy is slightly redder than normal elliptical galaxies.

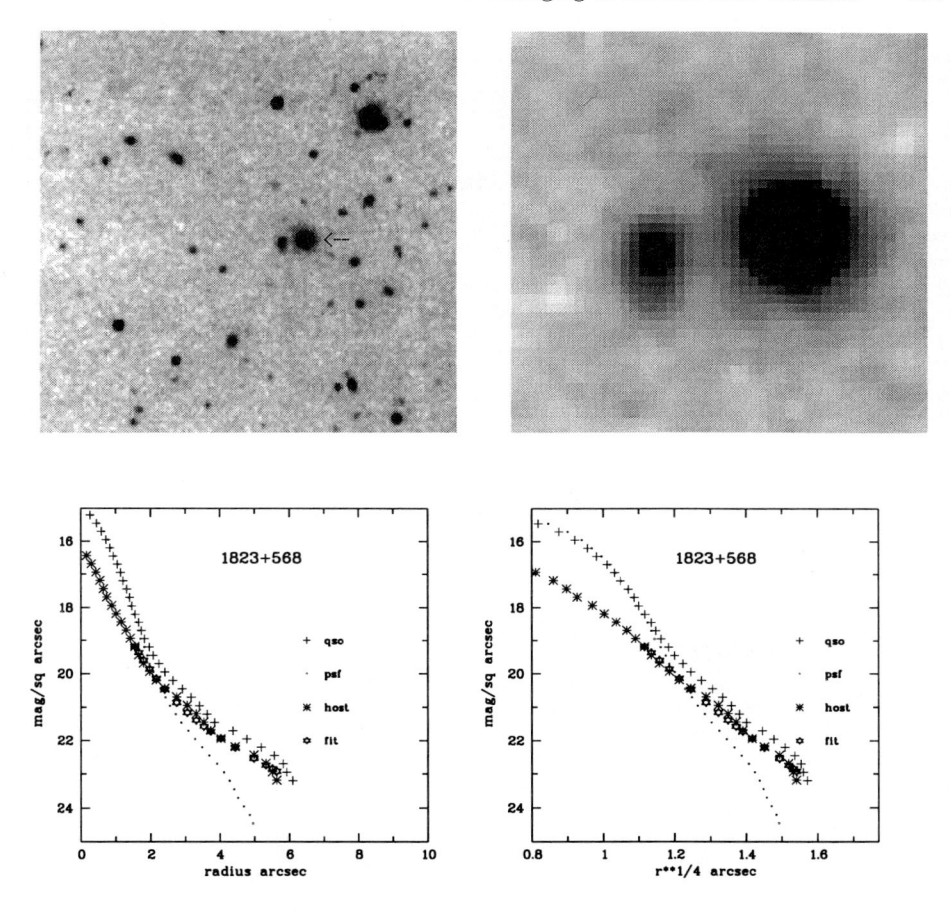

Fig. 2. Top) The field of S4 1823+658. North is up, east to the left. The image covers 90" on each side (left). S4 1823+568 and its close companion. The field is 15" wide. Note the asymmetric shape of the companion 5" to the east (right). Bottom) Results of the fits to the SBP of S4 1823+568. The left panel shows the results over the radius in linear scale, the right panel over $r^{1/4}$. The + signs give the observed SBP, the dots the scaled PSF, the asterisks the residuals after the subtraction of the PSF and the stars the fit to the residuals.

4 PKS 0139-097

The observations of PKS 0139-097 showed the most spectacular results obtained from the imaging survey so far. The redshift of this BL Lac is unknown, a lower limit of $z < 0.501$ based on a single Mg II absorption line has been given by Stickel et al. (1993). In previous observations it showed a stellar appearance (Falomo et al., 1990, Stickel et al., 1993). It was observed in K' for a total

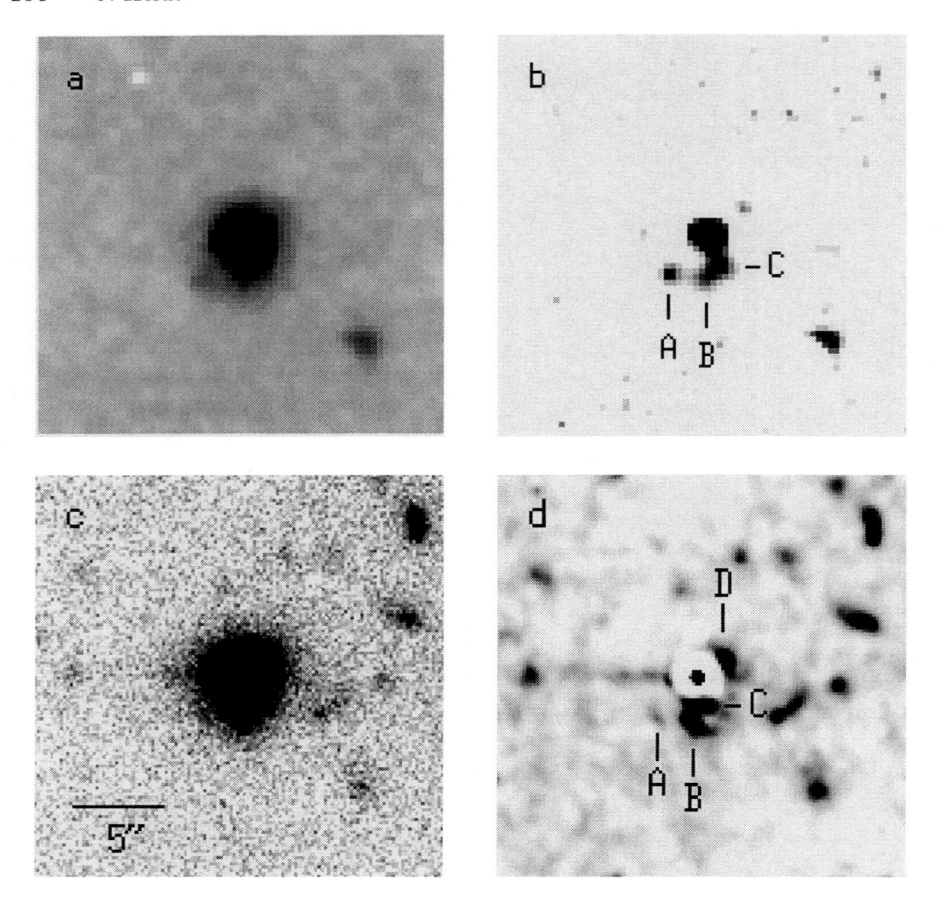

Fig. 3. Left: The original frame of PKS 0139-097 in the K' band (a) and in the R band (c). The frames cover 22.5" on either side. North is up and east to the left. PKS 0139-097 is already slightly elongated towards the south on these images. The feature A in K' can also be seen already in the unrestored image. Right: The restored images after deconvolution with the Lucy-Richardson algorithm in K' (b) and R (d). The white ring around PKS 0139-097 in the R image is an artifact produced by the deconvolution method used (Gibbs oscillations).

of 4 hours, the resulting limiting magnitude is ~ 20.5. The image is shown in figure 3a. For comparison, an optical image obtained at the NOT on La Palma is shown in figure 3c. This image is extremely deep (m(R) ~ 26.0) and taken under excellent seeing conditions (FWHM $< 0\rlap{.}''8$). PKS 0139-097 looks nonstellar in both colours, on the K' frame is also faint, close feature to the southeast visible. The K' and R images have been deconvolved using the Lucy-Richardson algorithm and the MEM (for full details, see Heidt et al., 1996). The results of the Lucy-Richardson deconvolution are shown in figures 3b and d.

There are at least 4 features within 3" from the center of PKS 0139-097 visible, 3 in both colours. The R-K colour indexes of the features are rather diverse. Whereas feature A is extremely red (R-K $\sim$ 6.8) and may be either a red quasar (Webster et al., 1995) or an unevolved early-type system at high redshift (z > 1), feature D is very blue and might be a faint late-type field galaxy (Glazebrook et al., 1995). Features B and C have magnitudes, colours and distances found in the range of Mg II absorbers towards QSOs (e.g. Bergeron & Boissé, 1991). Thus the latter features are most likely responsible for the Mg II absorption feature seen in the spectrum of PKS 0139-097. If this is the case, PKS 0139-097 is another BL Lac candidate, which might be affected by gravitational microlensing. It is also worth to note that neither in K' nor in R the host galaxy of PKS 0139-097 could be detected although especially the optical observations are extremely deep. However, as long as no redshifts of the BL Lac and the features are available, it can not be ruled out that they are at the same distance. Finally, on the deconvolved optical image (figure 3d) several faint galaxies to the west of PKS 0139-097 with peculiar morphologies can be seen, which may be part of a small cluster of galaxies. While the observations of PKS 0139-097 presented here have shown rather spectacular results, the real nature of the features around this BL Lac can only been studied by imaging using the HST or adaptive optics in combination with spectroscopy using the VLT.

Acknowledgements: This work was supported by the DFG (SFB 328). I would like to thank J. Fried for the analysis of the host galaxy of S4 1823+568.

References

Abraham R.G., McHardy I.M., Crawford C.S., 1991, MNRAS 252, 482
Bergeron J., Boissé, 1991, A&A 243, 344
Bruzual G., Charlot S., 1993, ApJ 405, 538
Elias J.H., Frogel J.A., Matthews K., Neugebauer, G., 1982, AJ 87, 1029
Falomo R., Melnick J., Tanzi, E.G., 1990, Nat 345, 692
Falomo R., 1996, MNRAS 283, 241
Falomo R., Urry C.M., Pesce J.E., et al., 1997, ApJ, in press
Glazebrook K., Peacock J.A., Miller L., Collins C.A., 1995 MNRAS 275, 169
Heidt J., et al., 1996, A&A 312, L13
Heidt J., 1996, In L.O. Takalo (ed.) Proc. Workshop on two years of intensive monitoring of OJ 287 and 3C 66A, Tuorla Observatory Reports, No. 176, p. 86
McHardy I.M., et al., 1994, MNRAS 268, 681
McLeod K.K., Rieke G.H., 1994, ApJ 420, 58
Neugebauer G., Matthews K., Soifer B.T., Elia J.H., 1985, Apj 298, 275
Bender R., Möllenhoff C., 1987, A&A 177, 71
Stickel M., Padovani P., Urry C.M., Fried J., Kühr H., 1991, ApJ 374, 431
Stickel M., Fried, J.W., Kühr H., 1993, A&AS 98, 393
Urry C.M., Padovani P., 1995, PASP 107, 803
Webster R.L., et al., 1995, Nat 375, 469
Wurtz R., Stocke J.T., Yee H.K.C., 1996, ApJS 103, 109

HST Imaging of Quasar Host Galaxies Selected by Quasar Radio and Optical Properties *

Eric J. Hooper[1], Chris D. Impey[1], and Craig B. Foltz[2]

[1] Steward Observatory, University of Arizona, Tucson, AZ 85721, USA
[2] Multiple Mirror Telescope Observatory, University of Arizona, Tucson, AZ 85721, USA

Abstract. The Planetary Camera on the WFPC2 instrument of the Hubble Space Telescope was used to image 16 quasars selected from the Large Bright Quasar Survey in the redshift range $0.4 < z < 0.5$. Host galaxy luminosity is positively correlated with the luminosity of the quasar nuclear component. The measured axial ratios of many of the host galaxies in the sample are small, which may indicate that they are inclined disk systems. Alternatively, bars or other distinctive morphological features, which are visible while the bulk of the underlying lower surface brightness components of the host galaxy are not, may account for the elongated appearance.

1 Sample Selection

The HST imaging targets were drawn from the Large Bright Quasar Survey (LBQS; Hewett, Foltz, & Chaffee 1995 and references therein), the largest homogeneous optical quasar survey to date, containing over 1000 quasars in the redshift range $0.2 < z < 3.4$. The LBQS has many advantages, the most pertinent for this work being quantifiable selection criteria and inclusion of candidates whose images on the discovery plates are resolved due to an underlying host galaxy. This last point is particularly important for studies such as this one. Many previous optically selected quasar surveys required that a candidate appear point-like. This seeing-dependent and often subjective criterion could potentially exclude quasars with bright host galaxies. Quasars with typical colors and emission line strengths at the redshifts of the sample selected for WFPC2 imaging satisfy the LBQS magnitude limit and selection criteria as long as the total luminosity of the host galaxy is approximately equal to or less than that of the quasar (Hooper et al. 1995; Hewett et al. 1995). This luminosity restriction would exclude very few known galaxies *a priori*, since the absolute magnitudes of the quasars in the HST sample fall in the range $-25 < M_B < -23$.

Several criteria were considered for selection of targets for WFPC2 imaging: (1) a radio luminosity span typical of optically selected quasars; (2) similar redshifts to avoid problems of interpretation due to evolution and to provide a uniform rest-frame spatial resolution scale; and (3) quasar absolute magnitudes

---

* Based on observations with the NASA/ESA Hubble Space Telescope, obtained at the Space Telescope Science Institute, which is operated by the Association of Universities for Research in Astronomy, Inc., under NASA contract NAS 5-26555

bright enough to exclude Seyfert galaxies, yet not so bright that the nucleus completely obscures the host galaxy. LBQS quasars with redshifts $0.4 < z < 0.5$ satisfied these criteria. All six known LBQS radio-loud quasars with these redshifts were selected. These objects have radio luminosities between 10^{25} W Hz^{-1} and 3×10^{26} W Hz^{-1}. Ten quasars with 8.4 GHz upper limits $\leq 10^{23.5}$ W Hz^{-1}, well below the radio-loud classification threshold, completed the sample. The radio-quiet quasars spanned the absolute magnitude range $-25 < M_B < -23$, an interval containing nearly all of the LBQS quasars with redshifts $0.4 < z < 0.5$.

2 Data Reduction and Results

The quasars were imaged through the F675W filter, which approximates Johnson R, near the center of the Planetary Camera (0.046 arcseconds per pixel) of WFPC2. Total integration times of 20 to 30 minutes were divided into 3 or 4 separate exposures to avoid saturation. The images were corrected for bias, dark current, and flat-fielding during the standard pipeline processing.

Optimal values of host galaxy flux, exponential scale length or effective radius, axial ratio, and position angle, and the quasar centroid and flux were determined by cross-correlating the reduced data with two-dimensional (2-D) galaxy and point spread function (psf) models. A description of the general 2-D cross-correlation method was presented by Phillipps & Davies (1991); Boyce, Phillipps, & Davies (1993) discussed its application specifically to quasar host galaxy analysis. The psf, generated with the "Tiny Tim" software (Krist 1996), plus disk and elliptical galaxy templates were repeatedly cross-correlated with the region around the quasar in the WFPC2 images while varying the psf and model galaxy parameters. The set of parameters which maximized the cross-correlation function was adopted. Fluxes were converted to Johnson R using Equation (8) in Holtzman et al. (1995). Absolute magnitudes were then calculated using k-corrections from Coleman, Wu, & Weedman (1980) and Cristiani & Vio (1990) for galaxies and quasars, respectively.

Table 1 contains the numerical results, listing for each target: (1) source name; (2) redshift from Hewett et al. (1995); (3) 8.4 GHz luminosity (W Hz^{-1}) from Hooper et al. (1995, 1996); (4) and (5) apparent and absolute R magnitudes for the quasar nuclear component, respectively; apparent (6) and (7)absolute R magnitudes for the host galaxy using an exponential disk template; (8) and (9) apparent and absolute R magnitudes for an $r^{1/4}$ model template; and (10) the axial ratio of the host galaxy. The average of the derived quasar magnitudes is listed in columns (4) and (5), since the quasar fluxes differ by typically only 2% or less between the elliptical and disk fits to the underlying galaxy. The integrated host galaxy fluxes, however, are typically 0.5 to 1 magnitude fainter for the disk model, and both are listed in the table. Axial ratios derived from the two models are identical in most cases, and the average is listed in the few that are not.

Table 1. Results of WFPC2 Analysis

Name	z	$\log L_{8.4}$ (W Hz^{-1})	R (quasar)	M_R (quasar)	R ($r^{1/4}$)	M_R ($r^{1/4}$)	R (disk)	M_R (disk)	axial ratio
(1)	(2)	(3)	(4)	(5)	(6)	(7)	(8)	(9)	(10)
0020+0018	0.423	< 23.50	19.52	−22.39	19.33	−23.42	20.17	−22.19	0.7
0021−0301	0.422	< 23.50	19.15	−22.75	19.87	−22.87	20.67	−21.68	0.6
0100+0205	0.393	< 23.45	17.81	−23.93	19.97	−22.54	20.70	−21.47	0.3
1138+0003	0.500	25.60	17.94	−24.35	19.33	−23.97	20.26	−22.54	0.9
1149+0043	0.466	< 23.66	17.37	−24.76	19.17	−23.91	19.77	−22.84	0.6
1209+1259	0.418	< 23.52	19.56	−22.32	20.20	−22.51	20.85	−21.48	0.6
1218+1734	0.445	25.33	18.36	−23.66	19.63	−23.28	20.33	−22.15	0.6
1222+1010	0.398	< 23.46	18.46	−23.30	19.36	−23.18	20.59	−21.60	0.3
1222+1235	0.412	25.19	17.86	−23.98	18.85	−23.81	19.55	−22.74	0.6
1230−0015	0.470	25.77	17.51	−24.64	18.96	−24.14	19.65	−22.98	0.4
1240+1754	0.459	< 23.47	18.13	−23.96	19.50	−23.50	20.40	−22.16	0.3
1242−0123	0.491	< 23.47	18.26	−23.99	19.51	−23.73	19.91	−22.84	0.4
1243+1701	0.459	< 23.41	18.71	−23.38	20.84	−22.24	20.92	−21.64	0.1
2214−1903	0.397	< 23.48	18.98	−22.77	19.18	−23.35	19.75	−22.43	0.5
2348+0210	0.504	25.56	17.18	−25.13	19.57	−23.75	20.25	−22.57	0.4
2351−0036	0.460	26.40	17.92	−24.18	20.07	−23.02	20.87	−21.70	0.4

The reliability of the analysis technique and detection sensitivity was estimated by applying the 2-D cross-correlation method to stars found in the Planetary Camera quasar fields. This check was complicated by the fact that the stars are typically substantially fainter than the quasars. The brightest star analyzed, in the field of 1218+1734, had $R = 20.02$, 0.5 mag fainter than the faintest quasars in the sample (0020+0018 and 1209+1259). The "host galaxy" magnitudes produced for this star were $R = 21.71$ and $R = 22.23$ for the elliptical and disk templates, respectively, in each case at least 1.4 mag fainter than the results for the quasars. One quasar, 1243+1701, had a very low axial ratio of 0.1 and an unphysically small effective radius and exponential scalelength of ≤ 300 pc. Either the host has an unusual morphology and low surface brightness, for which the standard templates are inadequate, or there is little or no detectable flux from the host galaxy. The psf subtracted image of this field contains some extended residual emission. Until additional analysis is performed, it is unclear whether a host has truly been detected for this quasar or others which have faint extended emission only in the near vicinity of the quasar nucleus. Therefore, the following discussion will consider a subset consisting only of the most clearly visible host galaxies in addition to the entire sample.

3 Discussion

Fig. 1 shows absolute R magnitude (M_R) of the quasar nuclear component plotted against host galaxy M_R derived from $r^{1/4}$ and disk templates. The radio-loud and radio-quiet quasars are denoted by filled circles and stars, respectively. Magnitudes derived from $r^{1/4}$ fits and from many of the disk fits are brighter than the

Schechter galaxy luminosity function's characteristic luminosity, $M_R^* = -21.8$ (Lin et al. 1996). While published values of L^* differ by up to 1 mag, the basic result holds, that the elliptical fits are mostly brighter than the characteristic luminosity, and the magnitude distribution derived from disk templates straddles L^*. Note in particular that there is not a dominant population of low-luminosity host galaxies in the LBQS sample. A Kendall's τ test indicates that host galaxy luminosity is positively correlated with quasar nuclear luminosity at the 99% confidence level for the $r^{1/4}$ and disk template magnitudes. This correlation is consistent with the existence of a minimum host galaxy luminosity which increases with quasar luminosity, a trend noted by McLeod & Rieke (1995a,b) using H-band host galaxy magnitudes in lower-redshift samples. The lower luminosity envelope of the LBQS host galaxies is about 0.5 (disk template) or 1.5 (elliptical template) mag brighter than the McLeod & Rieke lower bound, converted to R-band. The host galaxy magnitudes deviate from this trend, becoming roughly constant with quasar magnitude, for quasars fainter than $M_R \approx -24$. A similar break occurs at roughly the same magnitude, after adjusting for differences in passband and cosmology, in the McLeod & Rieke (1995a,b) data. The radio-loud LBQS quasars all lie in the brighter half of the range of quasar nuclear M_R. This was not an effect of the selection of the radio-loud quasars, as the original absolute blue magnitudes for the majority were in the fainter half of the range of M_B for the HST imaging sample. There is no apparent distinction between the host galaxy magnitudes of radio-loud and radio-quiet quasars with similar nuclear component luminosities for a single type of galaxy model. However, if the host galaxy morphologies differ between radio luminosity classes, such that radio-quiet quasars are in spirals and radio-loud nuclei reside in ellipticals, then the hosts of luminous radio sources would consistently be approximately one magnitude brighter.

The distribution of axial ratios is skewed toward low values, with a median of 0.5; only two objects have ratios > 0.6. This result may indicate that the LBQS sample contains mostly inclined disk systems. However, several other HST imaging studies have found that early type host galaxies are prevalent, even among radio-quiet objects. McLeod & Rieke (1994a,b) did not find any host galaxies with measurable axial ratios < 0.5 in a sample drawn from the PG quasar survey, even among galaxies which appeared to be spirals. They attributed the lack of inclined disk systems to the additional reddening of the active nucleus caused by this viewing angle, which would make the quasar too red to be included in the UV excess-selected PG. Increased extinction would also make some potential candidates too faint for the PG magnitude limit. However, very few quasars which are intrinsically too bright to be included in the PG would cross the magnitude selection boundary of the survey due to extinction from an inclined host, because the bright flux limit of the PG exceeds the observed fluxes of virtually all known quasars. The LBQS selection is based on spectral features as well as color, and its bright magnitude limit is roughly coincident with the faint limit of the PG, so it is reasonable to expect that spiral host galaxies with higher inclinations could be included in the LBQS. The axial ratios are not consistent with

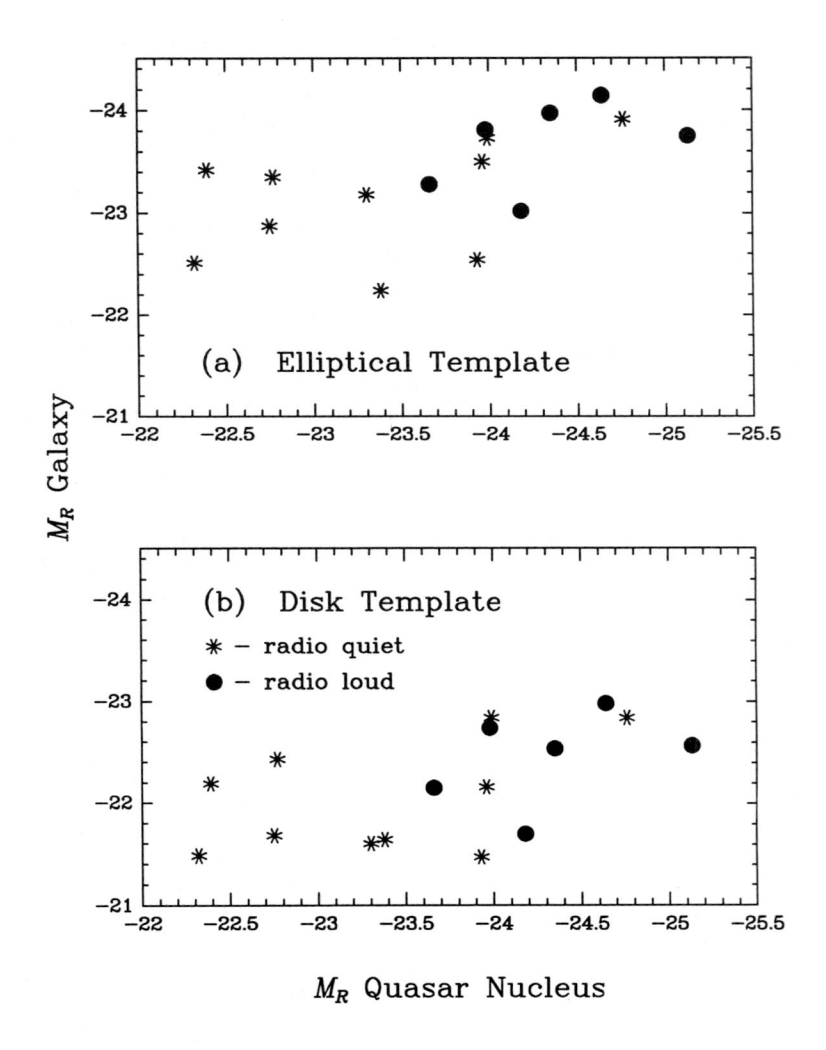

Fig. 1. Host galaxy absolute magnitude derived from (a) elliptical and (b) disk templates plotted against absolute magnitude of the quasar nuclear component. Filled circles are radio-loud quasars, and asterisks represent radio-quiet sources.

a typical population of elliptical galaxies, which generally appear less elongated. A Kolmogorov-Smirnov (KS) test indicated that the distribution of axial ratios in a sample of 171 elliptical galaxies from Ryden (1992) differs from the LBQS results at a confidence level > 99.99%. Fewer than 5% of the ellipticals have axial ratios < 0.6, where the bulk of the LBQS objects lie. Another possibility is that the detected portions of some of the host galaxies may be bars, particularly in the cases of very low axial ratios. These systems could have nearly face-on disks which are too faint to detect, leaving only the bar components visible.

The above analysis was repeated on half of the sample, those 8 quasars with the most obvious host galaxies, in order to determine whether the major results were distorted by the lower surface brightness host galaxies, for which only marginal emission is seen and which may not have been detected in some cases. The subsample contains 5 radio-quiet quasars (0020+0018, 0021−0301, 1149+0043, 1222+1010, 2214−1903) and 3 radio-louds (1222+1235, 1230−0015, and 2348+0210). The absolute magnitudes of the quasar nuclear components in these objects span the range of the full sample. A correlation with host galaxy luminosity is still suggested, especially among the more luminous quasars, but the trend is not strong enough to produce a significant result in the smaller sample (86% confidence). Radio-loud and radio-quiet quasars with similar central component magnitudes in the subsample have similar host galaxy luminosities when derived from the same class of galaxy template. Only one host galaxy in the restricted sample has an axial ratio > 0.6. The KS test comparing axial ratios was repeated, assuming that the host galaxies not included in the subsample had relatively round morphologies closely matching the axial ratio distribution of known ellipticals. The two distributions were different at $> 99\%$ confidence level, indicating that the LBQS sample contains a significant component of clearly flattened systems.

References

Boyce, P. J., Phillipps, S., & Davies, J. I. 1993, A&A, 280, 694
Coleman, G. D., Wu, C., & Weedman, D. W. 1980, ApJS, 43, 393
Cristiani, S., & Vio, R. 1990, A&A, 227, 385
Hewett, P. C., Foltz, C. B., & Chaffee, F. H. 1995, AJ, 109, 1498
Holtzman, J. A., Burrows, C. J., Casertano, S., Hester, J. J., Trauger, J. T., Watson, A. M., & Worthey, G. 1995, PASP, 107, 1065
Hooper, E. J., Impey, C. D., Foltz, C. B., & Hewett, P. C. 1995, ApJ, 445, 62
Hooper, E. J., Impey, C. D., Foltz, C. B., & Hewett, P. C. 1996, ApJ, 473, 746
Krist, J. 1996, The Tiny Tim User's Manual, version 4.1
Lin, H., Kirshner, R. P., Schectman, S. A., Landy, S. D., Oemler, A., Tucker, D. L., & Schechter, P. L. 1996, ApJ, 464, 60
McLeod, K. K., & Rieke, G. H. 1994a, ApJ, 420, 58
McLeod, K. K., & Rieke, G. H. 1994b, ApJ, 431, 137
McLeod, K. K., & Rieke, G. H. 1995a, ApJ, 441, 96
McLeod, K. K., & Rieke, G. H. 1995b, ApJ, 454, L77
Phillipps, S., & Davies, J. 1991, MNRAS, 251, 105
Ryden, B. S. 1992, ApJ, 396, 445

Optical Imaging and Spectroscopy of BL Lac Objects

R. Falomo[1], M-H. Ulrich[2]

[1] Osservatorio Astronomico di Padova, Italy
[2] European Southern Observatory

Abstract. High resolution images and spectroscopy of the close environment of several BL Lac objects are presented. The observations obtained with the NTT at ESO are part of an on-going programme aimed at studying the properties of the host galaxies and the immediate environments of BL Lac objects with different characteristics (radio vs X-ray selection, redshift, clustering of galaxies).

1 Introduction, Observations and Data Analysis

BL Lac objects are a special kind of AGN with extreme properties of non-thermal emission, polarization and strong and rapid flux variability. The basic current idea is that BL Lac objects are sources which have a relativistic jet closely aligned with the line of sight of the observer. This implies that there are objects intrinsically identical to BL Lacs but that have the jet pointing away from the observer's line of sight. From comparison of radio and optical properties it was proposed that radio galaxies of FR type I or a fraction of them (e.g. Ulrich 1989; Urry & Padovani 1995, and references therein) may represent these mis-aligned objects.

A direct test of this idea might come from the study of the host galaxies of BL Lacs and the environment properties as compared with those of parent objects. With subarcsec ground based imaging one is able to study the host properties up to $z \sim 0.5$. Beyond $z \sim 0.5$ imaging with HST capabilities is required and/or observing with excellent resolution in the IR band. At variance with the case of quasar hosts the study of BL Lacs hosts is hampered by the uncertainty of the redshift which is yet unknown or highly uncertain in many sources.

In the course of a program aimed at comparing the properties of host galaxies of BL Lacs at $z < 0.5$ from different samples we have acquired subarcsec optical images and spectroscopy of the nebulosity for selected targets. We present here preliminary results for two sources: 0301-24 and 1101-23. The observations were obtained using the 3.5m NTT at ESO with the direct imaging system SUSI and a R filter. Spectroscopy of the nebulosity was secured with EMMI and low resolution grisms in the spectral range 4000 to 8500 Å. We performed a surface photometry down to $\mu_R \sim 26$ mag./arcsec^2 in order to derive the morphological and photometric properties of the host galaxies. The luminosity profile was then fitted with a model consisting of a point source plus a galaxy (elliptical or exponential disk) convolved with the proper PSF.

2 Results

0301-243 Our image (seeing = 0.8″ ; exp =20min) of this BL Lac object clearly shows an extended nebulosity (ellipticity $\varepsilon = 0.3$) with complex morphology and close companions (figure 1). The immediate region around the object is rich with faint galaxies and there is a marked enhancement of the galaxy density within ~60 arcsec from the BL Lac object. Spectra of two galaxies (G1 and G2; see figure 1) at ~ 6″ and 20″ from 0301-243 indicate that they are at $z = 0.263$ suggesting a cluster of galaxies of Abell richness class 0 might be associated with the BL Lac source at this redshift (Pesce Falomo & Treves 1995). The radial profile is adequately well represented by a point source plus the elliptical model while the fit with an exponential disk is not acceptable. Figure 1 (right panel) shows the field after subtraction of the BL Lac model that reveals the faint galaxy ~ 3.5″ South of the nucleus. The surrounding nebulosity is very well centered on the nucleus within an accuracy of 0.2″ . Optical spectra of the nebulosity (slit off the nucleus by 2″), that are still dominated by the signal from the non-thermal source, show only one weak emission line at $\lambda = 6303$ Å. The most plausible identification is [O III] 5007 that yields a redshift of 0.26, identical to the redshift of the companion galaxies G1 and G2. At this redshift the absolute magnitude of the host galaxy is M_R = -24.2 (k-corrected and assuming $H_0 = 50$ km s^{-1} kpc^{-1} and $q_0 = 0$).

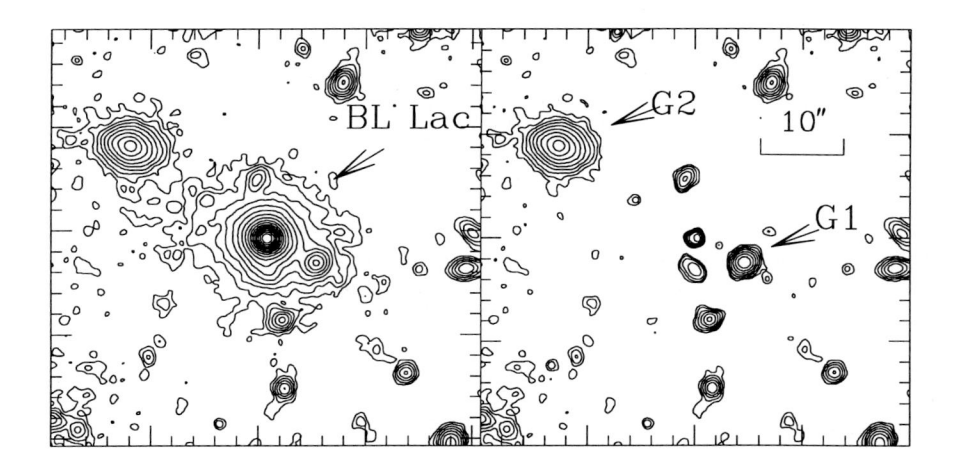

Fig. 1. Contour plot of the central part of 0301-24 before (left) and after (right) subtraction of the galaxy and nucleus. The spacing between isophotes is 0.5 magnitudes while faintest shown level is $\mu_R = 25.5$ mag/□″ . Galaxies G1 and G2 are at redshift z = 0.26.

1101-232 This is a BL Lac discovered from X-ray survey and is surrounded by a conspicuous rather elongated nebulosity (see Fig 2) at the proposed $z =$

0.186 (Remillard et al. 1989). Our optical spectroscopy confirms this redshift from absorption lines of the galaxy. The radial brightness profile extends to 15 arcsec along the major axis a with the ellipticity increasing up to 0.45. We found that the luminosity profile is well fitted by an elliptical galaxy model plus a point source. The absolute magnitude of this host galaxy is $M_R = -24.3$. External isophotes are found to be substantially *boxy*, possibly due to merging (*e.g.* Bender et al. 1989). This is the only clear evidence of substantial *boxy* isophotes found in BL Lacs hosts.

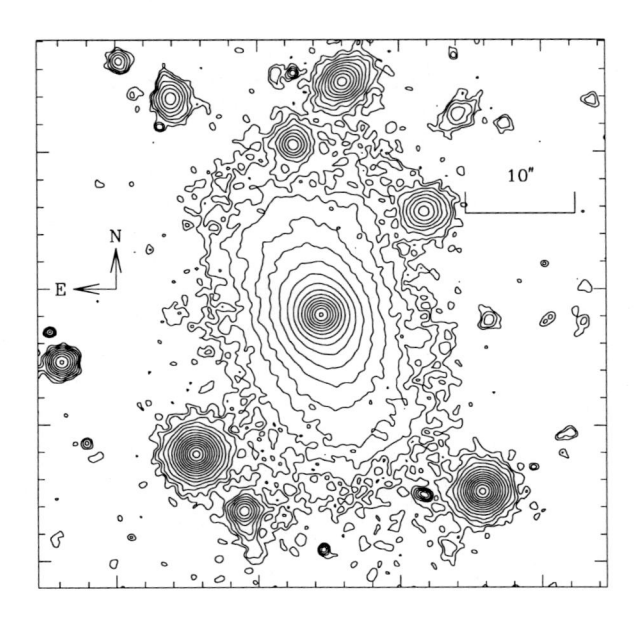

Fig. 2. Contour plot of the central part of the image of the 1101-23. The spacing between isophotes is 0.5 magnitudes while faintest shown level is $\mu_R = 25.5$ mag/$\square''$.

References

Bender, R., Surma, P., Döbereiner, S., Möllenhoff, C., & Madejsky, R., 1989, A&A, 217, 35
Pesce, Falomo and Treves 1995, A.J., 110, 1554.
Remillard , R.A., Tuohy, I.R., Brissenden, R.J.V., Buckley, D.A.H., Schwartz, D.A., Feigelson, E.D., and Tapia, S. 1989, ApJ , 345, 140.
Ulrich,M-H. 1989, in *BL Lac Objects*, Ed. Maraschi et al., p45.
Urry, C.M. & Padovani, P. 1995, PASP , 107, 803.

Host Galaxies of Radio-Loud AGN

Josef W. Fried[1] and Jochen Heidt[2]

[1] Max-Planck-Institut für Astronomie, Königstuhl 17, D-69117 Heidelberg, Germany
[2] Landessternwarte Heidelberg, Königstuhl 12, D-69117 Heidelberg, Germany

Abstract. We are studying the host galaxies and cluster environments of a sample of radio-loud AGN. The sample is large enough to compare sub-samples (quasars, FRI and FRII radio galaxies, RBL and XBL). First results show that elliptical host galaxies fall on the $\mu_e - r_e$ relation of bright elliptical galaxies.

1 Introduction

From deep direct R-band images of radio loud AGN we derive parameters of host galaxies and associated galaxy clusters. The sample is homogeneous and large enough (91 AGN) to extract and compare subsamples of different types of AGN. The aims of our study are to

(1) test the unified scheme for AGNs (Urry & Padovani 1995). Within the unified scheme we expect similar properties of host galaxies and cluster environments for quasars and FRII radio galaxies, BL Lacs and FRI radio galaxies,

(2) search for evolution of host galaxies (e.g. in absolute magnitude) with redshift,

(3) study the role of tidal interactions in feeding the monster (isophote twists and/or distorsions for $z \leq .1$, close companions for $z \geq .1$),

(4) search for evolution of cluster environment with redshift. Such an evolution has been found for radio-loud quasars (Ellingson et al. 1991, Fried 1992), and

(5) look for differences in environments between XBL and RBL, since these 2 groups differ in cosmological evolution (Stickel et al. 1991, Morris et al. 1991) and duty cycle (Heidt 1996)

2 The Sample

We have selected objects from the 1 Jy catalogue (Kühr et. al. 1981) with $z < 0.8$, the (optimistic) limit for spatial resolution of host galaxies and detection of associated clusters and $\delta > -25°$, the limit of observability from the Calar Alto observatory. Applying the same criteria, we have additionally selected XBLs from the EMSS sample (Morris et al. 1991). Our sample contains 21 radio galaxies (FRI, FRII), 26 quasars, 22 RBL and 22 XBL, in total 91 AGN.

3 Data Reduction

The field sizes are $> 1 Mpc$ projected linear size at the redshift of the AGN and reach $R_{lim} = 23.5$. The host galaxy images are derived by subtracting scaled PSFs from the AGN images, and then fitting seeing convolved galaxy profiles (i.e. de Vaucouleurs $\otimes$ seeing or exponential $\otimes$ seeing) to the galaxy profile. This procedure is repeated with different scaling of the PSF until best fit is reached. The resulting image is used to derive the host galaxy parameters e.g. M, r_e, μ_e. For the cluster analysis, all objects on the images are classified into stars and galaxies; the clustering is parameterized by the qso-galaxy covariance amplitude.

4 Results

Up to now we have obtained good data on 34 AGNs; first results for the host galaxies are:

1.) a host galaxy is detectable in 27 objects.

2.) 4 host galaxies are better fitted by disks rather than by ellipticals. All are hosts of XBL. For 3 of these Wurtz et al. 1996, too, had found that they are better fitted by disks than by ellipticals.

3.) the elliptical hosts of FRI, FRII, XBL, RBL follow very well the $\mu_e - r_e$ relation of bright elliptical galaxies

4.) a few hosts of quasars do not follow this relation. However, since all these are at redshifts $z \geq 0.5$ this is probably a seeing effect and requires further study

Full analysis of this ongoing project will be given elsewhere.

References

Ellingson, E., Yee, H. K. C., Green, R. F. (1991): ApJ **371**, 49
Fried, J.W. (1992): *Proc. "Physics of AGN"* (Springer, Berlin, Heidelberg), p.677
Heidt, J. (1996): *Proc. "Blazar Variability"* (Springer, Berlin, Heidelberg) in press
Kühr H., Witzel A., Pauliny-Toth I.I.K., Nauber U. (1981): AAS **45**, 367
Morris S.L., Stocke J., Gioia I.M., Schild R., Maccaro T., Ceca R. (1991): ApJ **380**, 49
Urry C.M., Padovani P. (1995): PASP **107**, 803
Stickel M., Fried, J.W., Kühr H., Padovani P., Urry C.M. (1991): ApJ **374**, 431
Wurtz R., Stocke J., Yee H.K.C. (1996): ApJS **103**, 109

Host Galaxies of Intermediate Redshift Radio-Loud and Radio-Quiet Quasars

Eva Örndahl, Jari Rönnback and Ernst van Groningen

Astronomiska Observatoriet, Box 515, S-751 20 Uppsala, Sweden

Abstract. We have obtained deep images for a sample of 96 radio-loud and radio-quiet quasars at intermediate redshifts (0.4–0.8). Our aim is to compare the properties of the host galaxies and the galaxy environments of the two groups of quasars in a systematic way. To facilitate a direct statistical comparison between radio-loud and radio-quiet objects, we selected pairs of objects with approximately the same redshift and V-magnitude. After analyzing about half the sample we find that there is no systematic difference in the host galaxy luminosities for the two classes of objects. We confirm that radio-loud quasars commonly occur in elliptical type galaxies, with some exceptions, while radio-quiet quasars are found in spiral disks or ellipticals. The optical colours of the host galaxies are bluer than can be expected from passive evolution. At this stage we cannot say if the blue colours are caused by a young stellar population or by light scattered by the ISM of the host galaxy.

1 Introduction

The global properties of the host galaxies of radio-loud quasars (RLQ) and those of radio-quiet quasars (RQQ) appear to differ to some extent (Smith et al. 1986, Véron-Cetty and Woltjer 1990, Hutchings and Neff 1992). For example, RLQ are generally found in elliptical hosts, whereas the RQQ most often are found in spiral host galaxies (although exceptions exist). This leads us to ask what connection there could be between the processes in the active centre and the processes that result in a certain galaxy morphology. At high redshifts ($z \geq 1$) both RQQs and RLQs are found in rich environments with many starbursting galaxies (Hutchings et al. 1995, Lowenthal et al. 1995, Hutchings 1995). Recent HST observations have resolved compact blue knots in the hosts of low-redshift quasars, indicating recent star formation (Hutchings and Morris 1995, Bahcall et al. 1995). Also, close companion galaxies seems to be a very common feature (Bahcall et al. 1995, Disney et al. 1995). It seems plausible that interacting and merging galaxies could trigger star formation and provide fuel for the active nucleus (e.g. Sanders et al. 1988).

In order to complement previous studies with intermediate redshift data we have carried out a campaign in which 96 quasars, at $0.4 \leq z \leq 0.8$, have been observed with several optical filters. For a presentation of the results for the first part of the sample, see Rönnback et al. (1996). In this paper we assume $q_0 = 0.5$ and $H_0 = 80$ km s^{-1} Mpc^{-1}.

2 Sample and Observations

The sample was selected from the catalogues of Véron-Cetty and Véron (1993) and Hewitt and Burbidge (1993), where the only selection criteria used were the redshift range ($0.4 \leq z \leq 0.8$) and the constraints imposed by the sky coordinates at the observation sites. From this subset we selected matched pairs of radio-loud and radio-quiet quasars with approximately the same redshift and V magnitude, in order to facilitate a direct statistical comparison between the two classes.

Observations were performed in 1994 at the ESO NTT (3 nights) and at the Nordic Optical Telescope (NOT, 20 nights). In total, 96 quasars were observed, primarily in the R band (all fields), but many of them also in V and/or I (33 fields). In this way we obtained data for 45 pairs. The mean and standard deviation of the redshift differences is -0.004 ± 0.022 and of the magnitude distribution -0.04 ± 0.33. At both the NOT and the NTT we were fortunate to get excellent seeing conditions: the median seeing value for the co-added frames is less than $0.8''$, with some frames having a seeing better than $0.5''$. On only 7 % of the images is the seeing worse than $1''$.

3 Host Galaxy Retrieval

3.1 PSF Construction

All object frames contain at least one unsaturated star, of comparable magnitude to the QSO, which can be used to obtain a model for the point spread function (PSF). Tests show that the variation of the PSF over the field in the frames is insignificant (the luminosity profiles of the test stars vary less than 2 per cent at any radial position, over a range of 8 magnitudes).

To increase the accuracy of the PSF subtraction and to measure the morphology of the residuals we need a better model of the two-dimensional PSF. We found that it was necessary to adopt different treatments for the core and the wings of the PSF. In the core (where the signal-to-noise ratio is high), small tracking errors, in combination with the excellent seeing, lead to a non-circular PSF, which cannot be adequately modeled. In the wings, where the PSF profile is as good as circular, the signal-to-noise of the data can become very low. To model the wings of the PSF, we used the profile discussed by Saglia et al. (1994) (using a routine kindly provided by Richard Hook). We constructed a composite PSF using empirical data in the core and model data in the wings. Some results are shown in Fig. 1. We switched to model data at a radius where the contours of the empirical PSF become sufficiently circular but before the noise starts to become significant. The advantage of this appro! ach is that no extra noise is added to the wing of the host-galaxy image when we subtract the PSF from the QSO, enabling us to see fainter levels.

3.2 PSF Subtraction

In order to obtain the image of the host galaxy, a properly scaled composite model PSF must be subtracted from the object frames. The QSO flux is not

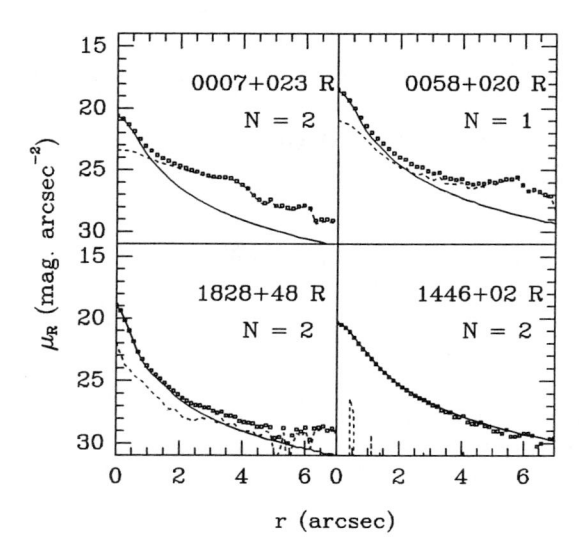

Fig. 1. Examples of azimuthally averaged luminosity profiles of some quasars (boxes) and composite PSFs (full lines). The dashed lines show the residuals after the subtraction of the PSF profiles from the quasar profiles. N indicates the number of stars which were used to generate the PSF profile.

known a priori, however. To measure it, we have used the results from a program based on a generalized Lucy-Richardson maximum entropy restoration method (Lucy 1994; Hook et al. 1994). The restored QSO flux is rescaled until we get a residual image where the luminosity profile is monotonically decreasing with radius. Magnitudes derived in this way must be regarded as upper limits since the method will oversubtract the QSO light, if the host galaxy has a very peaked profile (as in ellipticals and spirals with a bulge). The normalized PSF was scaled accordingly and subtracted from the QSO, and the azimuthally averaged luminosity profile for the residual computed. In case the profile was not decreasing monotonically with radius, the PSF was rescaled and the process repeated.

3.3 Reliability of the Method

We have used artificial data to estimate the accuracy of our method. Circular galaxy images with luminosity profiles corresponding to early-type and disc galaxies were created and convolved with a pure model PSF. A stellar image, including the sky background, was added on top of the model galaxy, and a tilted plane representing the sky level was subtracted. Then a composite model PSF was subtracted in the same manner as for real data. A grid of such simulations was produced to cover the range of observing conditions and expected properties of the hosts.

Except for the faintest models (with R = 21), the profile shapes are quite well reproduced. For the $r^{1/4}$ models the effective radii are systematically underestimated (with a factor $\gtrsim 0.2$ for R = 19, $\gtrsim 0.3$ for R = 20 and $\gtrsim 1$–2 for R = 21), while for the exponential discs the scale length is reproducible within 10–30 % over the whole grid. The final errors in the derived scale sizes are estimated to be 30–40 %.

Comparing the integrated fluxes of the PSF-convolved model galaxies and the PSF-subtracted QSO + host galaxy models, we find that they differ with 0.10±0.05 when R = 19 magnitudes, 0.20±0.10 when R = 20 and 0.50±0.10 when R = 21. These values are comparable to the results of Maraziti and Stockton (1994) in a similar study.

Additional possible complications which have not been included in the modeling above are that the nuclear point source need not be centered on the host galaxy, and that there may be deviations from a smooth luminosity profile, caused by spiral arms, H II regions or tidal arm structures. If the mis-centering of the QSO is less than a few pixels, as expected, the classification of host galaxies into ellipticals or spirals will not be significantly affected, although the derived structural parameters would change slightly. The observed host galaxies have typically R $\sim$ 20, and we are therefore quite confident that we have derived a highly probable result.

4 Results

4.1 Host Galaxy Magnitudes

We applied the above described technique of PSF subtraction for the 43 objects reduced and analyzed at the moment. In this (unmatched) sample we identify a host galaxy in 73% (77%) of the RLQs and in 67% (76%) of the RQQs, where the number in parenthesis include the uncertain cases. For the objects where the host galaxy is detected, the mean absolute R magnitude is −22.0±0.7 for the host galaxies in RLQs and −21.8±0.7 for the host galaxies in RQQs.

A subsample of 13 pairs matched in (z, V) yields basically the same result: for the cases where a host galaxy is detected we find an average absolute magnitude of −22.0±0.3 for the RLQs and −21.6±0.4 for the RQQs. Thus, both from the unmatched sample of 43 objects and the matched sample of 26 objects we conclude that there is no significant difference in the absolute magnitude of the hosts of radio-loud and radio-quiet QSOs in the redshift range of 0.4 to 0.8.

4.2 Host Galaxy Morphology

Based on the extracted luminosity profiles we conclude that RLQs commonly occur in elliptical type galaxies, with some exceptions, while RQQs are found in spiral disks or ellipticals. The mean of the metric scales of the scale lengths is about 4 kpc for elliptical types and and about 3 kpc for disk types. The data for the RLQs can be compared with the study by Romanishin and Hintzen (1989). We derive profiles with similar effective radii in the mean. Their mean R magnitude is about 0.2 magnitudes fainter than the one we derive. Thus, their main conclusion, that the host galaxies of RLQs are more compact than normal low redshift ellipticals, also holds for our data. The fact that the seeing results in systematically underestimated scale sizes could reduce this effect. The exact amount of this should be examined in more detail before any firm conclusions can be drawn, however. For the object in common in our study and Romanishin and Hintzen, PKS 0932+022, we derive the same effe! ctive radius and magnitude of the elliptical host galaxy, within the errors.

4.3 Host Galaxy Colours

The V−R colours of some of the hosts which were observed in several bands are displayed in Fig. 2. The colours of the template stellar populations shown in the figure were derived by integrating the redshifted filter profile convolved with spectral energy distributions adopted from Coleman et al. (1980). The colours were corrected for the different CCD quantum efficiencies in V and R. Although the errors are quite large, it is apparent that the colours on the whole are typical for present day late-type disks and irregular galaxies. It is remarkable that for three of these hosts we derive elliptical luminosity profiles.

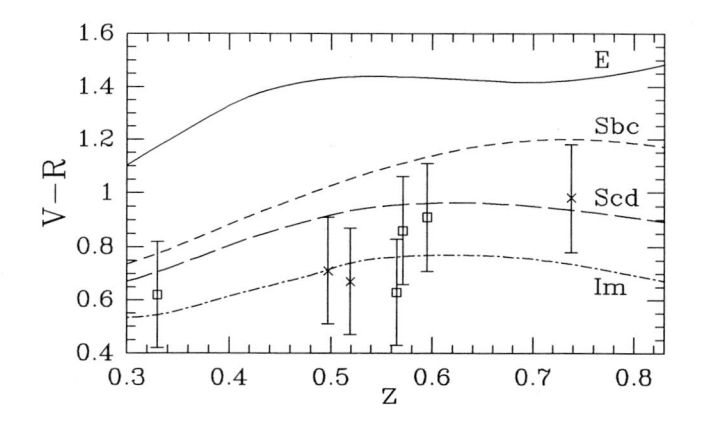

Fig. 2. The V−R colour as a function of redshift for non-evolving E (full line), Sbc (dashed), Scd (long dashed) and Im (dot-dashed) galaxies. Squares mark the positions of radio-loud quasar host galaxies and crosses mark radio-quiet quasar host galaxies. The error bars represent 1 σ, based on photon statistics.

What causes the blue colours of the hosts detected by us? One possibility is that the blue colours are produced by a mixture of stellar light and scattered QSO light. If the galaxies classified as ellipticals in reality are normal late-type galaxies, we can used the observed colour indices to estimate the fraction of scattered light from the QSO in the case of electron scattering. For a typical QSO with $M_R = -24.0$ and $V-R = 0$, where the host galaxy has $M_R = -22.2$ and $V-R = 0.8$, and an E galaxy has $V-R = 1.4$, we find that as much as $\sim 10\%$ of the nuclear light may be scattered. This light will contribute to the image of the host galaxy on the order of 50% or more. In the more probable case of dust scattering it is more difficult to compute the fraction of scattered light as a function of wavelength, since it then is dependent on the size, composition and distribution of the dust particles.

5 Conclusions

We have found host galaxies in about 70% of the observed radio-loud and radio-quiet quasars. RLQ host galaxies typically have elliptical luminosity profiles (although exceptions exist), with absolute R magnitude ~ -22.0, while the host galaxies of RQQs are large discs or ellipticals with R magnitudes ~ -21.7. The colours of the stellar population in the hosts are typical of late-type galaxies. The blue colours could be caused by a recent burst of star formation or a scattered light component from the QSO. In our continued study we will further address these issues.

References

Bahcall, J. N., Kirhakos, S., Schneider, D. P. (1995): ApJ **450**, 486
Coleman, G. D., Wu, C. C., Weedman, D. W. (1980): ApJS **43**, 393
Disney, M., Boyce, P., Blades, J., Boksenberg, A., Crane, P., Deharveng, J., Macchetto, F., Mackay, C., Sparks, W., Phillipps, S. (1995): Nat **376**, 150
Hewitt, A., Burbidge, G. (1993): ApJS **87**, 451
Hook, R. N., Lucy, L. B., Stockton, A., Ridgway, S. (1994): ST-ECF Newsletter **21**, 16
Hutchings, J. B. (1995): AJ **110**, 994
Hutchings, J. B., Morris, S. C. (1995): AJ **109**, 1541
Hutchings, J. B., Neff, S. G. (1992): AJ **104**, 1
Hutchings, J. B., Crampton, D., Johnson, A. (1995): AJ **109**, 73
Lowenthal, J. D., Heckman, T. M., Lehnert, M. D., Elias, J. H. (1995): ApJ **440**, 558
Lucy, L. B. (1994): *Proc. the Restoration of HST Images and Spectra* Eds. Hanisch, R. J., White, R. L., STScI, p. 79
Maraziti, D., Stockton, A. (1994): PASP **106**, 71
Romanishin, W., Hintzen, P. (1989): ApJ **341**, 41
Rönnback, J, van Groningen, E., Wanders, I., Örndahl, E. (1996): MNRAS **283**, 282
Saglia, R. P., Bertschinger, E., Baggley, G., Burstein, D., Colless, M., Davies, R. L., McMahan, R. K., Wegner, G. (1994): MNRAS **264**, 961
Sanders, D. B., Soifer, B. T., Elias, J. H., Madore, B. F., Matthews, K., Neugebauer, G., Scoville, N. Z. (1988): ApJ **316**, 584
Smith, E. P., Heckman, T. M., Bothun, G. D., Romanishin, W., Balick, B. (1986): ApJ **306**, 64
Véron-Cetty, M. P., Véron, P. (1993): ESO Sci. Rep. 13
Véron-Cetty, M. P., Woltjer, L. (1990): A&A **236**, 69

Some Examples of the Extremely Close Environments of BL Lac Objects

Aimo Sillanpää[1], Leo O. Takalo[1], Tapio Pursimo[1], Pekka Heinämäki[1], Kari Nilsson[1], and Jochen Heidt[2]

[1] Tuorla Observatory, Tuorla, FIN-21500 Piikkiö, Finland
[2] Landessternwarte Königstuhl, D-69117 Heidelberg, Germany

Abstract. Our ongoing programmes taking very deep images of different types of BL Lac objects with the 2.5m Nordic Optical telescope on La Palma have continued more than two years. In many cases we have observations with sub-arcsecond seeing conditions and in some cases also with almost sub-half-arcsecond seeing. Some of the objects show extremely close companions. Preliminary results of these environments are discussed.

1 Introduction

BL Lac objects are among the most violent and luminous objects in the Universe and especially among the objects called Active Galactic Nuclei (AGN). These properties mean that BL Lacs are maybe the best objects to test the Unified Scheme scenario for the AGN (Antonucci, 1993). To make this testing we have started a huge observing program with the 2.5m Nordic Optical Telescope on La Palma. This telescope gives the best possible ground-based resolution for our deep images which is extremely important when we are searching for hosts of the BL Lacs and also when we are searching for extremely close companions of our targets.

2 Our Programmes and Some First Results

To check if there are any differences and similarities between many different types of BL Lac objects and their environments we have choosen many subclasses to our observing programmes. The most important ones are:

1. "Random snapshots" of the suitable BL Lac objects taken from the Veron-Veron catalogue when we had "empty times" in our observing programmes. In this project we have observed about 25 objects which are both radio- and X-ray selected objects. Typical integration times in this sample have been about 900s in R-band.

2. The complete sample of the Einstein slew survey BL Lacs (Perlman et al., 1996) observable from the NOT (about 30 objects). So far we have observed 8

objects with typical integration times of around 1800s in R-band. All of these objects have been more or less typical elliptical galaxies harbouring BL Lac objects in their nuclei. The most interesting thing in these objects has been that in six of eight cases there seems to be a double nucleus in the center. Without any spectroscopic information it is, of course, impossible to say if these companions are real companions or stars and projected objects.

3. The complete sample of the 1 Jy BL Lacs (Stickel et al., 1991) observable with the NOT. This sample consists of 28 objects. Typical integration times have been around 5000-6000s. The first results of this sample have been already discussed in Heidt et al. (1996) and in Takalo et al. (these proceedings).

4. The sample of the BL Lac candidates which are both radio- and X-ray bright (ROSAT) objects (about 40 objects, Sally Laurent-Muehleisen, Wolfgang Brinkman and Stefan Wagner, private communication). Integration times in this sample have been about 2000s in R and we have also some colour and polarization information of these objects.

Totally our programmes include more than 120 BL Lac objects, being the largest sample of BL Lac objects ever observed with similar depth and resolution. So our sample gives extremely good tools when testing hosts and environments of different types of BL Lacs and also when testing for Unified Schemes for the Active Galactic Nuclei.

References

Antonucci, R., (1993): ARA&A **31**, 473-521.
Heidt, J., et al. (1996): A&AL **312**, L13-L16.
Perlman, E., et al. (1996): ApJS **104**, 251-285.
Stickel, M., Fried, J., and Kühr, H. (1993): A&AS **98**, 393-442.

The Optical Jet in 3C 371

L.O. Takalo, K. Nilsson, T. Pursimo, A. Sillanpää[1], and J. Heidt[2]

[1] Tuorla Observatory, Tuorla, FIN-21500 Piikkiö, Finland
[2] Landessternwarte Königstuhl, Heidelberg, Germany

Abstract. We present preliminary results of optical imaging of the BL Lac object 3C 371. Subarcsecond optical imaging at the Nordic Optical Telscope (NOT) has revealed an optical jet close to the nucleus of the elliptical galaxy in which the BL Lac object resides. This jet has four separate components (R = 21.2 to 23.5). The first two knots reside at the same location as component A in the radio jet (Wrobel and Lind 1990).

3C 371 is a nearby (z=0.051) BL Lac object with companion galaxies closeby. Previous optical imaging of this source (Stickel et al. 1993) has revealed it to be an interacting system with a tidal tail towards the neighbouring galaxy 75" to the southwest.

3C 371 is a strong radio source with a compact nucleus, a onesided radio jet and double radio lobes extending up to 30" from the nucleus in opposite directions (Wrobel and Lind 1990). It is one of the few BL Lac objects with this kind of radio structure.

The observations presented here were taken at the NOT on October 20th 1995 and June 16th 1996 under subarcsecond seeing conditions (0.85"). The total integration time in the coadded images is one hour in both the R-band and B-band. Observations have been reduced in the normal way, with bias and flat field corrections, using IRAF.

For the analysis we have modelled the object with an elliptical galaxy plus a point source in the core. This model has been subtracted from both B and R-band images. We have also deconvolved the B-band image, using MEM-algorithm. The result after 40 iterations is shown in Fig. 1. Four separate components can be seen towards the west from the core. In Table 1 we show the measured magnitudes for the components shown in Fig. 1. Component 1 coincides with the radio spot A (Wrobel and Lind 1990). Aligning the radio core (Akujor et al. 1994) with the optical nucleus we can see that the radio jet coincides with the optical components 1a and 1b. Note also that the optical radiation is more intense in the front edge of the jet.

Fitting a power-law to the observed fluxes from radio bands into the optical, for component 1, gives a spectral index of -0.77. This power-law would indicate that the radiation is due to synchrotron radiation.

Table 1. Magnitudes for the jet components

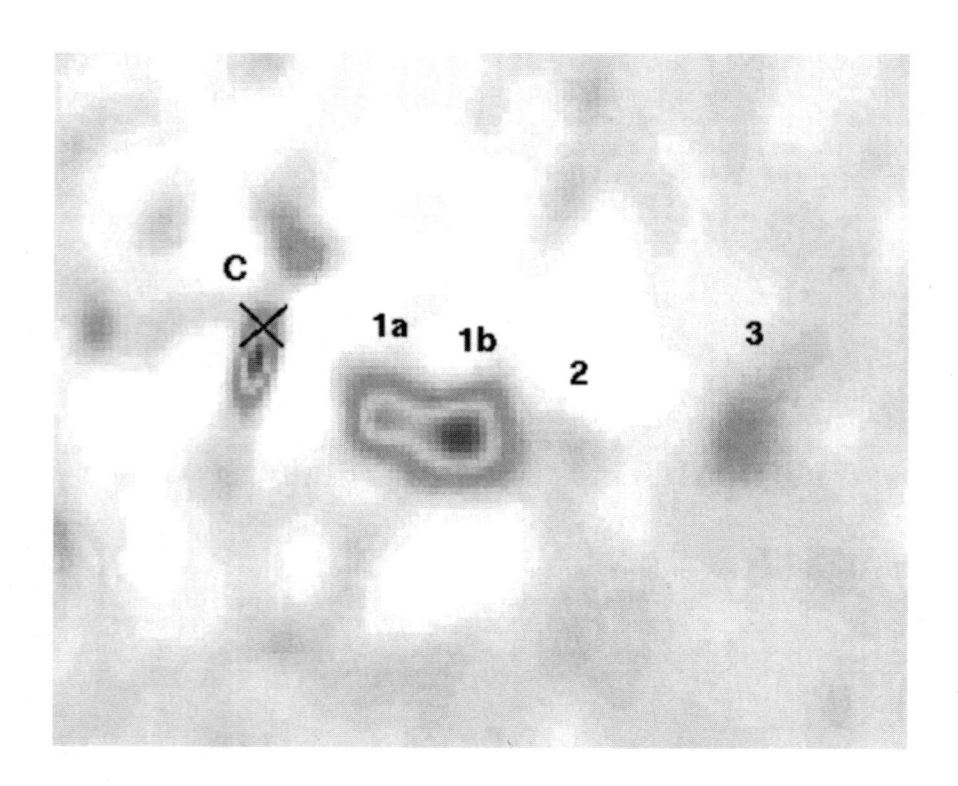

Component	B	R	B-R
1	21.5 ± 0.1	21.2 ± 0.1	0.3
2	–	23.5 ± 0.2	–
3	24.6 ± 0.2	23.1 ± 0.1	1.5

Fig. 1. The nuclear region of 3C 371 in the B-band after the model subtraction and deconvolution with MEM. After deconvolution the optical component 1 is split into two subcomponents 1a and 1b. The field size is 12.6"*10.6" and one pixel corresponds to 0.11". C marks the center of the BL Lac object.

References

Akujor, C.E., et al. 1994, AAS, **105**, 247
Stickel, M., Fried, J.W., and Kühr, H., 1993 AAS **98**, 393
Wroble, J.M., and Lind, K.R., 1990, ApJ, **348**, 135

Part 5

LOW REDSHIFT POPULATIONS

Quasar Host Astronomers touring Tenerife

Infrared QSOs

D.B. Sanders and Jason A. Surace

Institute for Astronomy, University of Hawaii, 2680 Woodlawn Dr., Honolulu, HI, 96822

Abstract. The *Infrared Astronomical Satellite* (*IRAS*) discovered a class of ultraluminous infrared galaxies (ULIGs) with "warm" mid-infrared colors whose total infrared luminosities are equivalent to the bolometric luminosities of optically selected quasi-stellar objects (QSOs). These "infrared QSOs" appear to represent a relatively short-lived phase during the merger of two gas-rich spirals, close in time to when the two nuclei merge. It is possible that most if not all QSOs begin their lives in a similar intense infrared phase.

1 Introduction

The fact that Seyfert galaxies and some QSOs can emit a substantial fraction of their bolometric luminosity at far-infrared wavelengths was suspected as early as the late 1960's, following the first sensitive mid-infrared observations of optically selected extragalactic sources (Low & Kleinmann 1968, Rieke & Low 1972). More detailed observations at mid-infrared wavelengths (e.g. Becklin *et al.* 1973, Stein *et al.* 1974, Rieke & Low 1975, Neugebauer *et al.* 1976, Rieke 1978, Telesco & Harper 1980) showed that the infrared spectrum of many optically selected AGN could plausibly be explained by models of thermal emission from dust (e.g. Rees *et al.* 1969, Burbidge & Stein 1970).

A more complete understanding of the infrared properties of extragalactic objects in general and AGN in particular was made possible by the first sensitive all-sky far-infrared survey carried out by *IRAS* (Neugebauer *et al.* 1984). In addition to detecting significant far-infrared emission from bright optically selected QSOs (e.g. Neugebauer *et al.* 1986, Sanders *et al.* 1989), *IRAS* uncovered a substantial population of previously uncataloged "infrared galaxies" (objects with f_{60}/f_B in the range 1 to 100 — e.g. Soifer *et al* 1984)[1], the most extreme of which had "ultrahigh" infrared luminosities equivalent to the bolometric luminosities of optically selected QSOs (Soifer *et al.* 1987, Sanders *et al.* 1988a). Of these "ultraluminous infrared galaxies" (ULIGs), perhaps the most intriguing were those objects with Seyfert 1 optical spectra, and a nearly pointlike appearance on the Palomar Sky Survey – but without the UV-excess characteristic of optically selected QSOs. The first of these "infrared QSOs" was reported by Beichman *et al.* (1986) and Vader & Simon (1987), followed by small lists of additional similar objects that had been identified in surveys of galaxies with "warm" ($f_{25}/f_{60} > 0.2$) mid-infrared colors (e.g. Low *et al.* 1988, Sanders *et*

---
[1] The quantities f_{12}, f_{25}, f_{60}, f_{100}, and f_B refer to the *IRAS* flux densities in Jy at 12, 25, 60, 100 μm and the Johnson B-band flux density at 0.44 μm respectively.

al. 1988b). Equally intriguing were those few warm ULIGs with infrared luminosities approaching $\sim 10^{13}$ $L_\odot$ (e.g. Kleinmann & Keel 1987, Hill et al. 1987, Frogel et al. 1989, Cutri et al. 1994) all of which are Seyfert 2s in direct optical emission, but which show broad (Seyfert 1) linewidths in polarized light (Hines 1991, Hines & Wills 1993, Hines et al. 1995) suggesting the presence of a hidden broad-line region.

The purpose of this brief review is to summarize the main evidence which suggests that infrared QSOs represent a transition stage in the evolution of ULIGs into optically selected QSOs. Following a brief introduction to ULIGs, we present this evidence by discussing multiwavelength observations that have been obtained for the complete sample of warm ULIGs identified by Sanders et al. (1988b). The discussion is then continued in a companion paper by Surace & Sanders, which presents important new *Hubble Space Telescope (HST)* images of the same sample of objects.

2 Ultraluminous Infrared Galaxies

The luminosity function of infrared selected galaxies is now relatively well established from several extensive redshift surveys (e.g. Saunders et al. 1990). A major result from these surveys is that at luminosities above $\sim 2 \times 10^{11}$ $L_\odot$ (i.e. $\sim 8\,L^*$), infrared selected galaxies appear to become the dominant population of extragalactic objects. In the luminosity range $10^{11} - 10^{12}$ $L_\odot$ the space density of infrared selected galaxies exceeds that of optically selected starburst galaxies and is approximately equal to the space density of optically selected Seyferts, while at the highest luminosities, the space density of ultraluminous infrared galaxies (ULIGs: $L_{\mathrm{ir}} > 10^{12}$ $L_\odot$)[2] exceeds by a factor of ~ 1.5 the space density of optically selected QSOs, the only other class of objects of comparable bolometric luminosity (e.g. Schmidt & Green 1983). The nature of luminous infrared galaxies in general, and ULIGs in particular, continues to be a subject of intense investigation.

2.1 Origin and Evolution

Ground-based observations of complete samples of luminous infrared galaxies (LIGs: $L_{\mathrm{ir}} > 10^{11}$ $L_\odot$) now reveal that mergers of molecular gas-rich spirals are responsible for the majority of the most luminous infrared objects — the general trend being that the more luminous objects are also the most advanced mergers with the largest nuclear concentrations of molecular gas (see review by Sanders & Mirabel 1996). Imaging of a few hundred ULIGs (Sanders et al. 1988a, Melnick & Mirabel 1989, Kim 1995, Murphy et al. 1996, Clements et al. 1996) indicates that > 90% are advanced mergers, while millimeter-wave interferometer observations of several tens of objects reveal that they all appear to contain enormous central

---

[2] $L_{\mathrm{ir}} \equiv L(8 - 1000\mu m)$, and is computed using the fluxes in all four *IRAS* bands following the prescription outlined in Perault (1987); see also Sanders & Mirabel (1996)

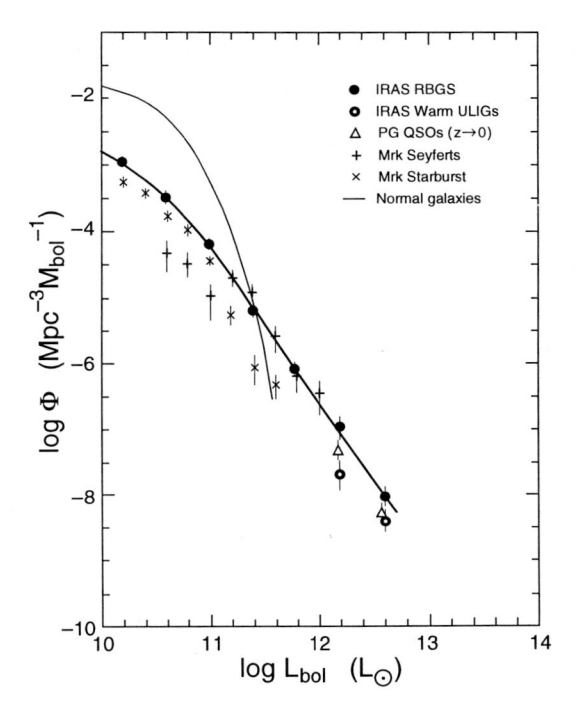

Fig. 1. The luminosity function for infrared selected galaxies compared with other classes of extragalactic objects, normalized to the same Hubble constant ($H_o = 75$ km s^{-1}Mpc^{-1}, $q_o = 0.5$) and plotted in units of bolometric luminosity (adapted from Sanders & Mirabel 1996). The data for PG QSOs are from Schmidt & Green (1983). The data for the *IRAS* RBGS and *IRAS* warm ULIGs are from Sanders *et al.* (1997) and Kim & Sanders (1997) respectively.

concentrations of molecular gas — typically $M(H_2) = 0.5 - 1 \times 10^{10}\ M_\odot$ at radii < 0.5 kpc (e.g. Scoville et al. 1991, 1997; Bryant 1996). These observations confirm predictions from numerical simulations (e.g. Barnes & Hernquist 1992), which show that mergers of molecular gas-rich spirals are extremely efficient at funneling a large fraction of the pre-existing gas from the individual disks into the merger nucleus. There is also a clear trend of an increasing percentage of Seyferts with increasing L_{ir}, from $\sim 5\%$ at $L_{ir} \sim 10^{11}\ L_\odot$ to $\sim 50\%$ at $L_{ir} > 2 \times 10^{12}\ L_\odot$, while the fraction of LINERS is both large ($\sim 40\%$) and relatively constant throughout this luminosity range (Veilleux *et al.* 1997a). Those *IRAS* galaxies with Seyfert-like optical spectra also are likely to have "warm" infrared colors, $f_{25}/f_{60} > 0.2$ (e.g. deGrijp *et al.* 1985).

For those objects that do eventually become ULIGs their ultimate fate is still not completely clear. However, both theory and observations suggest that ULIGs may already be in the process of expelling much of their circumnuclear gas and dust from the merger nucleus due to the combined forces of radiation pressure, supernova explosions, and stellar winds. Approximately 20% of ULIGs (principally those with "warm" f_{25}/f_{60} colors) appear to be at a stage where enough of the gas and dust has been expelled to allow a relatively unobscured look at the merger nucleus.

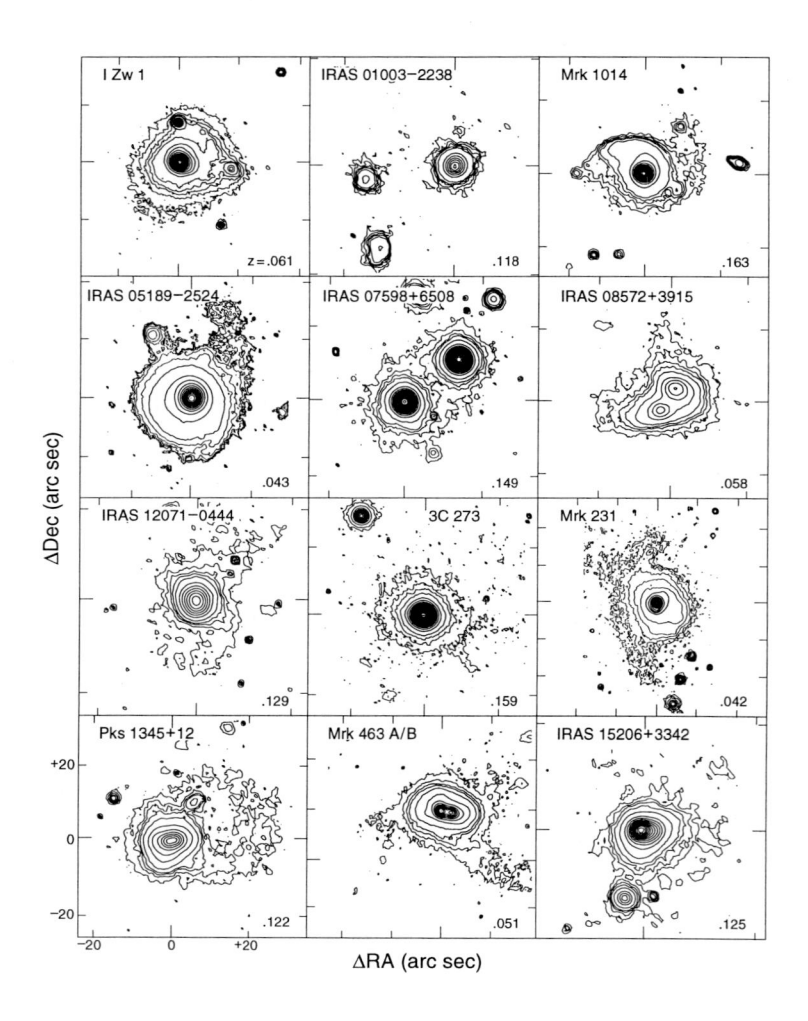

Fig. 2. Contour maps from CCD Gunn-r (6500Å) images of the complete sample of 12 warm ULIGs from Sanders et al. (1988b). The lowest contours are linear to emphasize faint large scale structure, while the higher contours are logarithmic. The object redshift is indicated in the lower right of each panel. The axis tick marks are at intervals of $20''$, offset from the *IRAS* source. (Note: The object $18''$ NW of IRAS 07598+6508 is a star.)

3 Warm ULIGs ↔ Infrared QSOs

Perhaps the best evidence to date that the dominant power source in all "warm" ULIGs is a dust-enshrouded AGN is provided by detailed observations of the complete flux-limited sample of 12 objects in Sanders *et al.* (1988b), whose images are shown in Figure 2. All of these objects have optical spectra characteristic of Seyferts (equally split between types 1 and 2), and they span a variety of classes of extragalactic objects, including four optically identified QSOs (three radio-quiet and one radio-loud), two powerful radio galaxies, and six objects not previously identified in optical catalogs. Near infrared spectroscopy (Veilleux *et al.* 1997b) shows that all of the optical Seyfert 2s show evidence of broad Seyfert 1 linewidths at Paα and/or [Si VI] emission, suggesting the existence of a genuine broad-line AGN in these objects as well.

The broad array of data that is now available for the objects in Figure 2 clearly supports a time evolution picture whereby warm ULIGs are more evolved versions of their cooler ULIG counterparts. All of the warm ULIGs in Figure 2 (with the exception of 3C273) have been detected in the CO(1→0) emission line indicating that they are rich in molecular gas, but with a mean H_2 mass approximately half that detected in "cool" ULIGs. Deep optical and near infrared ground-based images of the objects in Figure 2 show that they all appear to be advanced mergers, caught close to the time when the two nuclei actually merge. (The median projected nuclear separation for the warm ULIGs is ∼0.6 kpc compared to ∼2 kpc for cool ULIGs.) Compared to cool ULIGs, the warm objects show a much greater diversity of spectral types and infrared/optical flux ratios. This is illustrated in Figure 3 where the SEDs of the objects in Figure 2 are arranged in order of decreasing infrared/optical flux ratio. The progression is clearly such that those objects with the the warmest mid-infrared colors are closest in appearance to "classical" QSOs, i.e. have the most pointlike optical nuclei, the smallest infrared-to-optical flux ratios, and exhibit Seyfert 1 optical spectra.

The total time interval for the warm ULIG phase can be estimated from the fact that half of the sources in Figure 2 have single nuclei while the other half have an average nuclear separation of ∼0.8 kpc. Numerical simulations suggest that the average "time-to-merger" for these latter objects is $\sim 3.5 \times 10^7$ years, which would imply that the total time span represented by the full sample is only $\sim 7 \times 10^7$ years. Given that the observed ratio of cool to warm ULIGs is ∼6:1, this suggests that the cool phase lasts $\sim 4 \times 10^8$ years.

4 Future Work

If warm ULIGs are indeed the precursor phase of optical QSOs then it should be possible to detect the remnant signature of a gas-rich merger (e.g. large scale tidal debris, an aging circumnuclear starburst, remnant molecular gas, etc.) in the host galaxies of optically selected QSOs. An important new possibility has also been provided by new *HST* observations of the warm ULIGs in Figure 2 (Surace *et al.*

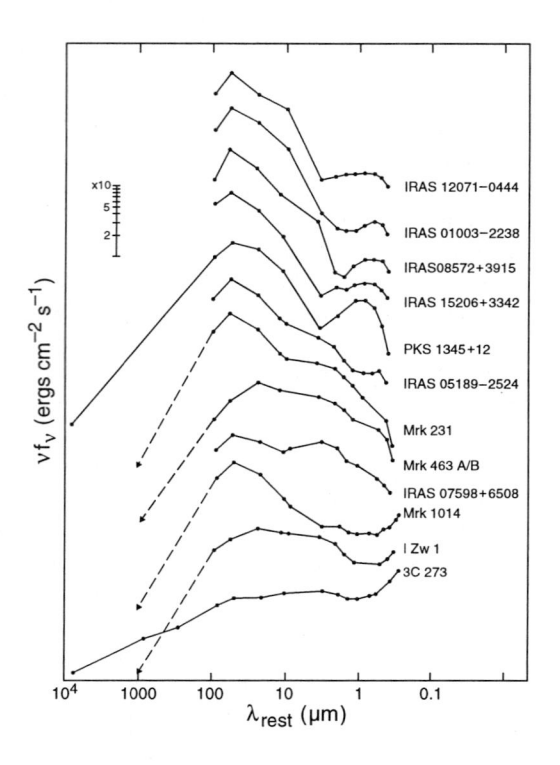

Fig. 3. SEDs for the complete sample of 12 warm ULIGs shown in Figure 2. Sources are ordered (top to bottom) in order of increasing f_{25}/f_{60} ratio and increasing optical-UV/infrared ratio. Dashed lines are an extrapolation beyond the figure boundary to the data point at $\lambda = 6$ cm.

1997) that are discussed in the following article by Surace & Sanders. The new *HST* data show that all of the warm ULIGs contain a significant population of extremely massive star clusters. Therefore, it should also be possible to detect an aging population of such clusters in optical QSOs at similar redshift, and then hopefully to use these clusters as a clock to age-date the infrared and optical phases.

Perhaps one of the more exciting new results to be presented at this workshop is that *HST* images of nearby optically selected QSOs (e.g. Bahcall *et al.* 1997, Boyce *et al.* 1997) appear to show that signs of strong interaction/merger (e.g. tidal features, putative star clusters) are common in QSO hosts. More detailed *HST* imaging and spectroscopy of both warm ULIGs and optical QSOs may indeed strengthen the evolutionary link between these two classes of objects.

Acknowledgments: DBS and JAS were supported in part by NASA through *HST* grant GO-05982, and NASA grant NAGW-3938.

References

Bahcall, J. N., Kirkahos, S., Saxe, D. H., Schneider, D. P. (1997): ApJ. in press
Barnes, J.E., Hernquist, L. (1992): ARAA. **30**, 705
Becklin,E.E., Matthews,K., Neugebauer,G., Wynn-Williams,G. (1973): ApJL. **186**, 69
Beichman, C.A., Soifer, B.T., Helou, G. et al. (1986): ApJL. **308**, 1
Boyce, P.J., Disney, M., Blades, J.C. et al. (1997): ApJ. in press
Bryant, P.B. (1996): PhD Thesis, California Institute of Technology
Burbidge, G.R., Stein, W.A. (1970): ApJ. **160**, 573
Clements, D.L., Sutherland, W.J., McMahon, R.G. et al. (1996): MNRAS. **279**, 477
Cutri, R.M., Huchra, J.P., Low, F.J. et al. (1994): ApJL. **424**, 65
deGrijp, M.H.K., Miley, G.K., Lub, J., deJong, T. (1985): Nat. **314**, 240
Frogel, J.A., Gillett, F.C., Terndrup, D.M., Vader, J.P. (1989): ApJ. **343**, 672
Hill, G.J., Wynn-Williams, C.G., Becklin, E.E. (1987): ApJL. **316**, 11
Hines, D.C. (1991): ApJL. **374**, 9
Hines, D.C., Schmidt, G.D., Smith, P.S. et al. (1995): ApJL. **450**, 1
Hines, D.C., Wills B.J. (1993): ApJ. **415**, 82
Kim, D.C. (1995): PhD Thesis, University of Hawaii
Kim, D.-C., Sanders, D.B. (1997): ApJ. submitted
Kleinmann, S.G., Keel, W.C. (1987): in Star Formation in Galaxies, ed C.J. Lonsdale-Persson (Washington DC:US GPO), p. 559
Low, F.J., Huchra, J.P., Kleinmann, S.G., Cutri, R.M. (1988): ApJL. **327**, 41
Low, J.J., Kleinmann, D.E. (1968): AJ. **73**, 868
Melnick, J., Mirabel, I.F. (1990): A&A. **231**, L19
Murphy, T.W., Armus, L., Matthews, K. et al. (1996): AJ. **111**, 1025
Neugebauer, G., Becklin, E.E., Oke, J.B., Searle, L. (1976): ApJ. **205**, 29
Neugebauer, G., Habing, H.J., vanDuinen, R. et al. (1984): ApJL. **278**, 1
Neugebauer, G., Soifer, B.T., Miley, G.K., Clegg, P.E. (1986): ApJ. **308**, 815
Perault, M. (1987): PhD Thesis, University of Paris
Rees, M.J., Silk, J.I., Werner, M.W., Wickramasinghe, N.C. (1969): Nat. **223**, 788
Rieke, G.H. (1978): ApJ. **226**, 550
Rieke, G.H., Low, F.J. (1972): ApJL. **176**, 95
Rieke, G.H., Low, F.J. (1975): ApJL. **200**, 67
Sanders, D.B., Mazzerella, J.M., Surace, J.A. et al. (1997): ApJ. submitted
Sanders, D.B., Mirabel, I.F. (1996): ARAA. **34**, 749
Sanders, D.B., Phinney, E.S., Neugebauer, G. et al. (1989): ApJ. **347**, 29
Sanders, D.B., Soifer, B.T., Elias, J.H., et al. (1988a): ApJ. **325**, 74
Sanders, D.B., Soifer, B.T., Elias, J.H. et al. (1988b): ApJL. **328**, 35
Saunders, W.S., Rowan-Robinson, M., Lawrence, A. et al. (1990): MNRAS. **242**, 318
Schmidt, M., Greene, R.F. (1983): ApJ. **269**, 352
Scoville, N.Z., Sargent, A.I., Sanders, D.B., Soifer, B.T. (1991): ApJL. **366**, 5
Scoville, N.Z., Bryant, P.B., Yun, M.S. (1997): ApJ. in press
Soifer, B.T., Rowan-Robinson, M., Houck, J.R. et al. (1984): ApJL. **278**, 71
Soifer, B.T., Sanders, D.B., Madore, B. et al. (1987): ApJ. **320**, 238
Stein, W.A., Gillett, F.C., Merrill, K.M. (1974): ApJ. **187**, 213
Surace, J.A., Sanders, D.B., Vacca, W.D. et al. (1997): ApJ. submitted
Telesco, C.A., Harper, D.A. (1980): ApJ. **235**, 392
Vader, J.P., Simon, M. (1987): Nat. **327**, 304
Veilleux, S.V., Kim, D.-C., Sanders, D.B. (1997a): ApJ. in press
Veilleux, S.V., Sanders, D.B., Kim, D.-C. (1997b): ApJ. in press

HST Images of Warm Ultraluminous Infrared Galaxies: QSO Host Progenitors

Jason A. Surace and D.B. Sanders

Institute for Astronomy, University of Hawaii, 2680 Woodlawn Dr., Honolulu, HI, 96822

Abstract. We present *Hubble Space Telescope* (*HST*) images obtained with the WFPC2 camera at B & I of a nearly complete sample of ultraluminous infrared galaxies (ULIGs) chosen to have "warm" mid-infrared colors. We find that all of these objects: (1) appear to be advanced mergers; (2) contain compact, luminous blue knots presumably powered by star formation, however the total luminosity of these knots is unlikely to be a major fraction of the total bolometric luminosity of the galaxy; (3) contain either one or two compact sources whose optical/near-infrared properties are similar to those of reddened QSOs. We believe that these warm ULIGs represent a critical transition phase between "cooler" ULIGs and optically selected QSOs, and as such represent the immediate progenitors of optically selected QSO hosts.

1 Introduction

The discovery by the *Infrared Astronomical Satellite* (*IRAS*) of a significant population of ultraluminous infrared galaxies (ULIGs: $L_{\rm ir} > 10^{12}\, L_\odot$) with "warm" mid-infrared colors ($f_{25}/f_{60} > 0.2$) has the potential for providing important clues for understanding the origin and evolution of QSOs (see Sanders & Surace, this volume). Ground-based observations of warm ULIGs have shown that nearly all exhibit AGN-like optical spectra, and that the trigger for the intense infrared emission in the majority, if not all, of these objects appears to be the merger of gas-rich spirals. Warm ULIGs span a wide range of types of extragalactic objects including radio-loud and radio-quiet optically selected QSOs, powerful radio galaxies, and luminous starbursts. It appears possible that they represent an important transition stage in the evolution of powerful circumnuclear starbursts into AGN. (A more extensive review of the properties of luminous infrared galaxies in general and warm ULIGs in particular can be found in Sanders & Mirabel 1996)

2 Background: Ground-Based Data

A sample of warm ULIGs that have proved particularly useful for study is the complete sample of 12 objects listed in Sanders *et al.* (1988). Ground-based images and spectral energy distributions (SEDs) for these objects are shown in Figures 1 and 2 respectively of Sanders & Surace (this volume). Morphologically, the majority of these objects have well-developed tidal tails and at least 3 (and possibly 5) objects appear to have double nuclei. Those objects with the largest mid-infrared excess are the most similar to optically selected QSOs (i.e. all have Seyfert 1 optical spectra and have the brightest and most compact optical nuclei). Additional observations of the molecular gas and dust distributions, as well as model simulations (e.g. Barnes & Hernquist 1992), suggest that the bulk of the luminosity is indeed generated in the inner few kpc of these objects; however, even this nearest sample of warm ULIGs is sufficiently distant ($z \approx 0.05 - 0.15$) that previously published ground-based optical and near-infrared observations are unable to resolve structure on scales smaller than ~1–2 kpc.

3 New HST Results

We have recently obtained *HST* images using the WFPC2 camera at B (F439W) & I (F814W) of the majority of objects (9/12) in our complete sample of warm ULIGs (Surace *et al.* 1997a)[1].

Figure 1 illustrates the new observations for two of the 9 warm ULIGs observed with *HST*. All of the warm ULIGs (except IRAS 01003–2238) clearly show tails, loops, and other features characteristic of the tidal debris associated with the advanced merger of two large spirals of relatively equal mass. No nuclei not already identified from ground-based images were revealed by *HST*: what were thought to be possible unresolved double nuclei in ground-based images (seeing $\sim 1''$) now appear to be single nuclei surrounded by "knots" of star formation.

All of the warm ULIGs have compact, blue knots distributed throughout their nuclear regions and in some cases also along what appear to be inner tidal features. These knots, with radii in the range 40–140pc, appear similar to the close packed groups of massive star clusters seen in *HST* images of more nearby luminous infrared mergers such as NGC 4038/39 (Whitmore & Schweizer 1995). Using spectral synthesis models (Figure 2), we derive typical ages for the knots of $\sim 6 \times 10^8$ years. However, there is a wide range in $B - I$ colors for the knots in any given system, suggesting that some of the knots may be as young as a few $\times 10^6$ years. Derived masses for individual knots are in the range $10^5 - 10^9$ $M_\odot$, with apparent weak gradients in mass and age as a function of galactocentric radius. These estimated values are upper limits; reddening can reduce the derived ages by factors of ~10, and can also decrease the derived masses typically by factors of 2–5. However, regardless of reddening, there are knots which must be extremely young, indicating that star cluster formation is still on-going in some of these galaxies. It is also interesting that there appear to be no young, blue knots with estimated masses as high as some of the old, red knots.

Despite the relatively large number of identified starburst knots in some of our objects (e.g. up to 20–30 in IRAS 15206+3342, and Mrk 463), the observed luminosities of these knots are such that even in total they appear not to be significant contributors to the high bolometric luminosity of any of these galaxies.

---

[1] New ground-based observations of these systems at H & K' using a fast tip/tilt image stabilizer at the UH 2.2m telescope (0.25″resolution) are also being obtained, and will be discussed in greater detail elsewhere (Surace & Sanders 1997b)

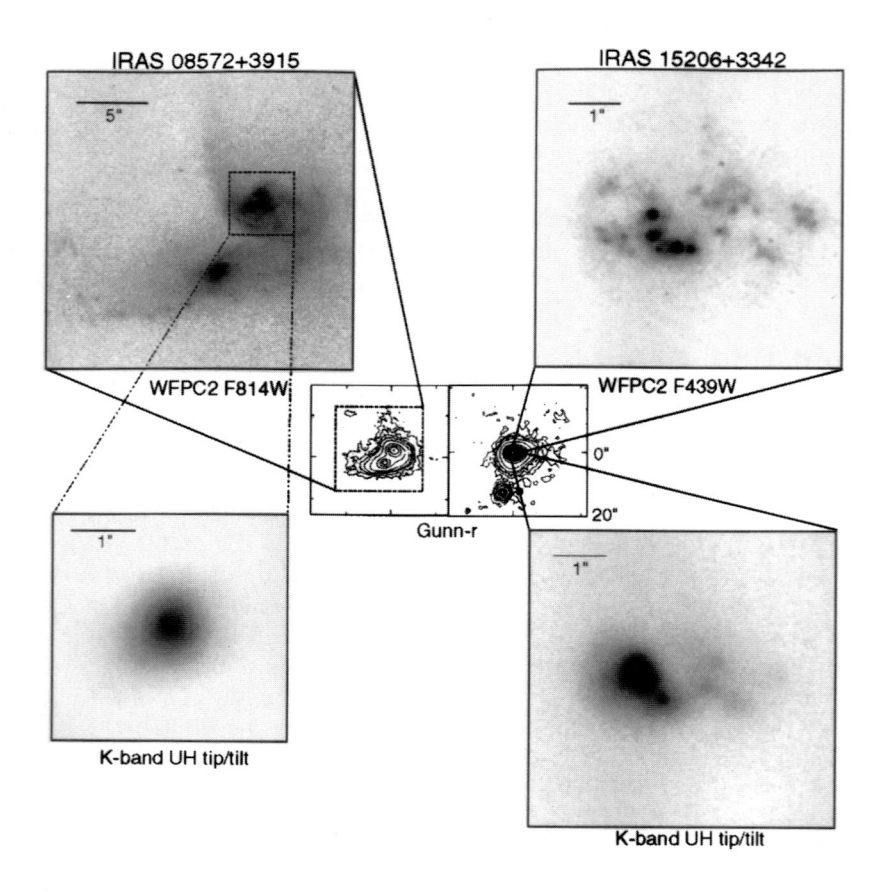

Fig. 1. Ground-based gunn-r, tip/tilt-K', and HST/WFPC2-$B\&I$ images of IRAS 08572−3915 and IRAS 15206+3342.

Furthermore, they are insufficiently dense to allow construction of an ultraluminous starburst from a hidden population of similar knots as the resulting ensemble would subtend nearly a kpc—such a large obscured region would have been detected in our images. Any ultraluminous starburst must have far more extreme properties than any of the starburst knots identified in our HST images.

We also have found that each of the warm ULIGs has at least one "knot" (or 2 "knots" in the case of 4 objects) whose luminosity and color (see Figure 3) would result in an extremely large derived stellar mass ($> 10^{10}$ $M_\odot$) inside an unreasonably small radius (often $<$ 30 pc). We suggest that these knots are the 'active' nuclei that are responsible for the dominant AGN-like optical spectrum (Seyfert 1 or 2) seen in ground-based optical spectra—e.g. Sanders et al. (1988), Veilleux et al. (1995). Compared to the other less luminous knots that are more likely to be powered by star formation, these putative nuclei are

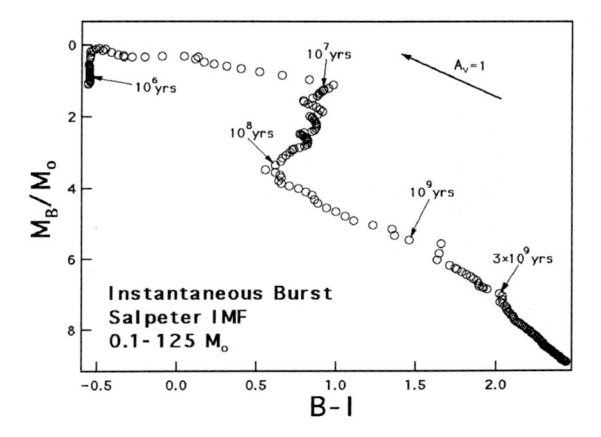

Fig. 2. Spectral synthesis model for an instantaneous starburst with a Salpeter IMF, normalized to $1 M_\odot$ (Bruzual & Charlot 1993). The vector (lower left) represents $A_V = 1$.

too luminous for their size and could only be attributed to the most extreme of starburst models. However, their luminosities and colors can plausibly be explained as AGN reddened by $A_V \sim 1$–4 mag. Those "nuclei" which appear radially symmetric lie very nearly on the quasar reddening line in Figure 3, while those that appear distorted (suggestive of patchy extinction and scattering) lie as expected slightly below and to the left of where they would be if $A_V = 0$.

Our ground-based near-infrared imaging (Surace & Sanders 1997b) also indicates that in all of the warm ULIGs the total luminosity is increasingly dominated by the putative nuclei at the longer wavelengths. The near-infrared images are also of sufficient resolution and depth to detect many of the starburst knots that we have identified in the *HST* data, and they seem to have the near-infrared colors expected for regions of massive star-formation. The nuclei, however, have very red colors and high K'-band luminosities, indicative of increasing contributions from hot dust. Additionally, the near-infrared images fail to detect any sizable population of knots sufficiently reddened so as to be hidden from optical detection yet still be detectable at K'. Any additional star formation knots must be hidden by substantially greater extinction than affects the more widespread starburst regions detected at optical wavelengths.

4 Future Work

Our *HST*/WFPC2 images of warm ULIGs support earlier conclusions based on ground-based imaging that these objects are indeed advanced mergers. A major new result has been the discovery of a population of luminous "knots" of star formation in the inner few kpc of all of these galaxies that accompany the putative AGN. Future high-resolution spectroscopy with *HST* is required to

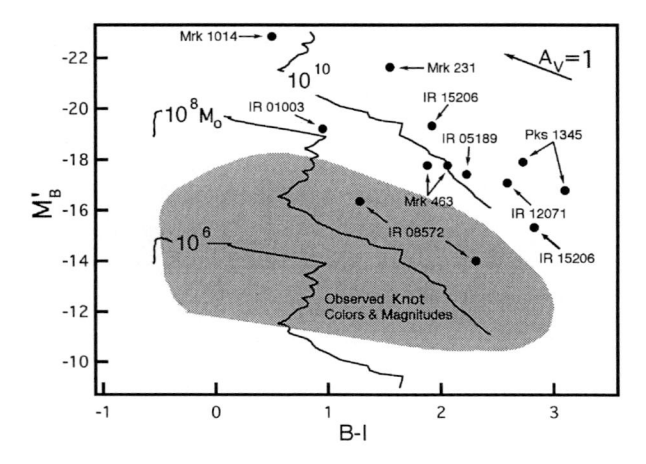

Fig. 3. Observed B-I colors and absolute B magnitudes of the galaxy nuclei (•) and star-forming knots (shaded region). The solid lines represent total cluster masses based on spectral synthesis modeling of starburst populations. Mrk 1014 (= PG 0157+001) is an infrared-loud, radio-quiet optically-selected QSO typical of the region populated by QSOs in the color-magnitude diagram. The vector (lower left) represents $A_V = 1$.

confirm the AGN identifications and to more precisely determine the ages and masses of the star clusters.

The morphology of the complete sample of warm ULIGs suggests that the warm infrared phase coincides with a relatively brief time interval (few $\times 10^7$ years) surrounding the actual merger of the two nuclei. This time interval is short compared to the time that it should take for the knots of star formation and larger scale tidal features currently observed in the warm ULIGs to completely fade from view (few $\times 10^8$ years). Thus it seems reasonable to expect that starburst knots and tidal features similar to those identified in our *HST* images of warm ULIGs, although fainter, should still be recognizable in *HST* images of many optically selected QSOs. Recent *HST* images of QSOs presented by Bahcall (this conference, and Bahcall et al. 1997) and by Boyce *et al.* (1997), would appear to confirm that a substantial fraction of optically selected QSOs may indeed have tidal features and knots similar what is observed in warm ULIGs. Deeper images and spectroscopy of optically selected QSO hosts may provide the additional evidence needed to establish whether the majority if not all QSOs begin their lives in an intense infrared phase.

References

Bahcall, J.N., Kirhakos, S., Saxe, D.H., Schneider, D.P. (1997): ApJ. in press
Barnes, J.E., Hernquist, L. (1992): ARAA. **30**, 705
Boyce, P.J., Disney, M., Blades, J.C. *et al.* (1997): ApJ. in press
Bruzual, G., Charlot, S. (1993): ApJ. **405**, 538

Sanders, D.B., Mirabel, I. (1996): Luminous Infrared Galaxies. ARAA. **34**, 749
Sanders, D.B., Soifer, B.T., Elias, J.H. *et al.* (1988): ApJ. **328**, L35
Surace, J.A., Sanders, D.B., Vacca, W.D. *et al.* (1997a): ApJ. submitted
Surace, J.A., Sanders, D.B. (1997b): ApJ. in preparation
Veilleux, S.V., Kim, D.-C., Sanders, D.B. *et al.* (1995): ApJS. **98**, 171
Whitmore, B.C., Schweizer, F. (1995): AJ. **109**, 960

ISO-SWS Results on Ultraluminous IRAS Galaxies

Dimitra Rigopoulou

Max-Planck-Institute für Extraterrestrische Physik, Postfach 1603, 85740 Garching, Germany

Abstract. We present first results on Infrared Space Onservatory (ISO) spectroscopic observations of Starbursts and Ultraluminous Galaxies. Observations of a wide range of ionic, atomic and molecular infrared lines enable us to explore in detail the physical conditions in the circumnuclear regions of these galaxies. With the help of theoretical modelling this multi-line spectroscopic database is used to further probe the central emission mechanism, discriminate between stellar and non-thermal processes and finally constrain the current models of galactic nuclei.

1 Introduction

The advent of the Infrared Space Observatory offers the possibility of sensitive spectroscopy of galactic nuclei. With observations in the wavelength regime between 1 to a few hundred μm it is possible to probe regions that are heavily obscured by line of sight dust. A number of spectral lines of key ions, atoms, molecules and characteristic signatures of various types of dust particles arise in this region. So it is possible to probe the spatial distribution, composition and dynamics of the interstellar matter in the immediate vicinity of the nucleus or the obscured disk star forming regions.

2 ISO Observations

The observations presented here were taken with the short wavelength spectrometer (SWS) onboard ISO (deGrauauw et al. 1996), during the calibration–verification (PV) phase and the first three months of the central program (CP) phase. The data obtained span the region between 2.5 – 45 μm. An example of a collection of lines obtained with the SWS for NGC 3256 is shown in Figure 1. More detailed discussion of the results can be found in the special issue of Astronomy and Astrophysics dedicated to ISO (Lutz et al. 1996, Rigopoulou et al.1996, Sturm et al.1996, Kunze et al.1996, Moorwood et al., 1996).

Here we concentrate on ISO observations of the first three Ultraluminous IRAS Galaxies (ULGs) that have so far been observed. We first calculate the extinction as this is now derived from the new ISO spectroscopic data and then we address the question of the energy source of the ULGs. Finally, we comment on their molecular hydrogen content.

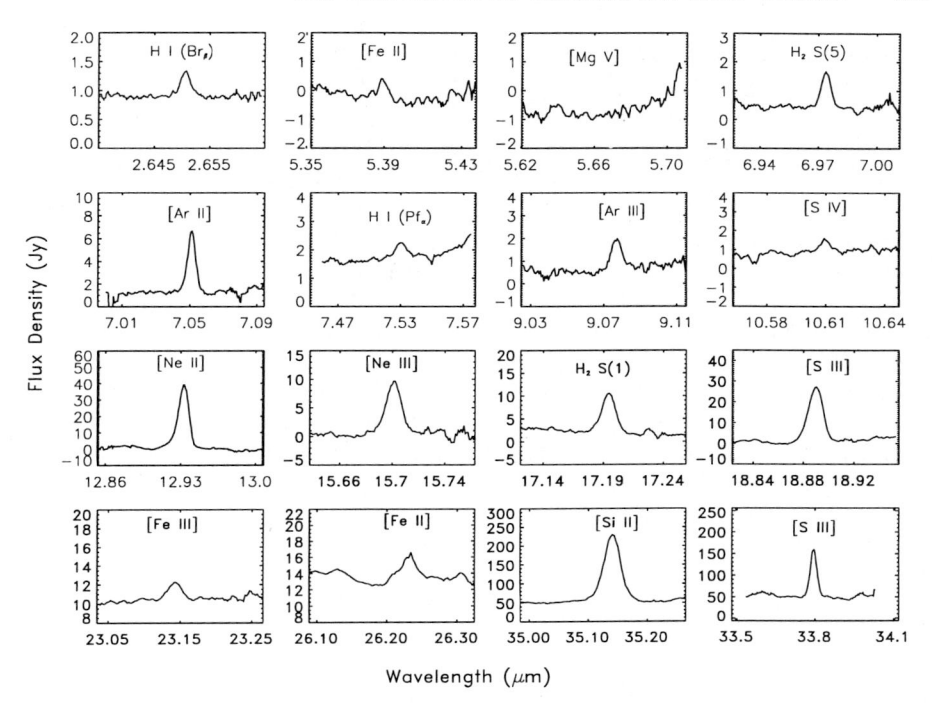

Fig. 1. ISO-SWS spectra for NGC 3256

3 Extinction

The extinction in galaxies is usually determined from the observed HI recombination lines. Assuming the validity of case B for the observed recombination spectrum, an A_V of ≈ 35 is infered for the starburst IR luminous galaxy NGC 3256.

In the case of Arp 220 the dust/gas mixed model, even at the high optical depth limit ($\tau_V \to \infty$), cannot reproduce the HI recombination lines observed by ISO. Thus, as pointed out by Sturm et al. (1996) it is possible that in Arp 220 the extinction is better described by a screen model. Using the two hydrogen recombination lines observed by ISO together with the Br$_\gamma$ flux from the literature, an A_V of 40 mag is derived. Using the two [SIII] lines an A_V of ≥ 59 is derived. The fact that ISO finds an extinction significantly higher than previous studies of Arp 220 removes one of the key constrains in previous attempts to model Arp 220 as a starburst.

4 What Powers Luminous Infrared Galaxies

The issue of the energy source of ULGs has so far been a matter of debate: Based on far-infrared, millimetre/submillimetre and radio observations it has been argued that star formation is the dominant process. On the other hand, the optical properties of ULGs imply a Seyfert-like nature of the ionizing source. With ISO it is now possible for the first time to use mid-infrared spectroscopy to probe the central parts of these galaxies where most of the luminosity originates. The ionic infrared emission lines towards galactic nuclei arise, like optical emission lines, predominantly from photoionized gas. Hence a powerful tool for investigating whether star formation or a central obscured AGN dominates is to study the excitation state of the mid- and far-infrared emission line spectrum. Of particular interest are high excitation 'coronal' lines that require much harder radiation field than can be delivered by stars, thus implying the presence of an obscured AGN nucleus.

Figure 2 is a plot of the (dereddened) [NeV]/[NeII] and [OIV]/[NeII] line ratios (or 3 σ upper limits) for 11 galaxies observed so far with SWS (Lutz et al. 1996). This includes several starburst and AGN templates as well as three typical (ultra-) luminous IRAS galaxies (arp 220, NGC 6240, NGC 3256). In all sources known to be powered by stars alone, the [NeV]/[NeII] and [OIV]/[NeII] ratios are ≤ 0.1, while these ratios are between 0.13 and 1.5 in the two AGNs. The three luminous IRAS galaxies have line ratios ≤ 0.1, strongly supporting the notion that a moderately extincted AGN cannot be the main source of their luminosity. One way out this constraint is to postulate that these sources contain AGNs that are hidden even at 15 to 30 μm requiring $A_V \geq 100$. Such a compact highly optically thick source would produce bright hot dust emission in the 5 to 20 μm regime which is not observed. Moreover, all three luminous IRAS galaxies emit most of their luminosities between 30 and 200 μm requiring minimum source sizes of several hundreds of pc. *It is thus possible that Arp 220, NGC 6240 and NGC 3256 are all powered mainly by stars, although one cannot exclude of course the possibility that an AGN contributes a small fraction of the bolometric luminosity.*

5 Starburst Models

Under the assumption that the source of luminosity in Arp 220, NGC 6240 and NGC 3256 is formation of massive stars then it is interesting to investigate the evolutionary state of the starburst. A number of near-infrared/optical studies addressing the issue for the three IRAS luminous galaxies can be found in the literature (e.g. Lester et al. 1988, Doyon et al. 1992, etc), all of them concluding that the observed far-infrared luminosity is not dominated by recently formed massive young stars. According to their conclusions if stars power these galaxies then the last period of star formation in these galaxies must have occured 10^8 yrs ago or stars more massive than 20 M$_\odot$ have not been formed recently.

However, the new ISO SWS data (see paragraph 1) indicate that the near-infrared/optical emission is much more affected by dust obscuration than previously thought. With the new extinction corrections as these were derived from the ISO SWS data and using the star cluster evolution code of Kovo and Sternberg (1996) Lutz et al. (1996) have calculated the number of stars of different

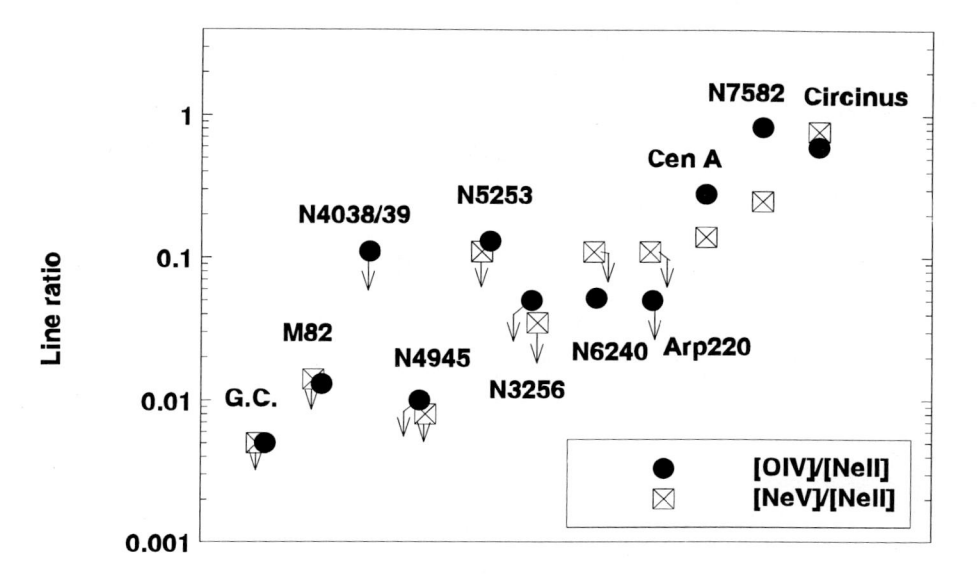

Fig. 2. Dereddened [NeV]/[NeII] and [OIV]/[NeII] ratios (or 3σ upper limits) for 11 galaxies observed with ISO SWS (Lutz et al. 1996). To the left are the starburst template galaxies, to the right three active galactic nuclei. The three (ultra-)luminous IRAS galaxies (NGC 3256, NGC 6240, Arp 220) are in the middle.

type and the global L_{Bol}/L_{Lyc}, L_K/L_{Lyc} and supernova rate to L_{Bol} ratios as a function of time in three different star formation histories: pure δ bursts ($\Delta t=10^6$ yrs), extended bursts ($\Delta t=2\times 10^7$ yrs) and constant star formation. Calculations have been carried out for Salpeter Initial Mass Function, solar abundances and upper mass cutoffs of 25, 50 and 100 $M_\odot$. The cluster averaged stellar spectrum was synthesized by coadding Kurucz atmospheres with the appropriate weighting of different stellar types. The results for the specific case of $m_u=100$ $M_\odot$ are plotted in Figure 3 which also shows the locations of luminous IR galaxies and starburst templates that were observed. *The best overall fits for both starburst templates and the luminous IR galaxies are for moderately extended bursts ($\Delta t \approx 1-2\times 10^7$ yrs) with mean ages ranging between 1 to 7×10^7 yrs and high upper mass cutoffs (50 to 100 $M_\odot$).*

6 Molecular Hydrogen

ISO offers the unique opportunity to study H_2 through its rotational lines and thus get direct estimates of the molecular hydrogen mass in galaxies. In NGC 3256 ISO detected three rotational lines S(1), S(2) and S(5), whereas in Arp 220

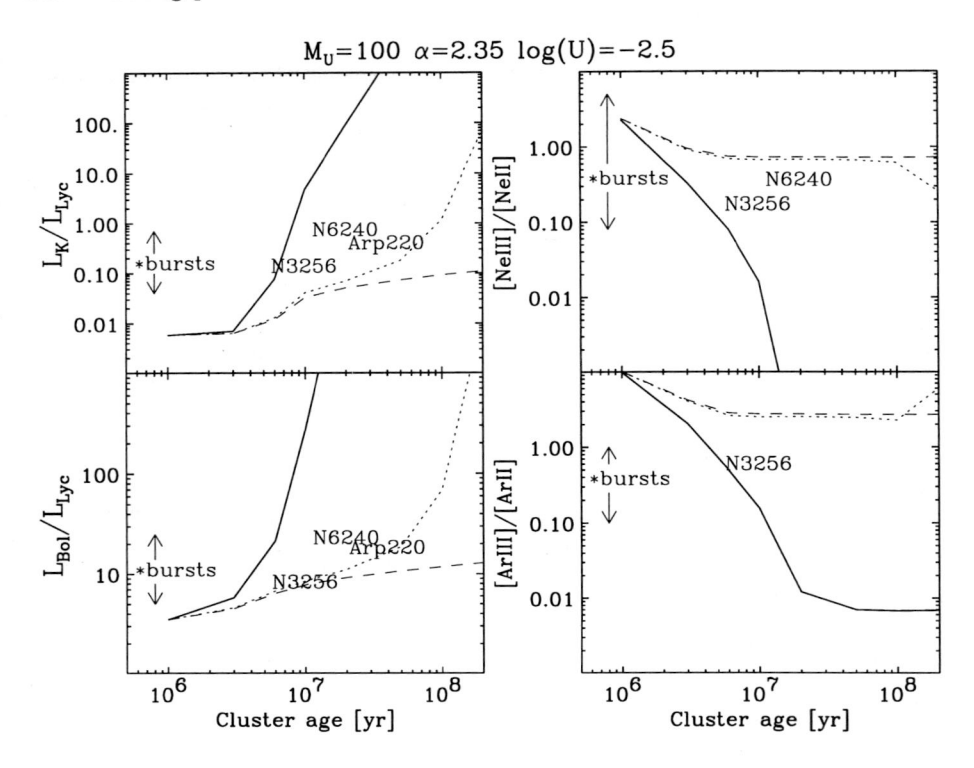

Fig. 3. $L_{bol}/L(Lyc)$, $L(K\text{-band})/L_{Lyc}$ and characteristic infrared line ratios for evolving star cluster models (from Lutz et al. 1996). Solid curves denote δ-bursts (star-formation rate$\approx\exp(-t/\Delta t)$ with $\Delta t=10^6$ yrs). Dotted curves denote extended bursts with $\delta t=2\times 10^7$ yrs, dashed curves are constant star formation models. In all cases solar abundances was used in the stars and the nebulae.

and NGc 4038/39 the S(1) and S(5) transitions were detected. The mass of the 'warm' gas where the lower transitions originate, is found to be of the of the order of few$\times 10^8$ $M_\odot$ or even up to 3×10^9 in the case of Arp 220. Comparing the current 'warm' mass estimates with those mass estimates derived from CO measurments we conclude that warm molecular gas is always several % of the total molecular mass as this is probed by CO observations.

References

deGraauw, T., Haser, L.N., Beintema, D., et al., 1996, A&A **315**, L49
Doyon, R., Puxley, P.J., and Joseph, R.D., ApJ **397**, 117
Kovo, O., and Sternberg, A., 1996, in preparation
Kunze, D., Rigopoulou, D., Lutz, D., et al., 1996, A&A **315**, L101
Lester, D., Harvey, P.M., and Carr, J., 1988, Ap.J., **329**, 614

Lutz, D., Genzel, R., Sternberg, A., et al., 1996, A&A **315**, L137
Moorwood, A.F.M., Lutz, D., Oliva, E., et al., 1996, A&A **315**, L109
Rigopoulou, D., Lutz, D., Genzel, R., et al., 1996, A&A **315**, L125
Sturm, E., Lutz, D., Genzel, R., et al., 1996, A&A **315**, L133

Quest for Type-2 Quasars with ASCA

Toru Yamada[1]

Astronomical Institute, Tohoku University, Aoba-ku, Sendai, 980-70, Japan

Abstract. The ASCA X-ray satellite brought us almost the first opportunity to systematically search for distant obscured (Type-2) quasars thanks to its high sensitivity over the wide energy band up to 10 keV. Here we summarize our efforts in surveying and investigating the type-2 quasar population with ASCA, namely, optical follow-up observations of ASCA surveys, observation of a distant type-2 QSO candidate, and observation of the quasar-cluster field. Many of these are still on-going projects, and the highlight presented here is the discovery of a radio-quiet type-2 quasar at z=0.9 in one of the deep survey fields.

1 Type-2 Quasars and Quasar Hosts

The concept of type-2 (Obscured) quasars is not new (e.g., see the review in Lawrence et al. 1991). On the analogy of Seyfert galaxies, it is natural to expect the type-2 counterparts of luminous quasars, too, since there is no clear gap between Seyfert 1 galaxies and quasars. Type-2 quasars are potentially very important objects in many fields, especially in understanding quasar hosts, although currently there are only a handful of examples or candidates, except for powerful radio galaxies (PRG) and ultraluminous far-infrared galaxies (ULFIRG) (see below).

First, let us re-define "type-2 quasars". They are objects, (1) as luminous as ordinary broad-line quasars at unobscured wavelength such as radio, FIR, and hard X-ray, or in emission lines from extended regions, (2) which show no broad lines at UV-optical wavelength, (3) but show clear evidences of obscured AGN, such as strong X-ray emission, optical high ionization and high excitation lines.

Of course, immediate candidates of type-2 quasars are powerful radio galaxies. They are as luminous as radio-loud quasars in radio emission and also in [OII] emission lines (e.g., Hes et al. 1993); as widely believed, PRG must be type-2 counterparts of *radio-loud* quasars. However, radio-loud quasars occupy only a minor fraction of entire quasar population, and thus type-2 counterpart of 95 % of quasars are little known. What we are interested is to search 'radio-quiet' type-2 quasars. Ultraluminous FIR galaxies detected by IRAS are another candidates of type-2 quasars since quasars are luminous in FIR and FIR emission little obscured. It is known that some ULFIRG show strong X-ray emission without broad optical-UV emission lines. Yet, entire ULFIRG population seems heterogeneous. There are some ULFIRG as luminous as quasars in FIR but show no sign of hidden AGN at all. Also, sensitivity of IRAS was still very limited in detecting high-redshift quasars. Thus current sample of ULFIRG are not representation of the entire type-2 quasars. There are also several type-2 quasars (and candidates) discovered serendipitiouslly; IRAS 10214+4724 (z=2.3), 1E0449-184 (Stocke et al. 1982, z=0.34), RX J13434+0001 (Almaini et al. 1995. z=2.5). I also note that the gravitationally lensed objects MG2016+112 (Lawrence et al. 1984, z=3.3) may be a type-2 quasar.

Why type-2 quasars are important ? Firstlly, they are important in testing the unified models of AGNs. If the simple viewing angle model is correct, there must be a good many type-2 quasars. The frequency of type-2 to type-1 quasars constrain opening angle of molecular-dust torus which obscure nuclei. Type-2 quasars are also important in understanding origins of Cosmic X-ray Background (CXB). A thirty-years enigma is the discrepancy between spectral indices of CXB and known quasars at 2-10 keV band; CXB has flatter spectrum than quasars. Type-2 obscured AGNs are good candidates of the sources which are responsible for hard band CXB.

In addition, we would like to emphasize here the importance of type-2 quasars in studying quasar host objects. It is very difficult to observe host objects of luminous high-redshift quasars, because of the large contrast between nucleus and host object in surface brightness. In type-2 quasars, however, UV to NIR light from the nucleus is obscured, and we have much more chance to see the host objects of type-2 quasars once they are identified. Since at least some quasars have strong connection to galaxy formation event (Yamada et al., this volume), studying host objects of high-redshift quasar may mean studying forming galaxies. In fact, the age of the stellar population in high-redshift PRG have been investigated extensively, in order to constrain early star-formation history in the universe (e.g., Chambers et al. 1990). Radio-quiet type-2 quasars are more useful in these studies, since radio-quiet quasars are major population of quasars and their photometric properties are not strongly affected by presence of radio jet.

2 Why ASCA ?

Why there are only small number of type-2 quasars or candidates while there are nearly 10000 quasars are known ? The answer is simply that type-2 quasars escape the selection criteria in searching type-1 quasars; they show little UV excess (in analogy of Seyfert 2 galaxies) and no broad lines, and they are faint in UV or soft X-ray emissions. Thus we have to search for type-2 quasars in other method, especially in wavelength less affected by obscuration. Unfortunately, in radio or FIR wavelength, quasars are not dominant population. Evolving population of star-forming galaxies strongly contaminate quasars. Since type-2 quasars may be as faint as galaxies, it seems difficult to identify type-2 quasars among a number of faint radio or FIR sources efficiently.

On the other hand, at hard X-ray wavelength, quasars are the dominant population. As photons at $\gtrsim$ 10 keV can penetrate even matters with column density of 10^{24} cm^{-2}, many type-2 quasars are expected to be as luminous as type-1 quasars at $\gtrsim$ 10 keV band. ASCA X-ray satellite has good sensitivity and imaging capability over the wide band from 0.5 to 10 keV, it is currently most suitable facility to systematically search for type-2 quasars. At z = 1, we can observe X-ray emission from 1 to 20 keV, and may have good chance to identify objects as luminous as quasars at hard band but less luminous at soft band, namely obscured AGN with hard X-ray spectra.

3 Quest for Type-2 Quasars with ASCA: Our Efforts

3.1 Optical Follow-Up Observations of ASCA Surveys

Unbiased and the most efficient way of searching for type-2 quasars must be deep X-ray sky surveys and the following observations for identification of optical counterpars. We are conducting optical follow-up program in the ASCA Large Sky Survey (LSS, Ueda et al. 1997, in preparation) fields and in the two (Lynx, SA57) of the ASCA Deep Survey fields (Ogasaka et al. 1997, in preparation).

ASCA LSS surveys 1×5 deg^2 field near the North Galactic Pole (next to the SA57 region), centered at 13^h15^m and $+31°30'$. Above the 4σ threshold about 120 sources are identified in either total or soft (0.7-2 keV) or hard (2-10 keV) band on SIS or GIS detectors. At 2-10 keV band, the threshold corresponds $\sim 1 \times 10^{-13}$ erg s^{-1}cm^{-2} and ~ 70 sources have been detected (Ueda et al. 1997). Interestingly, averaged value of the photon index for faintest objects comes close to the CXB value.

We have already obtained optical imaging data for these fields, and spectroscopic programs are on-going. The hardest source in the LSS field have been identified as a new Seyfert 2 galaxy at z=0.07. This object has photon index of $\Gamma = -0.2$ and observed X-ray luminosity is 10^{42} erg s^{-1}. Due to the relatively large uncertainty of the positional determination accuracy of the X-ray sources (typically 1 arcmin), whole reliable spectroscopic identification takes much time. In the next section, we introduce a highlight in the initial results, namely discovery of a type-2 quasar at z=0.9.

3.2 A Type-2 Quasar at z=0.9: AX J08494+4454

AX J08494+4454 is detected in the ASCA Deep Survey Lynx field (Ogasaka et al. 1996). This source clearly has a 'hard' X-ray spectra since almost no counts was detected below 2 keV while nearly a hundred counts above 2 keV. We have obtained optical spectrum for almost all the objects brighter than R $\sim$ 21 within 1 arcmin, typical positional uncertainty in ASCA surveys, from the X-ray source position, and found a galaxy at z=0.886 with strong and narrow emission line 7 arcsec from the X-ray source. There is no other plausible candidate in the field. In this case, moreover, the identification seems reliable since there is also a very faint X-ray source detected with ROSAT PSPC, whose positional uncertainty is much smaller than those of ASCA, a few arcsec from this object.

The clear evidences of radio-quiet type-2 quasar is seen in the following properties of this object;

- *Large luminosity.* X-ray luminosity is calculated to be 1-2 $\times 10^{44}$ erg/s at 2-10 keV band at rest ($H_0 = 50$ km s^{-1} Mpc^{-1}), already larger than those

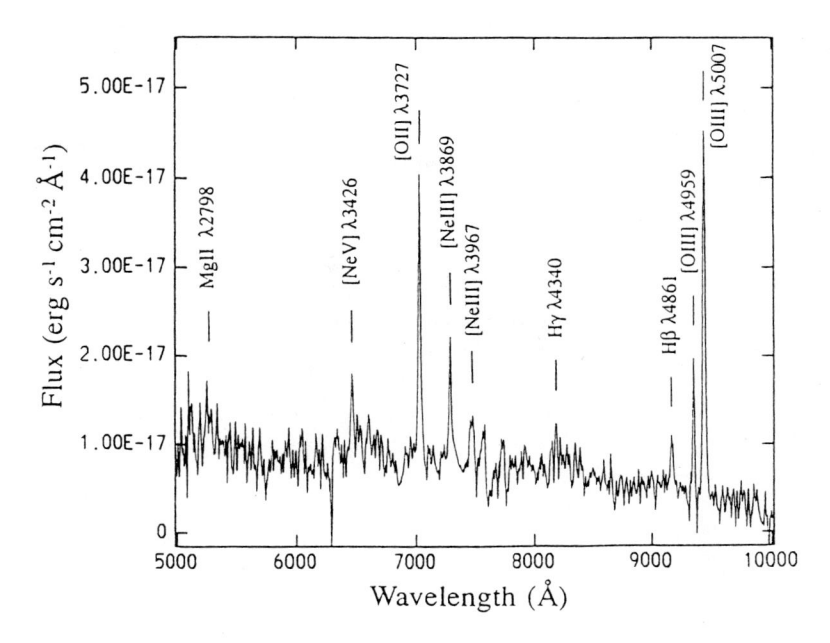

Fig. 1. Optical spectrum of the AX J08494+4454

of nearby Seyfert 1 galaxies, without any correction for the absorption. If we correct the extinction, the intrinsic X-ray luminosity would be as large as the characteristic luminosity of quasars at z ∼ 1.

- *Strong absorption.* Observed hard X-ray spectrum clearly indicate that.
- *Emission line width.* All permitted lines seen in our spectral range (Hβ, Hγ etc.) are narrow ($\lesssim$ 500 km s^{-1}) and there is no evidence of the broad component (Figure 1).
- *High ionization.* Obtained [OIII]/H$beta$ line flux ratio is sim8, which is only marginally consistent to those values for HII region but typical for AGN, indicating high ionization. Furthermore, detection of NeV λ3426 line, whose ionization potential is 97 eV, also indicate power-law photoionization as hard as AGN spectra.
- *Radio quit.* There is a weak radio source (0.21 mJy) at the position of the optical counterpart. Its radio power at 5 GHz is estimated to be $10^{23.5}$ W Hz6−1 which is more than order of magnitude smaller than typical radio power of the powerful radio galaxies or radio-loud quasars at high redshift. Radio to optical flux-density ratio, ∼ 10, is also far smaller than those of powerful radio galaxies.

Optical and NIR images were obtained with UH2.2m at the Mauna Kea in I and K band. The optical counterpart shows the two peaks; marginally resolved western object and faint extended eastern object. It is found that the emission

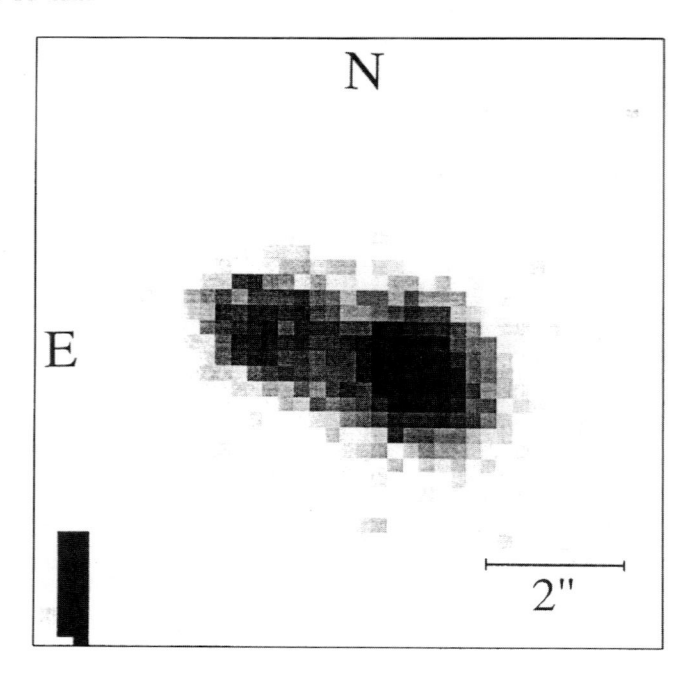

Fig. 2. Optical image of the AX J08494+4454. Emission line comes from eastern faint and extended object. The western object may be a foreground star.

lines mainly come from the eastern fainter object. It has $I \sim 20.5$ and $I-K \sim 3.5$. Detailed results are discussed in Ohta et al. (1996).

3.3 ASCA Observations of Quasar Cluster

A peak in the quasar cluster field discovered by Crampton et al. (1989) was observed with ASCA (100 ksec). Our motivation is; if the clustering of optically selected quasars are due to some physical process, e.g., environmental effect such as background large-scale density fluctuation, type-2 quasars must be harbored in the concentration, too. If this is the case, we can even determine the type-mix of quasars in this field.

Very interestingly, more than a few sources which have harder spectrum than those of usual quasars have been identified.

3.4 ASCA Observations of RX J13434+0001

RX J13434+0001 is a type-2 quasar candidate discovered by Almaini et al. (1995) through the optical identification of the ROSAT deep survey. This object shows only narrow Lyα emission line but the photon index in the ROSAT band is similar to that of normal quasar, 1.6. We have observed this object with ASCA in order to constrain the X-ray spectrum up to 30 keV at rest frame. ASCA observation was made in January 1996 and the source was detected but very weak, $\lesssim 10^{-3}$ cts/s/GIS. This result is consistent with the case that the photon index observed in the ROSAT band continues through the ASCA band. That is, the RX J13434+0001 seems to be nearly a normal quasar only partially absorbed in X-ray.

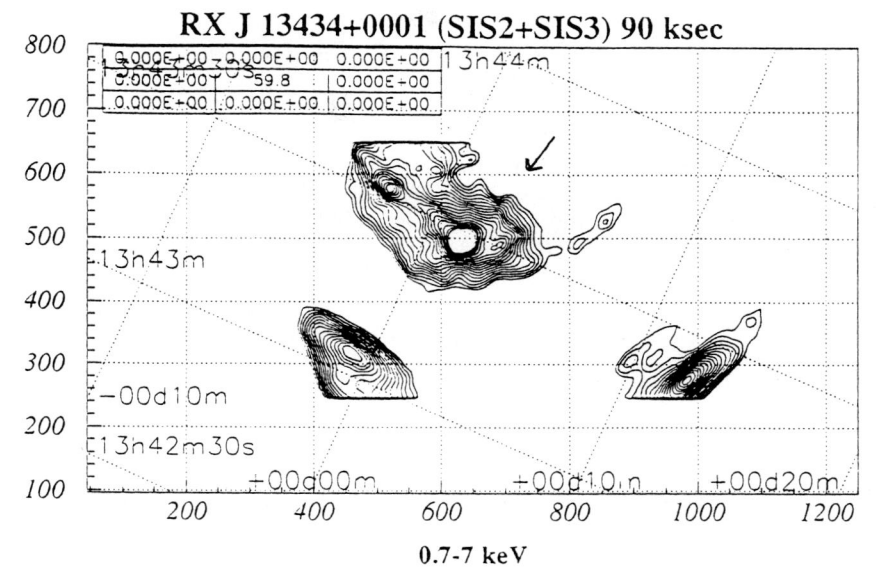

Fig. 3. ASCA SIS image of the field of RX J13434+0001. The object is indicated by an arrow.

Acknowledgement

The results presented here are production of a lot of fruitful collaboration with Kouji Ohta, Yoshihiro Ueda, Yasushi Ogasaka, Tsuneo Kii, Tad Takahashi, Takeshi Tsuru (ASCA surveys), Tatehiro Mihara, Nobuyuki Kawai, Massimo Cappi (RIKEN), and many colleagues in ASCA LSS team, ASCA deep survey team.

References

Almaini et al. (1995): Mon. Not. Roy. astr. Soc., **277**, L31..
Crampton, D. et al. (1989):Astrophys. J., **345**, 59.
Hes, R. et al. (1993):Nature, **362**, 326.
Lawrence, C.R. et al. (1984):Science, **223**, 46.
Lawrence, A. (1991): Mon. Not. Roy. astr. Soc., **252**, 586
Ohta, K. et al. (1996):AstrophysJ., **458**, L57.
Stocke, J. et al. (1982):AstrophysJ., **252**, 69.

The Local Luminosity Function of Quasars – Implications for Host Galaxy Studies

Thorsten Köhler and Lutz Wisotzki

Hamburger Sternwarte, Gojenbergsweg 112, D-21029 Hamburg, Germany
email contact: lwisotzki@hs.uni-hamburg.de

Abstract. We have investigated a new, flux-limited sample of bright low-redshift quasars and Seyfert 1 galaxies drawn from the Hamburg/ESO survey. We find much higher space densities than previous surveys, particularly of the most luminous QSOs ($M_B \lesssim -24$). At the faint end, the luminosity functions of Seyferts and quasars join smoothly, and we find no indication of structure in the combined luminosity function if the host galaxy contributions are properly subtracted. We suspect that present samples of low-redshift radio-quiet quasars may be heavily biased with respect to their host galaxy properties.

1 Introduction

The space density of luminous quasars in the local universe is not well known. Reliable determinations of the optical QSO luminosity function (QLF) exist almost exclusively for $z > 0.3$. The local ($z \approx 0$) QLF has to be constructed by extrapolation, using an *assumed* evolutionary model. This surprising fact can be explained by a simple lack of appropriate data – most optical QSO surveys discriminate against objects with extended morphology, causing severe incompleteness (and morphological bias) at low redshifts. Consequently, our knowledge of the local QSO population is incomplete, as are the parent samples where subsamples for statistical investigations are drawn from. Measurements of the luminosity function of Seyfert 1 galaxies based on dedicated galaxy surveys (e.g., CfA, Markarian) are obviously incomplete at *high* nuclear luminosities, where the AGN outshine the hosts, and the objects are not selected as galaxies anymore.

Within the Hamburg/ESO survey for bright QSOs, we have started to construct a large, optically flux-limited sample of low-redshift "Type 1 AGN", including QSO/Seyfert transition objects. In this paper we present a new determination of the optical luminosity function of these objects, for $z < 0.3$, and discuss the implications for existing and future samples of low-redshift quasars, in particular for host galaxy studies.

2 The Hamburg/ESO Survey

Since 1990 we conduct a spectroscopic wide-angle survey (the Hamburg/ESO survey; HES) based on objective prism plates taken with the 1 m ESO Schmidt telescope. An overview of the survey design, its prime scientific goals, and examples for the data material has been published by Wisotzki et al. (1996), and we summarise here only briefly some essential features. The survey material consists of ~ 400 objective-prism plates, to be digitised with the Hamburg PDS 1010G microdensitometer. The magnitude range of $12.5 \lesssim B \lesssim 17.5$ is well matched to the wide field, leaving essentially no bright cutoff for quasars and going deep enough to be able to build up samples of substantial sizes. The survey is intended to cover $\gtrsim 5000 \deg^2$ in the southern sky, of which, at the time of writing, about 60 % have been digitised and searched for QSOs.

The plate digitisation is followed by a series of largely automated reduction and analysis steps. Some features of this procedure were especially designed to minimise biases associated with low-redshift QSOs and other spatially resolved sources: (1) No morphological segregation is performed: Extended objects remain in the sample. (2) Fluxes are measured in an aperture of approximately the size of the seeing disk; the resulting sample is therefore almost limited by *nuclear* magnitude. (3) Candidate selection is performed by applying a multitude of selection criteria (UV excess, emission lines, etc.) simultaneously to the spectra; also a dedicated "Seyfert criterion" has been developed. Follow-up slit spectroscopy of all QSO and Seyfert candidates brighter than a certain (plate-dependent) magnitude limit is regularly performed with ESO telescopes.

3   The Local Luminosity Function

A first set of 33 ESO Schmidt fields has now been fully analysed, providing a flux-limited sample of QSOs and Seyferts over an effective area of $\sim 611 \deg^2$. We found in total 27 objects with $z < 0.3$ matching the completeness criteria, divided over two subsamples (for an exact definition of these subsamples and further procedural details see Köhler 1996): The "Sy 1 sample" contained 7 Seyfert 1 galaxies with $z < 0.07$; the "QSO sample" contained 20 QSOs and Seyfert 1 galaxies in the range $0.07 \leq z < 0.3$. The survey magnitude limit varied from field to field, typically $B < 17.0$–17.5.

In the following, quoted magnitudes are always based on the small aperture measurements made in the B band. For all sample objects, we obtained CCD images to improve the photometry, and to allow an estimate of the host galaxy brightness. The CCD magnitudes were then tied to the zeropoint of the corresponding photographic plates. To obtain unbiased magnitude estimates of the AGN components, the host galaxy contributions (i.e., the central surface brightness) had to be corrected. For the Sy 1 sample, this was done by subtraction of individually scaled point-spread functions from the CCD images. The subtraction was believed to be adequate when the residual galaxy surface brightness distribution did not show a central minimum. For the more luminous sources in the QSO sample, a standard template host galaxy with $M_{\rm gal} = -21$ was subtracted; the results are insensitive to the exact value of $M_{\rm gal}$.

Absolute magnitudes were computed for cosmological parameters of $H_0 = 50$ km s^{-1} Mpc^{-1} and $q_0 = 0.5$. Space densities were then derived using the $1/V_{\rm m}$ estimator, incorporating the full information about field-dependent flux limits. To obtain a reliable and unbiased estimate for the maximum redshift $z_{i,\max}$ at

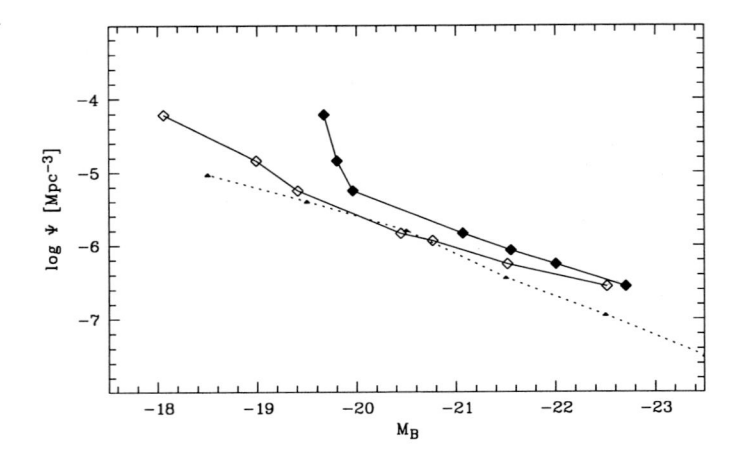

Fig. 1. Cumulative luminosity function of the "Sy 1 sample" ($z < 0.07$). Filled symbols: Uncorrected nuclear magnitudes; open symbols: Host galaxy contributions subtracted. For comparison, the Cheng et al. 1985 luminosity function is given by the dotted line.

which object i would still have been included in the survey, the available CCD images were used to simulate the mixture of AGN/host contributions to the nuclear magnitude as a function of redshift z.

The main results are presented in three diagrams: Figure 1 shows the *cumulative* luminosity function of the objects in the Sy 1 sample (i.e., the space density of Seyferts more luminous than a given M_B), for and after the correction for host galaxy contributions. We prefer the representation as cumulative relation as thus binning can be avoided, and the contribution of each object in the sample is apparent. The host galaxy correction was important particularly for the intrinsically faintest objects in the sample: Without a correction, the luminosity function shows a steep upturn at $M_B \simeq -20$; for these objects, even the central surface brightness is dominated by the host rather than by the AGN.

Although the sample is not large, the derived space densities are consistent with previous estimates (e.g., Cheng et al. 1985; Huchra & Burg 1992). However, none of these previous investigations has payed as much attention to an appropriate host galaxy correction as we have; in particular, the analysis of the CfA sample by Huchra & Burg was based on uncorrected Zwicky magnitudes, making a straightforward comparison impossible.

In Fig. 2 the cumulative luminosity function computed from the "QSO sample" is displayed. For comparison, we consider the only QSO sample with a well-defined flux limit covering a similar area in the (M_B, z) plane. The Bright Quasar Survey (BQS; Schmidt & Green 1983), part of the Palomar-Green UV excess survey, yielded 45 QSOs with $0.07 \leq z < 0.3$ over an area of 10 714 deg^2. We have plotted the cumulative LF of this sample into Fig. 2: The space den-

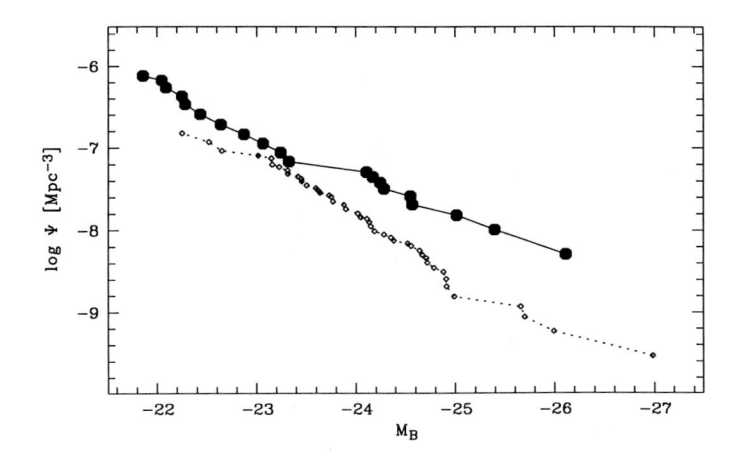

Fig. 2. Cumulative luminosity function derived from the HES "QSO sample" (0.07 < z < 0.3). Small symbols and dotted line: Corresponding relation for the Palomar-Green BQS. Magnitudes are uncorrected for host galaxy contributions.

sities found from our HES data are *much* higher than those derived from the BQS, with the discrepancy *increasing with luminosity*. For the most luminous sources, we find that there are almost an order of magnitude more quasars per unit volume than discovered by the PG survey: Our sample contains 5 QSOs with $z < 0.3$ and $M_B < -24.5$, while the BQS contains only 15 such sources in a more than 15× larger area.

In case of such a large discrepancy, one has to worry about systematic errors. A direct comparison of detected objects and photometric properties is not possible as the area covered by the present investigation shows almost no overlap with the PG area (this is going to change in the future, see below). However, we were not the first to detect significant incompleteness in the BQS: Already Wampler & Ponz (1985) showed that the photometric scale of the PG was possibly in error, and they suspected up to 50 % losses at certain redshifts. Later, Goldschmidt et al. (1992) found that the surface density of $B < 16.5$ quasars with $z > 0.3$ was underestimated by the BQS by a factor of ~ 3. Our own measurement for $z > 0.3$ agree well with the Goldschmidt et al. results, while the faint end of our number-magnitude relation matches perfectly with the values found in the LBQS (Hewett et al. 1995). We are thus confident that our data do not suffer from large systematic errors, and that the main reason for the discrepancy is substantial incompleteness in the BQS.

Around $M_B \simeq -22$ the luminosity functions of the two samples join smoothly. The combined local luminosity function of quasars and Seyfert 1 nuclei extends over almost $\Delta M = 8$ magnitudes (Fig. 3), and a single power-law for the differential LF, $\phi(L)dL \propto L^\alpha$ with $\alpha \simeq -2.2$, appears to be an acceptable parametric

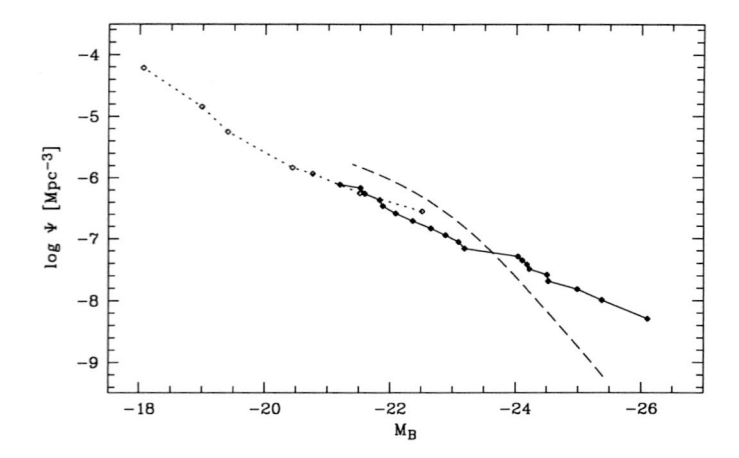

Fig. 3. The combined local ($z < 0.3$) luminosity function of Seyferts and quasars. Host galaxy contributions are subtracted. The dashed line gives the extrapolation of the Boyle et al. 1988 model.

description. Note that there is no sign of a break, or a significant change of slope, within the range of luminosities covered. This local QLF looks quite unlike the prediction of the 'pure luminosity evolution' model by Boyle et al. (1988) which we have also plotted into Fig. 3 for comparison.

4 Implications for Host Galaxy Studies

The statement that carefully selected and unbiased samples are essential to obtain statistically meaningful results is almost a commonplace, although there can be little doubt that much of the confusion and contradictory claims in the history of QSO host galaxy research is due to improperly selected samples.

Unfortunately, the Palomar-Green Bright Quasar Sample – up to now the prime source for low-redshift radio-quiet QSOs – is highly incomplete. Although it is hard, maybe impossible, to exactly pinpoint the reasons for this incompleteness, we think it likely that two effects have conspired: (i) General incompleteness by a factor of 2 or even 3, possibly caused by the high photometric errors of the original plates. (ii) A specific deficit of QSOs with extended host galaxies, caused by the imposed star-galaxy segregation.

It is thus not very far-fetched to suspect that samples of low-redshift QSOs drawn from the BQS may be not only sparse representations of the parent population, but also heavily biased with respect to host galaxy properties in the sense that those objects with the most luminous and most extended hosts are systematically avoided.

We also suspect another, more complicated bias: The apparently good agreement between HES and BQS around $M_B \approx -23$ could well be an artefact, caused by a systematic photometric overestimation of quasars with compact galaxy hosts. We know from own experience that such effects may occur on digitised photographic plates, and that it is exceedingly difficult to correct for it (this is why we used CCD images for the photometry); our suspicion – although we are presently unable to prove it – is that the BQS magnitudes of $M_B \simeq -23$ quasars may be systematically too bright, that the BQS may be overcomplete in this regime, and that thereby another morphological bias could have been introduced into the sample.

5 Conclusions and Prospects

We have analysed a new, flux-limited and well-defined sample of QSOs & Seyfert galaxies. Host-galaxy dependent selection and photometric biases, if not absent, are greatly reduced in comparison to other optically selected samples. We find that the derived space density of luminous QSOs in the local universe is much higher than previous results indicate. The local quasar luminosity function shows no break or significant change of slope over more than a factor of 1000 in luminosity, with classical Seyferts smoothly connecting to luminous quasars. If the determinations of the QLF at higher redshifts are correct, its shape must change with cosmological epoch, in contradiction to the "standard picture" of pure luminosity evolution (e.g., Boyle et al. 1988).

We have now acquired new plate material covering several fields common with the PG survey. When these fields will have been fully analysed, we shall be able to pursue the question of biases in the BQS by a direct comparison of objects, photometry, and morphological properties. As the Hamburg/ESO Survey progresses, we shall be able to increase the covered area by a factor of ~ 10, while maintaining our standards of completeness. Thus, within the next years, a substantial sample of $\gtrsim 300$ luminous low-redshift QSOs will become available to the community. This will allow, for a wide variety of applications, to define subsets of objects that are limited basically only by optical brightness, and yet contains enough objects to obtain statistically meaningful results.

References

Boyle B.J., Shanks T., Peterson B.A., 1988, MNRAS 238, 957
Cheng F.-z., et al., 1985, MNRAS 212, 857
Goldschmidt P., Miller L., La Franca F., Cristiani S., 1992, MNRAS 256, 65
Hewett P.C., Foltz C.B., Chaffee F.H., 1995, AJ 109, 1498
Huchra J., Burg R., 1992, ApJ 393, 90
Köhler T., 1996, Doctoral Thesis, Universität Hamburg
Schmidt M., Green R.F., 1983, ApJ 269, 352
Wampler E.J., Ponz D., 1985, ApJ 298, 448
Wisotzki L., Köhler T., Groote D., Reimers D., 1996, A&AS 115, 227

B-K Colours of Low-Luminosity Radio Quasars

J.I. González-Serrano[1], C. Benn[2], R. Carballo[1], S.F. Sánchez[1], and M. Vigotti[3]

[1] Instituto de Física de Cantabria, (CSIC-Univ. de Cantabria), Facultad de Ciencias, 39005 Santander, Spain
[2] Isaac Newton Group, Apto. 321, 38780 Santa Cruz de la Palma, Spain
[3] Istituto di Radioastronomia CNR, Via Gobetti 100, 40100 Bologna, Italy

Abstract. It has recently been suggested (Webster *et al.* 1995), on the basis of the observed broad range of $B - K$ colours of radio- selected QSOs, that a large fraction are heavily reddened by dust extinction in B, and that most QSOs may have been missed by optical surveys. We used the William Herschel Telescope on La Palma to obtain K images for 53 B3 QSOs, and find that although several are very red in $B-K$, most of these unusual colours can be attributed to an excess of light in K rather than a deficit in B; in many cases there is evidence for significant K light from stars in the host galaxy. Our data are consistent with there being no extinction-reddening (A_V less than 1 mag) of the QSO colours.

1 Introduction

This work was motivated by the claim by Webster *et al.* (1995) who show evidencies for a large undetected population of dust-reddened quasars. These authors present the $B - K$ colour distribution of a sample of flat-spectrum radio quasars selected from the Parkes survey. They compare their colour distribution with that of the LBQS (Large Bright Quasar Survey) which were optically selected at a limiting magnitude of $B \sim 19$. The main result of this work is that radio selected quasars show a colour distribution in the range $1 \lesssim B-K \lesssim 8$ compared with the narrower range for the optically selected QSOs ($1 \lesssim B - K \lesssim 4$). The authors interpreted the scatter in terms of dust-reddening implying that 80% of QSOs have been missed by optical surveys to a given magnitude limit. If true this would have important implications for the space density of quasars and its evolution with redshift.

In order to confirm or reject the Webster *et al.* claim we have made NIR observations of a radio-selected sample of QSOs.

2 The Sample

The sample of radio quasars has been selected from the B3 survey which has a limiting flux density of 0.1 Jy at 408 MHz. Vigotti *et al.* (1989) observed 1050 B3 sources using the VLA in A and C configurations at 1460 MHz. Of these, 30% were identified on the POSS plates and 172 starlike sources were found and selected (irrespective of their colour) as quasar candidates. These

172 identifications were observed spectroscopically and 125 were confirmed as quasars (Vigotti *et al.* 1996), the remainder being stars, galaxies or BL Lac objects. This sample is what we call the B3VLA QSO sample. The fraction of B3 quasars fainter than the POSS limit is small if we select sources brighter than 0.5 Jy at 408 MHz. From Fig. 11 of Vigotti *et al.* (1989) it can be estimated that $\sim 90\%$ of these sources have optical identification. There are 53 quasars matching the flux limit of 0.5 Jy in the range $07^h - 14^h$ of Right Ascension. Therefore we estimate that we miss ~ 5 quasars due to the optical detection limit.

3 Observations and Data Analysis

Infrared imaging of the 53 B3VLA QSOs was done at the 4.2m William Herschel Telescope (WHT) at the Observatorio del Roque de los Muchachos (La Palma). We used the WHIRCAM 256×256 camera with a pixel size of $0.''27$. We used the broad-band filter K_s which is basically the same as the standard Johnson K filter. Exposure time for the target sources was 600s. Flux calibration was done using several standard stars. Photometric errors were less than 0.1 mag and the 3σ limiting magnitude was 19 for points sources. One of the quasars was not detected. Details of the observational procedure and data reduction are given in Sánchez *et al.* (1996) and in Sánchez *et al.* in these proceedings. The results of applying a deconvolution-fitting method to detect and study host galaxies associated with the quasars are also presented in those papers.

Optical B and R magnitudes were obtained from the APM catalogue generated from the POSS plates (Automatic Plate-measuring Machine, Irwin *et al.* 1994). The original plate magnitudes should be converted to the standard B and R bands. The final optical magnitudes for our sources, after Galactic absorption correction, have errors of ~ 0.3 mag.

4 Results

4.1 Colours

In Fig. 1 we show the $B - K$ colour distribution of the B3 QSO sample together with the line corresponding to the B-band POSS limit.

Our colour distribution is similar to that presented by Webster *et al.* (1995) although we do not find quasars with $B - K$ redder that 6. It is possible that we are missing objects with $B - K > 6$ due to the POSS limit, but as we have estimated, these represent a small fraction of the sample.

4.2 Host Galaxies

Radio-selected QSOs are hosted by giant elliptical galaxies with absolute magnitudes $M_V \sim -23$ (e.g. Benn *et al.* 1988, Dunlop *et al.* 1993). At redshifts higher than ~ 0.4, which is the case for the B3VLA QSOs, these host galaxies are

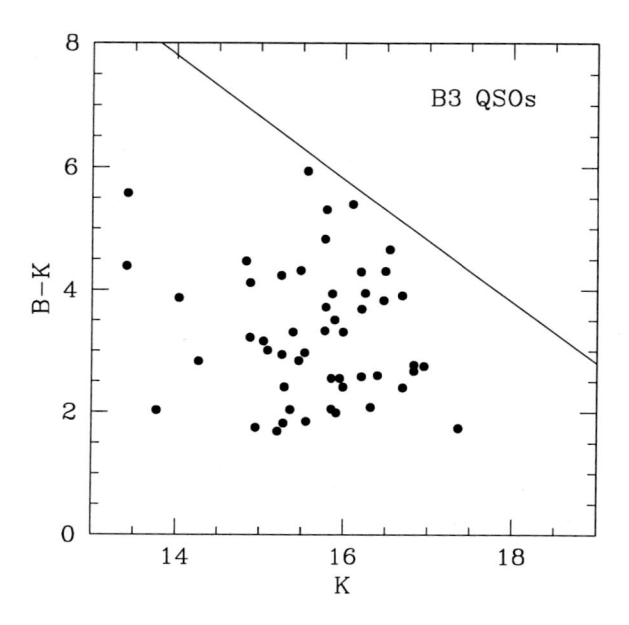

Fig. 1. $B - K$ colours of the B3 quasars. The solid line is the POSS limit in the B band.

undetectable in the POSS plates. At wavelengths $\gtrsim 1\mu m$ however, the spectral energy distribution of the galaxy reaches the maximum whilst the nuclear source has a local minimum (Sanders et al. 1989). In fact, near-infrared surveys find host galaxies with $M_K \sim -26$ (e.g. McLeod & Rieke 1994a, 1994b) which means apparent magnitudes $K \lesssim 18 - 19$.

In Sánchez et al. (1996) and in the poster by Sánchez et al. (these proceedings) it is presented the detection of host galaxies in a large fraction of the quasars of our sample. Of the 52 sources we analyzed 65% due to focus problems in one of the nights. Of these 34 sources $\sim$ 50% have associated galaxies brighter than $K = 17.9$. The mean absolute magnitude of these host galaxies is $\langle M_K \rangle = -25.9$ with a dispersion of 1.1 mag. We detect galaxies with redshifts in the range $0.6 - 2.3$, filling the gap between the low-redshift sample of Dunlop et al. (1993) and the 6 high-redshift QSOs of Lehnert et al. (1992). The host galaxies detected by us follow the $K - z$ relation found by Lilly & Longair (1984) for radio galaxies and by Dunlop et al. (1993) for the host galaxies of low-redshift QSOs. We confirm and extend this result for low-luminosity radio quasars up to a redshift of 2.3.

5 Discussion

If we assume that the colour dispersion is due to reddening along the line-of-sight of the quasars then high extinctions are derived. However, before concluding high reddening we should consider other alternatives. First is variability since we are combining 1950 POSS magnitudes with 1996 K magnitudes. Typical dispersions in $B-K$ colours due to variability are some tenths of a magnitude (Di Clemente et al. 1996; Cristiani et al. 1996), although changes of $1-2$ magnitudes on timescales of years are not rare in low-luminosity QSOs (Elvis et al. 1994) and extreme $B-K$ colours can be observed.

Second alternative is that pointed out by Serjeant & Rawlings (1995). These authors argue that flat-spectrum QSOs can have unusual red colours, probably due to a red-light extension of the non-thermal spectrum observed in the radio. Our sample contains mainly steep-spectrum radio sources and we do not expect this effect to be important. However, the sources observed by Webster et al. (1995) were selected to have flat radio spectra with $\alpha > -0.5$ ($S_\nu \propto \nu^\alpha$) and their colours could be partially explained in this way.

Third possibility is contribution from the host galaxies of the quasars to the total K flux. As we have noted above, QSOs are hosted by luminous galaxies which emit most of their light in the NIR. We have detected galaxies associated with the B3 QSOs that follow the $K-z$ relation and have high luminosities. In the optical, the contribution of these galaxies to the total QSO flux is small ($\sim 15\%$) but could be important in the K band. This would redden $B-K$ by about 1 mag.

In Fig. 2 we show an histogram of the $B-K$ colours of our sample where extended and flat-spectrum objects are marked separately. In this histogram we have represented the most clear extended sources ($\sim 40\%$ of the analyzed QSOs) which means galaxies contributing more than 30% to the K flux. Two of the extended sources are flat-spectrum sources with $B-K$ colours 3.51 and 4.47 but are marked only as extended in the histogram. This distribution is consistent with that found for a sample of steep-spectrum Parkes QSOs by Dunlop et al. (1989).

The main result from the histogram is that $\sim 60\%$ (8 out of 13) of the sources with $B-K$ colours higher than 4 are either extended or flat (in fact one is extended and flat). Although our sample contains few flat-spectrum QSOs these tend to have red $B-K$ colours (5 out of 6 have colour higher than 3) supporting the suggestion by Serjeant & Rawlings (1995).

The mean $B-K$ colour of the whole sample is 3.2 with dispersion 1.0. Considering only the extended+flat sources the colour is 3.8 (dispersion 1.2) whilst the mean colour for the non-extended QSOs is 2.8 (dispersion 0.8). The extended+flat population is therefore 1 mag redder than the point-like QSOs. Applying a t-test to these two distributions we find that this excess is significant at the 99% level. The difference in colours is also consistent with that expected considering the typical absolute magnitudes of host galaxies. Therefore it appears that the extended+flat quasars are responsible for a fraction ($\sim 60\%$) of the red tail of the $B-K$ observed distribution. Similar numbers result when considering only the extended sources.

The point-like sources may still have some K excess due to the host galaxy. If we decontaminate the K flux in the extended QSOs, the mean 'nuclear' $B-K$

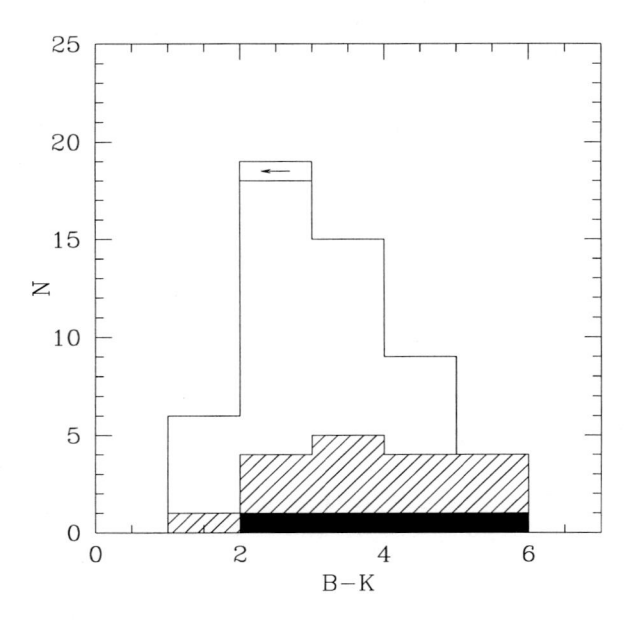

Fig. 2. Distribution of $B - K$ colours for the radio-selected B3 QSOs. The shaded regions represent sources where we detect a host galaxy. Black regions are flat-spectrum B3 quasars ($\alpha > -0.3$). The upper limit corresponds to one QSO undetected in K.

colour is then 2.2 (dispersion 1.2) compared with 2.8 of the point-like QSOs. There is a difference of 0.6 mag in colour which is significant at the 90% confidence level.

Taking into account the K flux after decontamination of the host galaxy, we can recalculate the extinction from the $B - K$ dispersion. Using the extinction law by Calzetti et al. (1994) the extinction due to dust is less than $A_V \sim 1.6$ at redshift 0.5 and less than $A_V \sim 1.0$ at redshift 2. These estimations are consistent with dust extinction implied by other observations. Schmidt (1968) derived rest-frame $A_V < 0.8$ from the $U - B$ and $B - V$ colours of 3CR QSOs out to redshift 2. Netzer et al. (1995), using $Ly\alpha/H\beta$ line ratios of 3CR and radio-loud QSOs, find $A_V < 1.2$. Boyle & di Matteo (1996), using X-ray-selected QSOs, estimate dust extinctions less than 1 mag. These limits are consistent with the dust extinction (~ 0.3 mag) obtained from the optical-UV colours of optically-selected QSOs (Rowan-Robinson 1995).

6 Conclusions

In summary, from our analysis of the $B - K$ colours of B3VLA QSOs, we can conclude the following:

1. There is no need for large amounts of dust extinction to explain the dispersion in colours. High $B - K$ colours in QSOs are most likely due to the host-galaxy contribution to the K flux.
2. The claim that optical surveys miss 80% of the quasars due to high extinction is probably wrong. We believe that the difference in $B - K$ distributions of radio- and optically-selected samples is due to the effects of apparent-magnitude and colour selection on the latter. The LBQS sample has a limiting magnitude of 19 (B band) and QSOs were selected by their blue colour.

References

Benn C.R. et al. 1988, MNRAS, 230, 1
Boyle B.J., di Matteo T. 1996, MNRAS, 277, L63
Calzetti D., Kinney A.L., Storchi-Bergmann T. 1994, ApJ, 429, 582
Cristiani S. et al. 1996, A&A, 306, 395
Di Clemente A. et al. 1996, ApJ, 463, 466
Dunlop et al. 1989, MNRAS, 274, 428
Dunlop J.S., Taylor G.S., Hughes D.H., Robson E.I. 1993, MNRAS, 264, 455
Elvis M. et al. 1994, ApJS, 95, 1
Irwin M., Maddox S., McMahon R.G. 1994, Spectrum (UK Royal Observatories) 2, 14
Lehnert M.D. et al. 1992, ApJ, 393, 68
Lilly S.J., Longair M.S. 1984, MNRAS, 211, 833
McLeod K.K., Rieke G.H. 1994a, ApJ, 420, 58
McLeod K.K., Rieke G.H. 1994b, ApJ, 431, 137
Netzer H. et al. 1995 ApJ, 448, 27
Rowan-Robinson, M. 1995, MNRAS, 272, 737
Sánchez et al. 1996, MNRAS, submitted
Sanders D.B. et al. 1989, ApJ, 347, 29
Schmidt M. 1968, ApJ, 151, 393
Serjeant S., Rawlings S. 1995, Nature, 379, 304
Vigotti M. et al. 1989, AJ, 98, 419
Vigotti M. et al. 1996, A&AS, in press.
Webster R.L. et al. 1995, Nature, 375, 469

Cygnus A: Host Galaxy of a Nearby Quasar?

F. Cabrera-Guerra[1], I. Pérez-Fournon[1], J.A. Acosta-Pulido[1,2,3], A.S. Wilson[4,5], and Z.I. Tsvetanov[6]

[1] Instituto de Astrofísica de Canarias, E-38200 La Laguna, Tenerife, Spain
[2] Max-Planck-Institut für Astronomie, Königstuhl 17, D-69117 Heidelberg, Germany
[3] ISO Science Operation Center, VILSPA, E-28080 Madrid, Spain
[4] Astronomy Department, University of Maryland, College Park, MD 20742, USA
[5] Space Telescope Science Institute, Baltimore, MD 21218, USA
[6] Department of Physics and Astronomy, Johns Hopkins University, Baltimore, MD 21218, USA

Abstract. According to the unification scheme of FRII radio galaxies and radio loud quasars Cygnus A should harbour a hidden quasar in its nucleus. Cygnus A presents observational features, such as a bright Extended Narrow Line Region (ENLR), which can be explained by the interaction of the radio jets with the ISM of the host galaxy and/or by photoionization by the hidden active nucleus. Similar processes are most probably present in higher-redshift radio galaxies and radio quasars. However, due to its proximity and to the fact that the active nucleus is hidden from our line of sight, we can study these processes in much more detail in Cygnus A. We review the observational evidence for a hidden quasar-like nucleus and present new estimates of the intrinsic luminosity of the nucleus from a study of the extended gas.

1 Introduction

Cygnus A is the best example of a powerful radio galaxy at low redshifts. In the radio it is far more luminous than any other radio galaxy at similar redshifts (Barthel and Arnaud 1996). While the redshift of Cygnus A is only 0.05562 the next FRII radio galaxy with a similar radio power is 3CR280, which has $z \sim 1$. Then, it is possible to study this object with a spatial resolution at least ten times better than for any other object of this type. For $H_0 = 75$ km s^{-1} Mpc^{-1} one arcsec corresponds to about one kpc. With HST and the VLA one can resolve the structure at the scale of 100 pc.

Studies in different spectral bands have tried to obtain evidence for the predicted hidden quasar nucleus. Barvainis and Antonucci (1994) and McNamara and Jaffe (1994) have failed to detect CO absorption. The detection of a relatively broad (FWHM ~ 270 km s^{-1}), HI 21cm absorption line by Conway and Blanco (1995) can be interpreted as evidence for a dusty, atomic torus. From infrared, optical, and X-ray data Ward (1996) obtained very high values of the extinction to the nucleus (up to $A_V = 185 \pm 35$). From the analysis of optical emission lines Tadhunter *et al.* (1994) have estimated the number of ionizing photons necessary to explain the observed line ratios. Contrary to the expectations, the derived photon luminosity $q \geq 1.3 \times 10^{54}$ *photons* s^{-1} sr^{-1} is low for a quasar of this radio power. However, this value is only a lower limit.

Perhaps the most important evidence for the presence of a nucleus comparable to those of quasars is the detection of broad MgII (Antonucci *et al.* 1994). Finally, the X-ray data can be explained by two components (Arnaud 1996): 1) a point source with a power-law spectrum of energy index $\alpha = -1$ and a high column density that obscures the nucleus in the optical and soft X-rays, and 2) an extended component associated with the intracluster gas.

We present new results from our spectroscopic observations of the extended emission line regions in Cygnus A which provide new constraints to the luminosity of the hidden nucleus.

2 Observations

Long-slit spectroscopic data were taken under excellent seeing conditions, between 0.7 and 1 arcsec, with the ISIS spectrograph on the WHT. A total of five spectra were obtained with the blue and red channels of ISIS with the slit positioned parallel and perpendicular to the radio jet axis. Fig. 1 shows the slit positions superimposed on a broad-band F622W HST WFPC2 image. The contours correspond to a VLA image at 8 GHz (Perley and Carilli 1996).

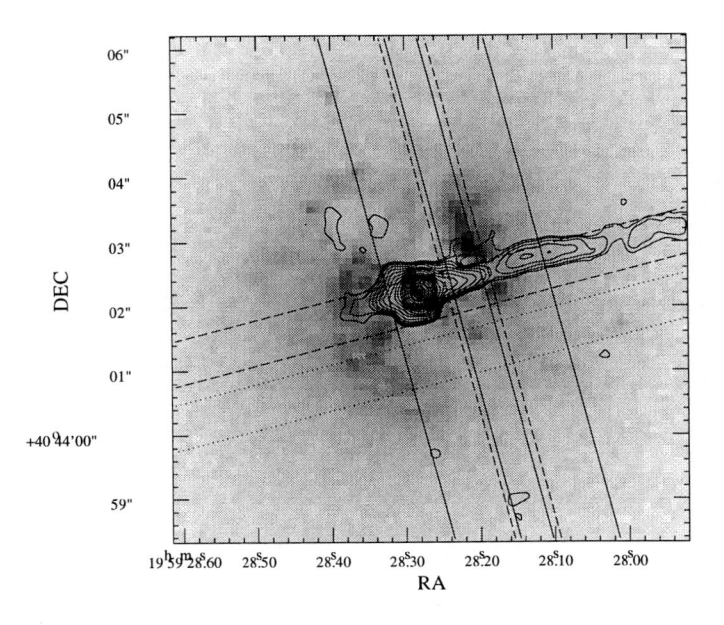

Fig. 1. Slit positions on an HST WFPC2 image with contours of a VLA image at 8 GHz.

3 Models

Several models have been proposed to explain the line emission from ENLRs in AGNs. Until recently, photoionization by the nuclear continuum was considered the most plausible interpretation. However, new models of autoionizing, high-velocity shocks have been proposed (e.g., Dopita and Sutherland 95, 96). The physical processes that produce such shocks are expected in objects with evidence for jet/cloud interactions (Pérez-Fournon et al. 1996, Cabrera-Guerra et al. 1996). Alternative models combining the line emission from matter- and radiation-bounded photoionized clouds can explain the observed line emission from nearby Seyfert galaxies and NLRG (Binette et al. 1996). In the following, we describe the grid of models which has been used to compare with the observed data for several regions on the ENLR of Cygnus A.

3.1 Photoionization of Radiation-Bounded Clouds

The photoionization code CLOUDY was used for several ionizing spectral shapes (Blackbody, $T = 2, 1 \times 10^5$ °K; Power law, $\alpha = -1, -2$), and for each spectral shape the ionization parameter varies from log U=-1. to log U= -3.5

3.2 Photoionization of a Combination of Matter-Bounded and Radiation-Bounded Clouds

These models were obtained by Binette et al. (1996) using the photoionization code MAPPINGS in order to explain the line ratios observed in Seyferts and NLRG. The main characteristics of the models are: 1) the ionizing spectral shape is a power law with spectral index $\alpha = -1.3$, 2) the ionization parameter of the matter-bounded (MB) clouds is 0.04 and they absorb 40% of the ionizing radiation, and 3) the ionization-bounded (IB) clouds see the spectrum filtered by the MB ones. The main parameter of the sequence of models is $A_{m/i}$, the ratio of the solid angles subtended by the MB clouds and the IB ones.

3.3 Autoionizing, High-Velocity Shock Models

These models were proposed by Dopita and Sutherland (1995, 1996) using the code MAPPINGS II. For this type of shocks, the shock region produces a strong flux of high-energy photons that ionize the precursor. The main parameters of these models are the shock velocity, the magnetic field and the presence or absence of a precursor region. Morse et al. (1996) have discussed in detail these models.

4 Results

We have extracted average spectra from several regions (nucleus, north-west lobe at different distances from the nucleus, and region outside of the "ionization cone"). In Fig. 2 we present the observed values and the predictions of the considered models for several line ratios. For all diagnostic diagrams, excluding the one involving the [NeV] and [NeIII] lines, there is one model of the three types considered that can explain well the observed values for regions inside the "ionization cone".

The models of photoionization of radiation-bounded clouds by the nuclear continuum produce a good fit to the spectra of regions inside the "ionization cone" for a power law of $\alpha = -1$ and $U = 3.2 \times 10^{-3}$. The derived spectral index of the ionizing continuum is consistent with the one obtained for the nuclear source from X-ray observations (Arnaud 1996). For this model, we have estimated the flux of ionizing photons produced by the nuclear source by two methods: 1) $q = U\,c\,d^2\,n$, where U is the ionization parameter obtained from the [OIII]/[OII] ratio (Penston et al. 1990), n is the density obtained from the ratio of [SII] lines and d is the distance of the clouds to the nucleus. The value obtained is $q \sim 1.8 \times 10^{54}$ $photons\ s^{-1}\ sr^{-1}$; 2) the second method calculates the flux of ionizing photons necessary to produce the observed line luminosity. For reasonable values of the covering factor and the assumed geometry we obtain $1.1 \times 10^{54} \leq q \leq 2.3 \times 10^{55}$ $photons\ s^{-1}\ sr^{-1}$ These values correspond to a quasar of medium power. The main problem of this model is that it predicts higher values for the [OIII]λ5007/[OIII]λ4363 ratio than the observed ones.

The models of high velocity shocks with precursor region produce a good fit to the observed line ratios, including the ratios of the [OIII] lines, for a shock velocity of 500 km s^{-1} and a magnetic parameter of 2 μG cm$^{3/2}$. Such high shock velocities are consistent with the large line widths observed in positions close to the jet axis (Cabrera-Guerra et al. 1996). The ionization parameter of the high-energy radiation produced by the shock is $U_{shock} = 8 \times 10^{-3}$ for this model, higher than the ionization parameter deduced above for the "standard" photoionization models. If one assumes that the photoionization of the observed regions within the cone is due to the photons produced by high-velocity shocks, one can then put constraints to the ionizing luminosity of the nucleus in the direction of the ENLR clouds. The ionization parameter of the photon field from the nucleus seen by the ENLR clouds cannot be much higher than U_{shock}, which is only a factor 3 larger than the one obtained for the standard photoionization model. Otherwise, this external photon field would dominate the one produced locally at the shocks and the expected line ratios would not fit the observed ones.

5 Conclusions

Although there is some evidence for a hidden quasar nucleus in Cygnus A, the luminosity of the nucleus appears to be low, contrary to the expectations from unified schemes. Cygnus A is the prototype of powerful radio galaxies. However, it presents some peculiarities that could be due to very particular conditions in this object. Barthel and Arnaud (1996) have argued that the high radio power is due to strong radiation losses in a dense environment rather than to an intrinsic high-luminosity quasar. Bahcall (these proceedings) has shown that quasars oc-

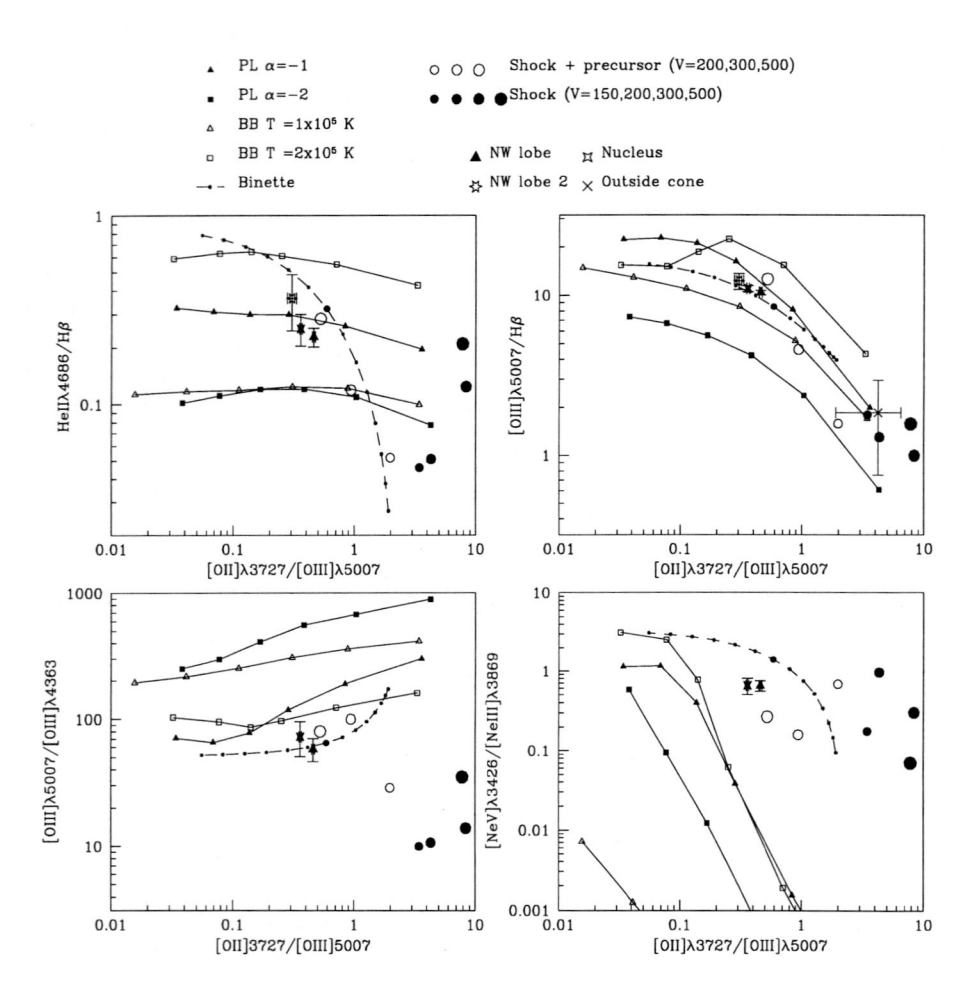

Fig. 2. Diagnostic diagrams for several line ratios. The meaning of the symbols is indicated on the top of the figure. The solid lines join the standard photoionization models with a separation in log U of 0.5 (U increases to the left). The dashed line represents the $A_{m/i}$ sequence ($0.063 \leq A_{m/i} \leq 25$), with the symbols separated by 0.2 dex in $A_{m/i}$ ($A_{m/i}$ increase to the left; the big dot corresponds to $A_{m/i} = 1$). The shock models are for a magnetic parameter of $2\mu G cm^{3/2}$. The observational data are from spectra extracted on the nucleus, the brightest region of the north west lobe (NW), another region of the north west lobe inside the ionization cone (about 1.8 arcsec from the nucleus) and a region outside the ionization cone south of the nucleus.

cur in a wide variety of hosts galaxies and environments. Even if Cygnus A is not representative of all types of radio quasars, from the detail study of this object we can learn about the processes that most probably ocurr in more distant radio quasars and radio galaxies. For instance, the ionization of the extended gas appears to be due to a combination of photoionization by the nuclear source, and high-velocity shocks associated with the jet. However, it is not clear if shocks are present in all types of extragalactic radio sources. It may well be that only sources where the radio jet interacts strongly with the interstellar medium of the host galaxy present observational features comparable with those of Cygnus A.

In order to test unification models for the hosts galaxies of FRII radio galaxies and radio loud quasars, a detail study, including spectroscopy of the extended line regions, of complete samples of both types of radio sources is needed. An important test to determine the importance of shocks in the ENLRs of radio galaxies and quasars is to look for differences in the profiles of lines produced in the precursor and those produced in the shocked region since the corresponding velocities are different.

References

Antonucci, R., Hurt, T., Kinney, A. 1994 Nature, **371**, 313.
Arnaud, K. 1996, in Cygnus A - Study of a Radio Galaxy, p. 51, eds. C.L Carilli and D.E. Harris, CUP.
Barthel, P. D., Arnaud, K. A. 1996, MNRAS, **283**, L45.
Barvainis, R., Antonucci, R. A. 1994, AJ **107**, 1291.
Binette, L., Wilson, A. S., Storchi-Bergmann, T. 1996 A&A, **312**, 365.
Cabrera-Guerra, F., Pérez-Fournon, I., Acosta-Pulido, J. A., Wilson, A. S., Tsvetanov, Z. I. 1996, in Cygnus A - Study of a Radio Galaxy, p. 23, eds. C.L Carilli and D.E. Harris, CUP.
Conway J. E., Blanco, P. R. 1995 ApJ **449**, L131.
Dopita, M. A., Sutherland, R. S. 1995 ApJ, **455**, 468.
Dopita, M. A., Sutherland, R. S. 1996 ApJS, **102**, 161.
McNamara, B. R., Jaffe, W. 1994 A&A, **281**, 673.
Morse, J. A., Raymond, J. C., Wilson, A. S. 1996 PASP, **108**, 426.
Penston, M. V., et al. 1990 A&A, **236**, 53.
Pérez-Fournon, I., Cabrera-Guerra, F., Burgos-Martin, J., Acosta-Pulida, J. A., Wilson, A. S., Tsvetanov, Z. 1996. Vistas in Astronomy, **40**, 51.
Perley, R. A., Carilli, C. L. 1996, in Cygnus A - Study of a Radio Galaxy, p.168, eds. C.L Carilli and D.E. Harris, CUP.
Tadhunter, C. N., Metz, S., Robinson, A. 1994 MNRAS, **268**, 989.
Ward, M. J. 1996, in Cygnus A - Study of a Radio Galaxy, p. 43, eds. C.L Carilli and D.E. Harris, CUP.

X-Ray Extended AGN?

Petra Nass[1,2]

[1] Max-Planck-Institut für extraterrestrische Physik, Garching, Germany
[2] Hamburger Sternwarte, Hamburg, Germany

Abstract. Normally, AGN appear as pointlike sources in X-ray surveys. However, some of the newly identified AGN from the ROSAT All-Sky Survey (RASS) are extended, similar to clusters of galaxies, which might be a hint for clusters around these AGN; thus we searched the RASS for such objects. We present and discuss the first results from these environmental studies of our sample of X-ray extended AGN.

1 Basic Idea

Clusters of galaxies and AGN are known to be powerful X-ray emitters, and because of this, X-ray surveys are a very efficient means to identify new objects. AGN are pointlike X-ray sources, whereas the X-ray emission coming from clusters of galaxies is extended.

1.1 Why Should We Search for AGN in Clusters?

First, the presence of AGN in clusters or groups is important, as galaxy-galaxy interactions or merging events could occur, interaction with the intracluster gas might be possible, accretion of gas can increase the luminosity, and the formation of the AGN's activity might be triggered by interaction. If the AGN is not a member of the cluster, but behind it, lensing effects could be expected. These processes might help to identify the poorly understood triggering mechanism that generates the AGN phenomenon.

Second, the aim of identification programs of surveys like for example X-ray surveys is to identify new objects as well as to derive number counts of different types of objects. Both have a strong influence on physical models that are used to explain the observational results. To recognize sources which belong to more than one type of object is very important. It is possible to miss an AGN in a cluster especially if it is a faint object or to overlook a cluster around a high-z AGN. But the identification programs mostly end up with a number density for the objects, presented as a log N-log S diagram. Now, an unknown AGN in (or behind) the cluster will significantly increase the amount of X-ray emission apparently coming from the cluster and, of course, this would also mean that this AGN is missing in the log N-log S for AGN from this survey. A wrong log N - log S distribution could then lead to wrong assumptions about the underlying physical model.

If we can prove that a certain class of AGN is always located in a cluster this might be an easy way to find clusters at high redshifts. Then, it would only be necessary to search for an AGN at a particular redshift to start with the search for the surrounding cluster (e.g. LeFevre et al. 1996).

2 The Project

During environmental studies of AGN from the RASS, we found some unusually extended X-ray counterparts. Their X-ray extension was similar to that detected from clusters of galaxies. Thus, we decided to search for AGN which appear as X-ray extended sources in the RASS. Our first AGN sample was taken from an identification program at the Hamburger Sternwarte where the RASS was correlated with an area of 6000 square degrees from the Hamburg Quasar Survey (HQS). Follow-up spectroscopy led to the discovery of several new AGN (Bade et al. 1995, Nass et al. 1996). Among the AGN or AGN candidates in this region, 120 sources with cluster-like extended X-ray emission were found. The advantage of this sample is that many groups are involved in different identification programs. Some are searching for AGN, others for clusters. Therefore, some of these X-ray sources are appearing in newly identified cluster samples, others in AGN samples, and some in both! This is a unique sample of double identifications, as most other identifications start with the exclusion of objects which are already identified.

Another approach was the correlation of the AGN from the Veron-Cetty & Veron catalogue (1996) with all 70 000 RASS II sources (Voges et al. 1996). 57 objects fulfilled our selection criteria for extension. This subsample has the advantage that a lot of data on these objects can be found in the literature.

3 The Observations and Results

The total candidate sample derived from both samples described above consists of all types of AGN, of different luminosities, and different redshifts. Candidates from the first sample are only located on the northern sky whereas the second sample covers the whole sky. We selected the most promising candidates for our observations according to their observability and their luminosity and discovered very interesting objects. Some of them will be presented in the figures and discussed in the figure captions.

X-ray observations were and will be done with the ROSAT HRI during AO6 and AO7 for about 20 objects. Many of our X-ray observed candidates are indeed BL Lac objects, since more of the BL Lac objects than of the other AGN passed the proposal committee. Except for one source (which was probably a detection error in the RASS), all sources were detected with the HRI. The X-ray emission was always centered on the AGN's position, taking into account that offsets of up to 10 arcsec between X-ray and optical images can be explained by attitude uncertainty. For almost all sources, we had problems with the standard source detection in EXSAS because of source confusion. Arbitrarily selected AGN from the ROSAT archive did not cause such problems and it should be checked again when other data sets from A06 become public. To correct for errors in the attitude of the satellite, we split the observations into the single intervals and aligned the single maps onto the X-ray peak. However, the extended component did not vanish. Subtraction of the ROSAT HRI point spread function (PSF) revealed an excess emission, containing about 40 percent of the original counts. Because of the attitude uncertainty, this should be checked in detail again. Further comparison with ROSAT HRI data from the archive will be done (e.g. to determine a star's psf for another subtraction). From the ROSAT archive, we can also obtain data for similar objects, because there are already a few known AGN-cluster combinations or candidates (e.g. Hall et al. 1995). Additionally, some of our objects were observed with the PSPC at an earlier date and we will also compare these data sets (Nass et al. 1997).

Optical spectra of the AGN and some direct images were obtained with the 2.2m and 3.5m telescopes at Calar Alto during the identification of the AGN (Bade et al. 1995, Nass et al. 1996) and direct imaging and spectroscopy of possible companions was also done during runs for this project (Nass et al. 1997). The direct images showed enhancements of galaxies around the candidates which are probably clusters of galaxies or possible groups around our observed candidates. The spectra obtained of some companions were mostly those of faint galaxies, except for one object whose companions proved to be foreground stars. Only one BL Lac object was found in our sample to lie behind a group or cluster while the other determined redshifts were the same as the AGN's z. The combination of our X-ray data with more optical data and data from archives will clarify the connection between an AGN and its potential companion or neighbours.

Some of the objects in this sample were included in Nass et al. 1996, and radio data were obtained for them at the VLA in May 1994. Radio catalogues were also used for this study.

4 Conclusions

By selecting extended AGN from the RASS, some extraordinary objects could be discovered like the ones mentioned above (Nass et al. 1997). The X-ray data from the RASS indicate possible extension of the X-ray emission which is confirmed by the ROSAT HRI data. For some sources, this extended part is not homogeneously distributed and further studies should be done to check for possible X-ray emitting sources near the AGN. We will now search for companions near BL Lac objects, because many of the BL Lac objects in our sample have possible neighbours visible on the direct plates.

The reliability of the ROSAT HRI PSF should be checked again before starting with attempts to separate and calculate the AGN's and the clusters' contributions to the total X-ray flux as measured with ROSAT. However, the first results of this study show that it is very difficult to conclude from the X-ray extension of an X-ray source whether a cluster, an AGN or both are present. There are some intermediate extended X-ray sources, where a cluster or an AGN might be the source of the X-ray emission and there are some X-ray sources maintained by two emitters: AGN in or behind clusters or groups of galaxies.

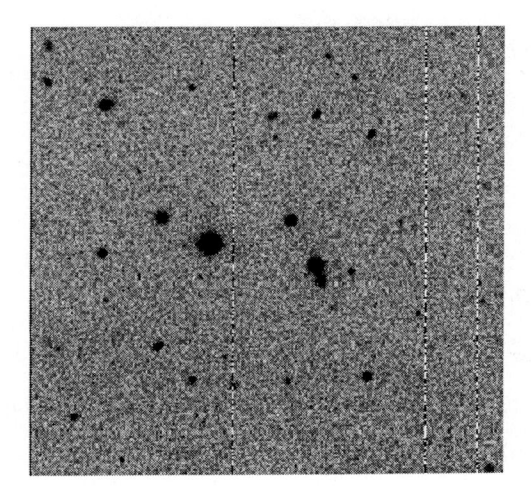

Fig. 1. Optical image from Calar Alto observation of RXJ0020+1034 (N to the top, E to the left; 1cm corresponds to 35 arcsec). We took three spectra of this object and all of them showed extremely broad double-peaked emission lines. Similar objects are only a handful of radio galaxies, but RXJ0020+1034 has no radio emission above the threshold of any currently existing catalogues. Today it is three magnitudes brighter than in 1950, but during the last 3 years its optical brightness remained the same - at least during our observations. The optical observations showed that a group is more likely than a cluster (one object southwards and faint object northwards). Recent X-ray observations are not delivered yet.

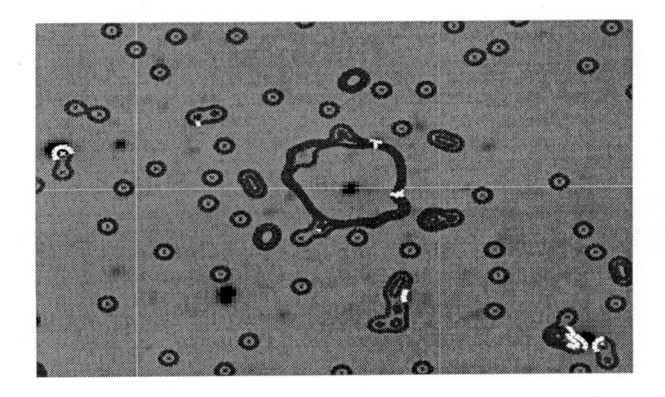

Fig. 2. Optical image of the BL Lac object RXJ0227+0201 from Calar Alto observation overlayed with X-ray contours (2,4,7 sigma above background) from ROSAT HRI observations; 1cm corresponds to 15 arcsec. We identified the X-ray source RXJ0227+0201 as a BL Lac object with z=0.45 whereas another group found a cluster at z=0.185. The contour image shows clearly that the X-ray emission peaks at the BL Lac's position.

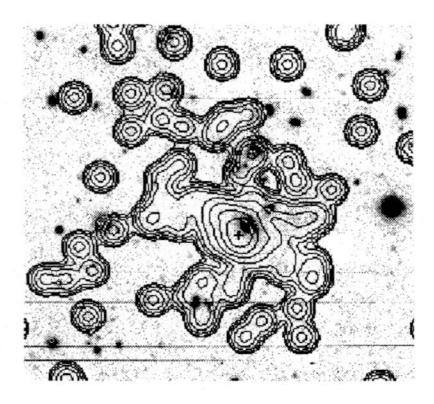

Fig. 3. Optical image of the BL Lac object RXJ0930+4950 from Calar Alto observation overlayed with X-ray contours from ROSAT HRI observations; 1cm corresponds to 30 arcseconds. For RXJ0930+4950 the extended X-ray component seems to be associated with companions visible on our deep direct image.

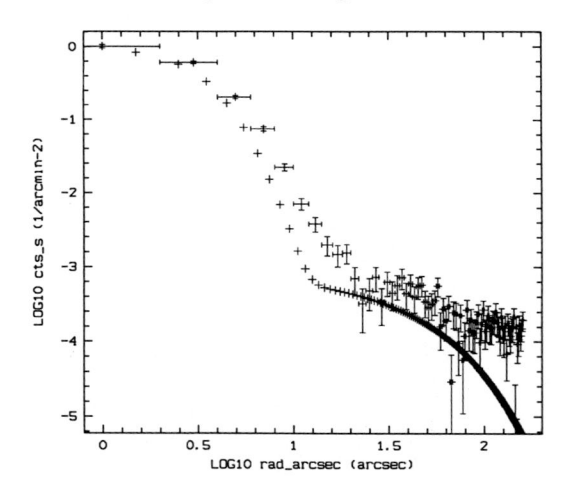

Fig. 4. Radial profile of RXJ0930+4950 and the HRI point spread function which was later on subtracted. After the subtraction, about 40 percent of the detected photons were left.

References

Bade, N., Fink, H.,Engels, D. et al.(1995): A&AS **110**, 469–511
Hall, P., Ellingson, E., Green, R., Yee, H. (1995): AJ,**110** 513–521
LeFevre, O.,Deltorn, J., Crampton, D., Dickinson, M.(1996): ApJ **471**, L11–L14
Nass, P., Bade, N., Kollgaard, R.I. (1996): A&A **309**, 419–430
Nass, P., Böhringer, H., Bade, N., Voges, W. (1997): in preparation
Veron-Cetty, M., Veron, B. (1996): *7th edition*
Voges, W., Boller, T., Dennerl, K. et al.(1996): MPE Report **263**,637–638

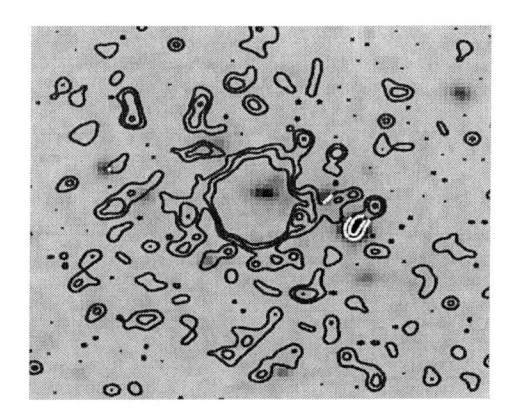

Fig. 5. Optical image of the BL Lac object RXJ1341+3959 from the digitized POSS overlayed with X-ray contours (2,4,7 sigma above the background) from ROSAT HRI observations; 1 cm corresponds to 20 arcseconds. RXJ1341+3959 is located in the Abell cluster A 1774.

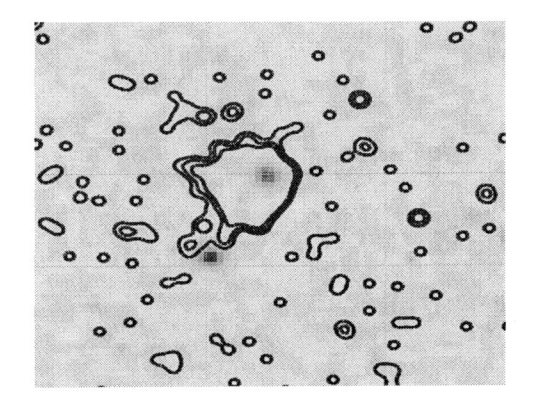

Fig. 6. Optical image of the BL Lac object RXJ1456+5048 from the digitized POSS overlayed with X-ray contours (2,4,7 sigma above background) from ROSAT HRI observations; 1cm corresponds to 35 arcseconds. RXJ1456+5048 has a neighbouring 207 mJy radio source (47 arcsec away) without visible or X-ray detected counterpart. The X-ray contour map of RXJ1456+5048 shows signs of an extension towards the radio source (see extension to NE). RXJ1456+5048 itself could not be detected in the radio, only an upper limit of 3.7 mJy could be determined for it.

A Search for Galaxies with a Variable Nucleus

Dario Trèvese[1], Matthew A. Bershady[2], and Richard G. Kron[3]

[1] Istituto Astronomico, Università di Roma "La Sapienza", via G.M. Lancisi 29, I-00161 Roma, Italy
[2] Department of Astronomy and Astrophysics, Pennsylvania State University, University Park, PA 16802, U.S.A.
[3] Fermi National Accelerator Laboratory, Box 500, Batavia, IL 60510, U.S.A.

Abstract. We have selected a sample of candidate AGNs with extended images on the basis of variability, to B=22.5 in Selected Area 57. Most of the objects have colors typical of AGNs. The brightest three of them have been observed spectroscopically and show emission lines typical of Seyfert I galaxies. Variability, as a detection criterion, allows us to fill a region of the redshift-luminosity plane containing intrinsically faint QSOs or high redshift Seyfert I galaxies.

1 Introduction

Quasar can be detected by various techniques. In the past we have used non-stellar colors, lack of proper motion and variability to detect quasars to faint limits ($B = 22.5$) in the Selected Area 57 (Kron and Chiu 1981, Koo, Kron, and Cudworth 1986, Koo and Kron 1988, Trèvese et al. 1989 [T89], Trèvese et al. 1994 [T94]). As shown in T89 and T94, the QSO sample selected by variability strongly overlaps the the color selected of SA 57 sample placing limits on the incompleteness of color-selection technique (see also Majewski et al. 1991). The analysis of T89 and T94 was restricted to point-like sources. An extension of the variability criterion to diffuse objects is highly desirable because: i) the color-criterion cannot be applied to diffuse objects since most galaxies appear as non-stellar in color space; ii) the distinction between QSOs and galaxies with an active nucleus is arbitrary both in terms of absolute luminosity of the nucleus and in terms of visibility of the host galaxy; iii) the visibility of the host galaxy depends not only on the intrinsic nucleus-to-host luminosity ratio, but also on the seeing, on the detector, on the observing band and thus on the redshift; iv) the cosmological evolution of the intrinsically faint part of the QSOs luminosity function is particularly interesting and poorly known; v) last but not least, the identification of complete samples of QSOs with detectable host galaxies would provide a basis for statistical studies of the hosts and their relation to the active nucleus. In the following we present a sample of diffuse objects selected in SA 57 on the basis of their variability.

2 Data and Detection Procedure

The data are derived from a homogeneous collection of 9 IIIa-J plates obtained at the prime focus of the Mayall 4-m telescope at Kitt Peak National Observatory

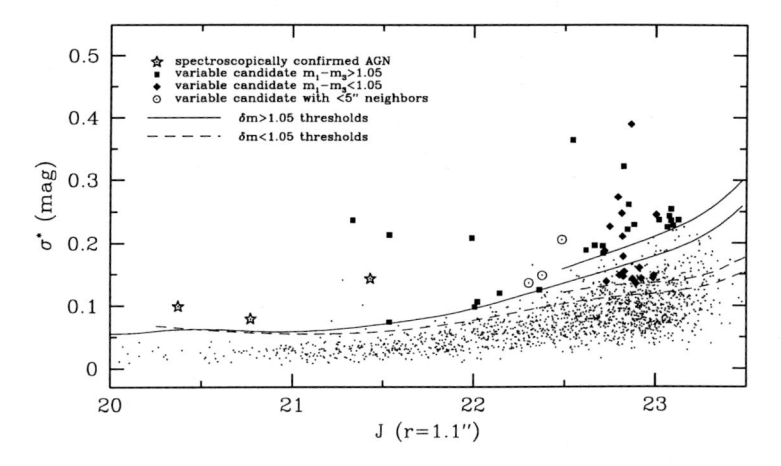

Fig. 1. Weighted root-mean-square deviation of the magnitude respect to the mean of the 9 observations, as a function of the B_J magnitude m_4. Solid and dashed lines represent $\sigma_N^* = 2.5$ and $\sigma_N^* = 3.5$, for objects with concentration $\delta > 1.05$ and $\delta < 1.05$ respectively. The lower threshold is adopted for $m_4 < 22.4$ to define the primary sample. The higher threshold is adopted for $m_4 > 22.4$ to define the secondary sample.

during 11 years. Microphotometric plate scanning, automatic object detection, classification and photometry are described in T89 and T94. The main difference with the previous analysis is that the aperture which provides the optimal signal-to-noise ratio depends on both the apparent magnitude and the compactness of the object. Thus we measure magnitudes in different apertures m_1, m_2, m_3, m_4 (from the central 9 pixels to 1.1 arcsec) and we use m_4 and $\delta \equiv m_1 - m_3$ as a measure of the apparent brightness and image concentration respectively. The parent sample considered in the present analysis does not contain the point like sources analyzed in T89 and T94, and consists of 1637 objects. As a variability index σ^* we adopt, for each object, a weighted r.m.s. deviation around the mean value of the light curve. The plot of σ^* versus m_4 for the entire parent sample of 1636 objects is shown in Figure 1. For most of the objects, which are not variable, σ^* is simply determined by the photometric noise. The ensemble average $\langle\sigma^*\rangle$ of the variability index, computed as a function of m_4 and δ, is a measure of the photometric accuracy and represents a bias in the variability estimate, while the uncertainty on σ^* is measured by its the r.m.s. deviation $\sigma(\sigma^*)$. Thus we can define a normalized variability index $\sigma_N^* = (\sigma^* - \langle\sigma^*\rangle)/\sigma(\sigma^*)$ and select as variable the objects with σ_N^* exceeding a given threshold. Since the dependence of the σ^* distribution on δ is weak, we simply split the sample in two bins: $\delta < 1.05$ and $\delta > 1.05$. We consider a prime sample selected according to the condition $\sigma_N^* > 2.5$ for $m_4 < 22.5$. A secondary sample is defined for $m_4 > 22.5$ using a threshold $\sigma_N^* > 3.5$.

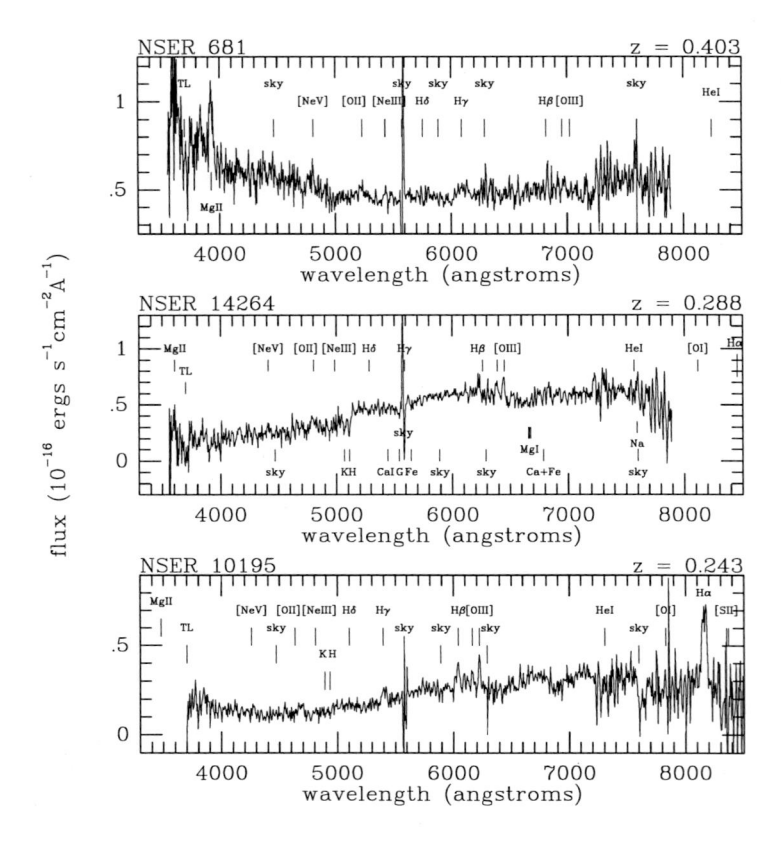

Fig. 2. Spectra of three variable galaxies, obtained with Nessie fiber spectrograph at KPNO 4m telescope. The AGN nature of these objects is confirmed by the presence of broad emission lines of $MgII$ for NSER 681 and NSER 14264, and H_α for NSER 10195. The object NSER 681 was also observed with the GoldCam spectrograph at the KPNO 2.1m telescope (Bershady Trèvese and Kron 1991) and shows also $H|beta$ and $[OIII]$ lines.

3 Results

The selected objects are marked with different symbols in Figure 1. The prime sample consists of 14 objects. Considering the different selection technique, this number can be considered roughly consistent with the 24 of Seyfert I galaxies expected at the same magnitude limit on the basis of the CfA counts of AGNs (Burg 1987). The secondary sample contains 34 objects. For three bright objects we have spectra, from KPNO 4m Nessie fiber spectrograph, shown in Figure 2, confirming their AGN nature. In Figure 3 the colors, taken from the photometric catalog of Koo (1986), are plotted as a function of m_4. On average the objects of the primary sample are bluer than the parent population as it would be expected

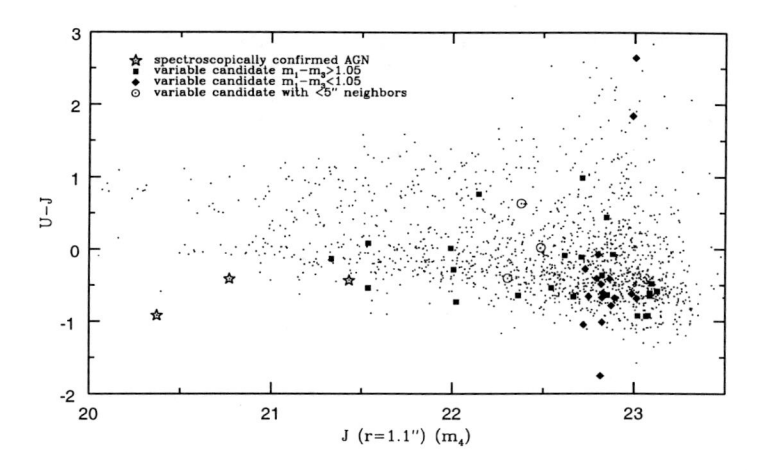

Fig. 3. $U - B_J$ color for the same sample of Figure 1.

for AGN. Most of the objects of the primary sample are likely to be bona fide AGNs. At fainter magnitudes the color distribution is not clearly distinct from the parent population. This could mean that either the secondary sample is less reliable or that the bluer nucleus is less dominant in fainter Seyfert galaxies. However, an analysis of the distribution of the secondary sample in the $U - B_J$ vs. δ plane shows that also the secondary sample contains a disproportionate number of objects with blue and compact images (see Bershady, Trèvese and Kron 1997) and could be useful to select candidates for spectroscopic verification.

References

Bershady, M.A., Trèvese, D., Kron, R.G. (1991): *Variability of Active Galactic Nuclei* eds. Miller, H.R., Wiita, P.J. (Cambridge University Press), 67
Burg, R. (1987): thesis, Massachussets Institute of Technology
Koo, D.C. (1986): ApJ **311**, 651
Koo, D.C., Kron, R.G., Cudworth, K.M. (1986): PASP **98**, 285
Koo, D.C., Kron, R.G. (1988): ApJ **325**, 92
Kron, R.G., Chiu, L.T. (1981): PASP **93**, 397
Majewski, S.R., Munn, J.A., Kron, R.G., Bershady, M.A., Smetanka, J.J., Koo, D.C. (1991): *Workshop on the Space Distribution of Quasars* ed. D. Crampton (Provo: ASP), 21, 55
Trèvese, D., Pittella, G., Kron, R.G., Koo, D.C., Bershady, M. (1989): AJ **98**, 108 (**T89**)
Trèvese, D., Kron, R.G., Majewski, S.R., Bershady, M.A., Koo, D.C. (1994): ApJ **433**, 494 (**T94**)

Proper Motions in the Center of the Galaxy

A. Eckart and R. Genzel

Max-Planck Institut für extraterrestrische Physik, 85740 Garching, Germany

Abstract. We report the first measurements of stellar proper motions in the innermost core of the Galaxy. From high resolution near-infrared imaging over the last 4 years we have determined proper motions for 39 stars between 0.03 and 0.3 pc from the compact radio source SgrA*. Proper motion and radial velocity dispersions are in very good agreement indicating that the stellar velocity field on average is close to isotropic. We have detected significant changes in the structure of the innermost complex of stars in the immediate vicinity of SgrA*, suggesting stellar motions of $>10^3$ km/s within 0.01 pc of the compact radio source. Taking radial and proper motion data together we find a $2.45(\pm 0.4) \times 10^6$ $M_\odot$ central dark mass within < 0.015 pc of SgrA*. The density of this central dark mass therefore is in excess of 10^{12} $M_\odot \text{pc}^{-3}$, implying that the central mass concentration is likely a single massive black hole.

1 Introduction

Probably the most unambiguous way to answer the question of whether or not galactic nuclei contain central massive black holes is from dynamical measurements (see Kormendy and Richstone 1995 for a review). The result of these investigations is of considerable importance for the understanding of active galactic nuclei and the evolution of galaxies. The Galactic center is a unique target for searching for a central massive black hole as it is so close (8 kpc, 1"≈0.039 pc, Reid 1993) and detailed measurements of gas and stellar dynamics are now possible down to a scale of about 0.01pc. In a recent study of the radial velocities of 222 stars within 1 pc from the dynamic center, Genzel et al. (1996 see also references therein) found that a central compact (core radius <0.06pc) dark mass of 2.2 to 3.2×10^6 $M_\odot$ (for a distance of 8.0 kpc) is required with fairly high significance if the stellar motions are isotropic. For determining the mass distribution from stellar dynamics without the a priori assumption of isotropic motions, it is necessary to determine both radial and proper motions of the stars. In contrast to external galaxies stellar proper motions in the Galactic center can realistically be measured in a few years time if relative stellar potions can be determined to about 10 milliarcseconds (mas). To achieve this goal we have been carrying out since 1991 a program of high resolution 2.2μm (K-band) imaging at the 3.5 m New Technology Telescope (NTT) of the European Southern Observatory (ESO) in La Silla, Chile. The first results of this program based on 7 epochs of images spanning four years have been summarized in detail in Eckart & Genzel (1996, 1997).

2 Observations and Analysis

Using a high resolution camera developed specifically for this purpose (SHARP, Hofmann et al. 1993), we have applied speckle imaging techniques to obtain diffraction limited resolution. The observations were carried out at the ESO 3.5m NTT at a wavelength of 2.2μm. Techniques and results from the near-infrared imaging have been discussed in several earlier papers (Eckart et al. 1992, 1993, 1994, 1995). For the present study we analysed 65 independent speckle image reconstructions (data sets) from 7 observing runs in 1992.25, 1992.65, 1993.65, 1994.27, 1995.6, 1996.25 and 1996.43. For determining accurate relative stellar positions with our imaging system (to 1/20..1/10 of the diffraction limited beam) first - for each data set - a final PSF was constructed by averaging the five brightest stellar images. Then relative pixel offsets from IRS16NE were determined for each data set from cross correlating this PSF and the data in the central 5x5 pixels around the peak of each star. For single stars the dispersions between different data reduction methods range between 3 and 20 mas.

In the second step all instrumental imaging parameters up to second order were extracted for each individual data set with respect to a list of N reference-stars taken from images of our 1994 epoch. A detailed description of this procedure is given in Eckart & Genzel (1997). From this analysis we found that the fitted pixel scales and the very small second order distortion parameters (0.001 of the first order parameters) differed by <0.5% between different data sets so that the main fit parameters were the base position of the N stars and the camera rotation angle. After correction for the instrumental parameters the final position fit errors ranged between 8 and 20 mas per data set for the brighter, isolated stars. These formal fit errors are in fact a good representation of the experimental (non-systematic) errors. This can be seen from Fig.1 which shows the final derived relative x- and y- positions of 5 stars as a function of time. These coordinates correspond approximately to RA and Dec. Each black dot in these panels represents an individual independent data set. The scatter in the dots for a given epoch is an empirical measure of the experimental position uncertainties. They agree quite well with the fit errors giving confidence in our data reduction method. The proper motion solutions turned out to be very stable in terms of selection of the N reference stars, as long as N>>20 to 30 >N_{min}=6. Almost independent from the reference-star-list the systematic errors in our velocity estimates are typically ±60 km/s.

3 Results

In the present analysis we have determined positional changes as a function of time for 39 stars between 0.9" and 8.8" and 11 stars <0.9" from SgrA*. Fig.1 shows the proper motion vectors of stars in the central few arcseconds, superposed on a 2μm image. Our data show that the relative positions of many of the stars in the central 0.1pc appear to change monotonically with time. Of the 39 stars motions at the >5σ level are detected in at least one coordinate for 10 cases and at the >4σ level for a total of 19 cases.

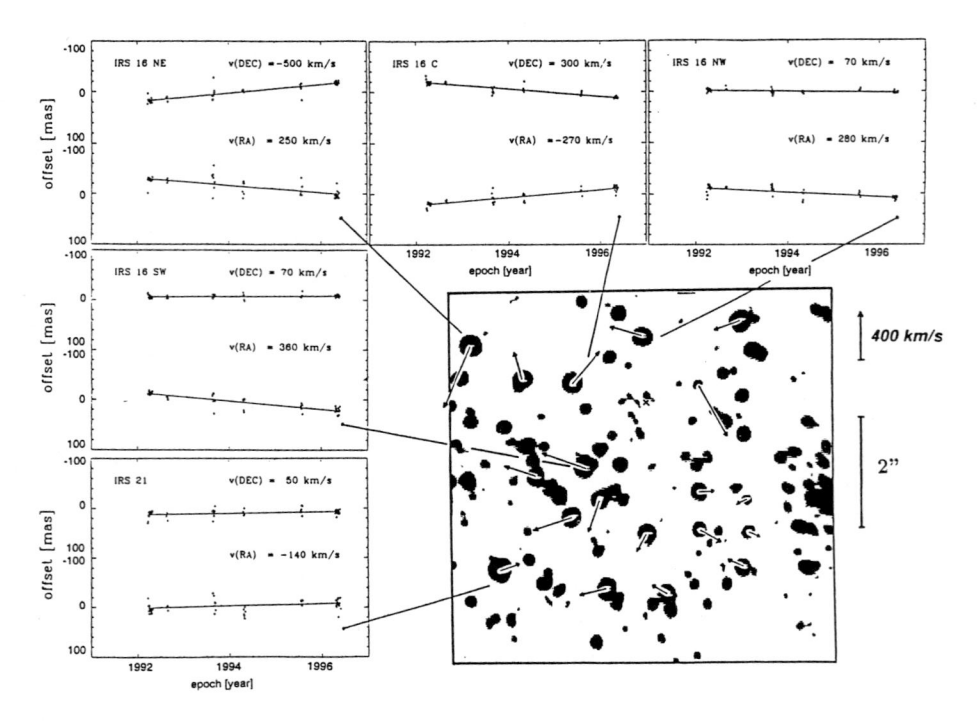

Fig. 1. Proper motion of stars in the central few arcseconds.

A simple consideration shows that the motions we have detected cannot be due to orbital motion in multiple star systems that are not resolved at 0.15" resolution. It can also be shown that the motions in the very center are not due to gravitational lensing (Eckart & Genzel 1997). We therefore conclude that we have detected orbital proper motions of the stars in the central gravitational field. An immediate important conclusion is that the velocity dispersion is very similar in all three coordinates and hence any anisotropy of the stellar motions must be small. This can be best seen from Fig.2 where we have plotted proper motion and radial velocity dispersions (from Rieke and Rieke 1988, McGinn et al. 1989, Sellgren et al. 1990, Lindqvist et al. 1992, Genzel et al. 1996, Haller et al. 1996) as a function of projected radius p from SgrA*. Taken together the measurements for p>0.053 pc now show a very significant Keplerian ($s(p) \propto p^{-0.5}$) falloff of the stellar velocities with distance from the dynamic center (assumed to be at SgrA*), demonstrating that there must be a central, compact and dark (cf. Genzel et al. 1996) mass concentration. We find that the late type stars that are projected within a few arcseconds of SgrA* have consistently smaller proper motions (and radial velocities) than the early type stars (He I stars) in the same area. This is consistent with the well established fact (Sellgren et al. 1990, Haller et al. 1996, Genzel et al. 1996) that the distribution of brighter late type stars shows a central hole.

The proper motion data thus strongly suggest that the dark mass concentration is centered within <0.009 pc on the compact radio source SgrA*. The SgrA* location relative to the near-infrared stellar distribution has recently been determined by Menten et al.(1996) to ±30 mas. SgrA* lies approximately at the center of a hammer-shaped concentration of fainter stars (SgrA*(IR), Eckart et al. 1995). In all our best images we obtain between 1994 and 1996 we were able to resolve the central SgrA*(IR) cluster into about a dozen components. The same structure has recently been confirmed with the Keck 10m telescope (Klein et al. 1996) and via adaptive optics with the CFHT (R. Doyon, privat com.). A close inspection of our images shows that those from the two epochs in 1996 are very similar while there are fairly obvious structural changes between 1994 and 1996. The morphology of the 1995 epoch maps ranges somewhere between those of the 1994 and 1996 data. Several of the stars in SgrA*(IR) appear to move with total proper motions of 1000 to 1600 km/s (Fig.3). The most convincing case is S1 (relative to S2/S3) which is within 0.2 (0.008 pc) of SgrA*. S1 appears to move with 1600±400 km/s with respect to the overall base position and with 1950±500 km/s relative to the centroid of S1/S2. For the SgrA*(IR) cluster we derived a velocity dispersion of 560±90 km/s at p=0.01 pc (corrected for measurement bias).

We fit to the projected velocity dispersion data in Fig.2 a model with a central point mass plus an extended isothermal cluster of dispersion at large radii of 50 km/s (providing an excellent fit to stellar mass data at projected radii p greater than a few parsecs, see Genzel et al. (1996) for details). Proceeding as in Genzel et al. (1996) we have also calculated enclosed masses from the Bahcall-Tremaine and virial theorem projected mass estimators. Jeans equation mass modelling as in Genzel et al.(1996) of the combined radial and proper motion data at projected radii of p>0.053 pc also is best fit by a combination of a dark point mass plus the extended cluster of stars radiating in the near-infrared. This is shown in Fig.5 in Eckart & Genzel (1997) where we plot the derived enclosed masses as a function of true radius R, along with two mass models.

The best fit is obtained as a combination of a 2.45±0.25 (statistical) ±0.4 (total) × 10^6 $M_\odot$ point mass and a visible (isothermal) stellar cluster of core radius 0.38 pc and a core stellar density of 4x10^6 $M_\odot pc^{-3}$, representing the stellar cluster radiating in the near-infrared with a M/L($2\mu m$) ratio of about 2 (Genzel et al. 1996). Mass modelling with a dark cluster shows that its density then would have to be >10^{12} $M_\odot pc^{-3}$ and its core radius <0.0062 pc (Fig.5 in Eckart & Genzel 1997). The second one is the combination of the same visible stellar cluster with a second, dark cluster. To fit the data, the central dark cluster must have a central density of at least 6.5x10^9 $M_\odot pc^{-3}$ and a core radius less than 0.035pc.

4 Conclusions

The Galactic center contains a dark central mass concentration with densities above 10^{12} $M_\odot pc^{-3}$. A similarly high mass concentration has been found for the

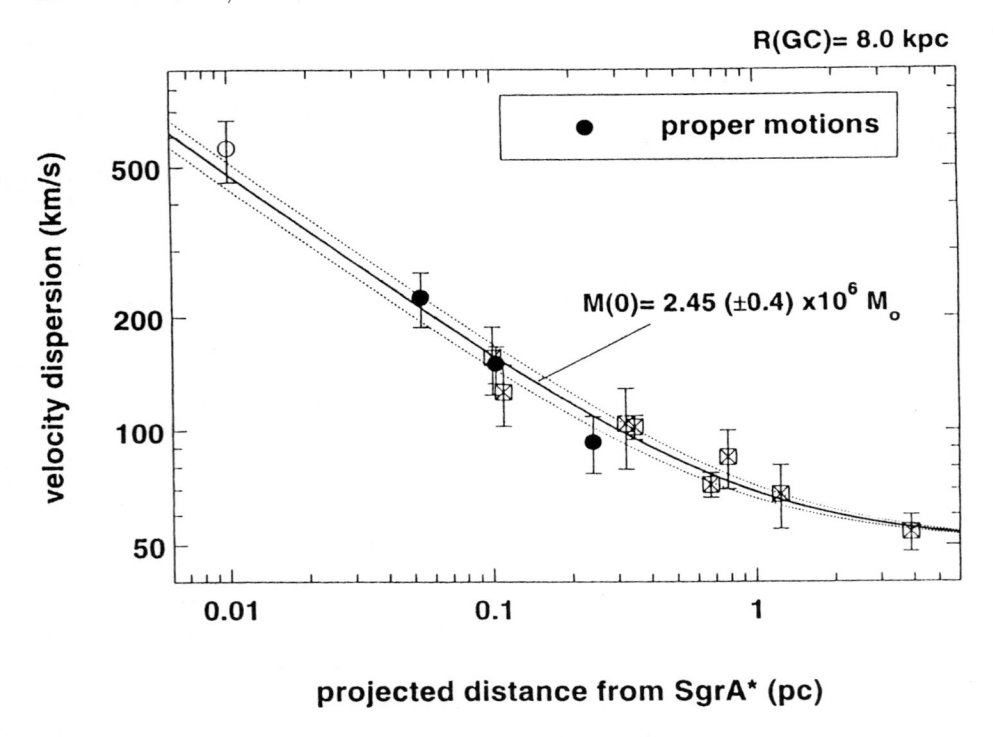

Fig. 2. Proper motion and radial velocity dispersions as a function of projected radius p from SgrA*.

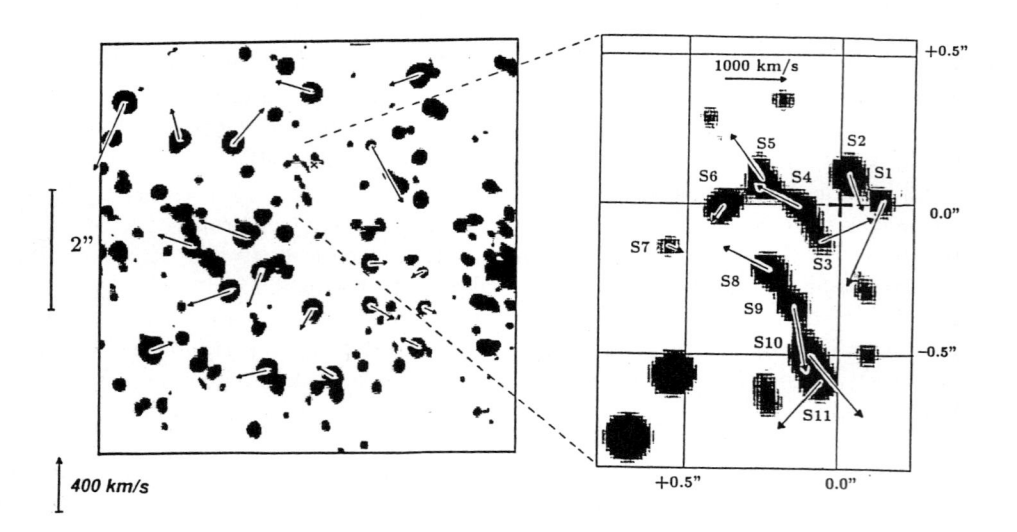

Fig. 3. Proper motions in the SgrA*(IR) cluster.

mega-maser galaxy NGC 4258 (Greenhill et al. 1995, Myoshi et al. 1995). What is the nature of this dark mass? Given that the dark mass in the Galactic center is now constrained to be at least ten times more concentrated and therefore one thousand times denser than the visible stellar cluster we can now exclude with some confidence that it consists of a cluster of solar mass remnants (neutron stars or white dwarfs, see the detailed discussion in Genzel et al. 1996). Considering the stellar dynamics outside R$\approx$0.05 pc, the remaining possibilities are a core collapsed cluster of 10-20 $M_\odot$ stellar black holes (Morris 1993, Lee 1995), or a single massive black hole, or a combination thereof. The proper motion data on the SgrA*(IR) cluster indicate that this must be a massive black hole.

Acknowledgements: A number of people at MPE and ESO have been involved in making this experiment possible and carrying it out. We would like to especially thank N.Ageorges, S.Drapatz, R.Hofmann, A.Krabbe, B.Sams, L.E.Tacconi-Garman and H.van der Laan.

References

Eckart, A., Genzel, R., 1996, Nature Vol. 383, 415.
Eckart, A., Genzel, R., 1997, M.N.R.A.S 284, 576.
Eckart, A., Genzel, R., Krabbe, A., Hofmann, R., van der Werf, P.P. and Drapatz, S. 1992, Nature 355, 526
Eckart, A. Genzel, R., Hofmann, R., Sams, B.J. and Tacconi-Garman, L.E. 1993, Ap.J. 407, L77
Eckart, A. , Genzel, R., Hofmann, R., Sams, B.J., Tacconi-Garman, L.E. and Cruzalebes, P. 1994, in The Nuclei of Normal Galaxies, eds. R.Genzel and A.I. Harris, Kluwer (Dordrecht), 305
Eckart, A. Genzel, R., Hofmann, R., Sams, B.J. and Tacconi-Garman, L.E. 1995, Ap.J. 445, L26
Genzel, R., Thatte, N., Krabbe, A., Kroker, H. and Tacconi-Garman, L.E. 1996, Ap.J. in press
Greenhill, L. et al. 1995, Ap.J. 440, 619
Haller, J.W., Rieke, M.J., Rieke, G.H., Tamblyn, P., Close, L. and Melia, F. 1986, Ap.J. 456, 194
Hofmann, R., Blietz, M., Duhoux, P. , Eckart, A., Krabbe, A. and Rotaciuc, V. 1993, in Progress in Telescope and Instrumentation Technologies, ed. M.H. Ulrich, ESO Report 42, 617
Klein, B.L., Ghez, A.M., Morris, M., Becklin, E.E., 1996, in: The Galactic Center ASP Conference series, Vol. 102, 1996, Roland Gredel, ed.
Kormendy, J. and Richstone, D. 1995, Ann.Rev.Astr.Ap.33, 581
Lee, H.M. 1995, Mon.Not.Roy.Astr.Soc. 272, 605
Lindqvist, M., Habing, H. and Winnberg, A. 1992, Astr.Ap. 259, 118
McGinn, M.T., Sellgren, K., Becklin, E.E. and Hall, D.N.B. 1989, Ap.J. 338, 824
Menten, K.M., Eckart, A., Reid, M.J. and Genzel, R. 1996, Ap.J. submitted
Morris, M. 1993, Ap.J. 408, 496
Myoshi, M., Moran, J.M., Hernstein, J., Greenhill, L., Nakai, N., Diamond, P. and Inoue, M. 1995, Nature 373, 127
Reid, M. 1993, Ann.Rev.Astr.Ap. 31, 345
Rieke, G.H. and Rieke, M.J. 1988, Ap.J. 330, L33
Sellgren, K., McGinn, M.T., Becklin, E.E. and Hall, D.N.B. 1990, Ap.J. 359, 112

The Seyfert Nucleus in the S0 Galaxy NGC 5252

Jose Acosta–Pulido[1], Baltasar Vila–Vilaró[2], and Ismael Pérez–Fournon[3]

[1] ISO Science Operations Centre, Astrophysics Division of ESA, Vilspa, Apdo. Correos 50727, E-28080 Madrid, Spain
[2] Nobeyama Radio Observatory, Nagano 384-13, Japan
[3] Instituto de Astrofisica de Canarias, E-38200 La Laguna, Spain

Abstract. We present a spectroscopic study (Acosta–Pulido *et al.*, 1996) of the S0 galaxy NGC 5252, which harbours a Seyfert 2 nucleus. Our spectra sample the nucleus and part of the extranuclear emission line regions, and for all cases we find emission–line ratios characteristic of photoionization by an AGN–like continuum source. The emission varies strongly with the distance from the nucleus. Very close to the nucleus, two different kinematical components are found, plus a broad Hα component. This last component is indicative that NGC 5252 contains a partially hidden Seyfert 1 nucleus. Outside the nucleus the velocity field of the gas is very complex, and cannot be explained as mere rotation in a disk.

1 Overview

The influence exerted by an active nucleus on the interstellar medium of the host galaxy has provided very useful information about the innermost structure of the nucleus. One of the best examples comes from the detection of cones of extended very highly ionized gas. Those cones are most easily recognized in early type galaxies, where no contamination by normal HII regions is expected. NGC 5252 appears like a typical S0 galaxy in continuum light. However it shows a *biconical morphology* in the emission line [OIII]λ5007 image, showing a characteristic shell morphology (Tadhunter & Tsvetanov, 1989). The subarcsecond structure is dominated by the nucleus and two bright knots along PA 35° as seen in the Hα+[NII] HST image (Tsvetanov *et al.*, 1996). Radio emission is formed by two weak jets which closely align with the conical structure (Wilson & Tsvetanov, 1994). Neutral hydrogen is detected outside the ionized gas (Prieto & Freudling, 1996), suggesting that both components may form a single structure.

2 Observations and Data Processing

We have carried out long–slit spectroscopy at two different positions: one through the nucleus aligned with the axis of the ionization cone, and the other at a distance of 15" SE of the nucleus but oriented perpendicularly w.r.t. the previous position. We have used the ISIS double spectrograph on the 4.2m WHT. Details of the observations and the data processing can be found in Acosta–Pulido *et al.*

(1996). We were able to measure very weak emission lines in the nucleus, after subtraction of the dominant stellar emission. Outside the nucleus the spatial variation of several emission lines was traced.

3 Results

3.1 Gas Excitation

The source of gas ionization was investigated by using diagnostic diagrams: the line ratios were typical of AGN–like ionizing continuum for all the regions investigated. A clear distinction between the nuclear and extranuclear emission was found, which is attributed to a different population of emitting clouds: optically thick clouds dominate at the nucleus, whereas outside the optically thin ones do. This hypothesis is further supported by a correlation between the line brightness and gas ionization and good agreement with theoretical modelling of the ENLR using clouds of different optical depth.

3.2 The Nucleus

The emission lines appear resolved very close to the nucleus. We can decompose the line into the following components:

- Two relatively narrow components (FWHM$\sim$ 270 km/s) separated by 200 km/s, which may be related to the bright knots present in the subarcsecond structure observed in Hα (Tsvetanov et al., 1996).
- A broad Hα component (FWHM$\sim$ 2485 km/s). No similar component was detected for Hβ, albeit the spectral resolution was lower at this wavelength range.

The presence of a hidden Seyfert 1 nucleus is confirmed by the detection of broad Hα in the nuclear spectrum. Furthermore, the nuclear luminosity needed to produce the observed ENLR luminosity is at least 10 times higher than that observed directly.

3.3 Gas Kinematics

Outside the nucleus the gas velocity is roughly antisymmetric about the nucleus ($\Delta v \simeq 200$ km/s), and tends to be sistemic in the outer parts. There are velocity gradients along the shells and no smooth trend between consecutive ones. This argues strongly against a simple planar geometry of the gas, which has been confirmed by recently obtained HI maps (Prieto & Freudling, 1996).

References

Acosta–Pulido J.A., et al.(1996), *ApJ,* **297**, 621.
Prieto M.A., & Freudling W. 1996, *MNRAS,* **279**, 63.
Tadhunter C.V., & Tsvetanov Z.I. (1989), *Nature,* **341**, 422.
Tsvetanov Z.I., et al.(1996), *ApJ,* **458**, 172.
Wilson A.S., & Tsvetanov Z.I. (1994), *AJ,* **107**, 1227.

NGC 3393 — Broad(er) Lines in a Nearby Seyfert 2 Galaxy

Andrew Cooke[1], Jack Baldwin[2], Gary Ferland[3], Hagai Netzer[4], Bev Wills[5], and Andrew Wilson[6]

[1] Institute for Astronomy, University of Edinburgh, Blackford Hill, Edinburgh EH9 3HJ, United Kingdom
[2] Cerro Tololo Interamerican Observatory, Casilla 603, La Serena, Chile
[3] Department of Physics and Astronomy, University of Kentucky, Lexington, KY 40503, USA
[4] Department of Physics and Astronomy, Tel Aviv University, Tel Aviv 69978, Israel
[5] McDonald Observatory and Astronomy Department, University of Texas at Austin, Austin, TX 78712, USA
[6] Department of Astronomy, University of Maryland, College Park, MD 20742, USA

Abstract. The Narrow-Line Region (NLR) of the Seyfert 2 galaxy NGC 3393 is dominated by a symmetric structure which appears as S-shaped arms in HST images. These arms apparently border a linear, triple-lobed radio source. Ground-based long-slit and Fabry–Perot imaging spectroscopy suggests that there is both a narrow and a broad component to the optically emitting gas. The broader line and general structure appear to be influenced by the radio outflow, but the gas ionization is probably dominated by radiation from an anisotropic central source.

1 Introduction

NGC 3393 is one of the nearest (v=3700 km/sec) and brightest Seyfert 2 galaxies after NGC 1068 with a spatial scale of $180h^{-1}$ pc/arcsec.

2 Data

HST PC images reveal a small-scale (3″) S-shaped emission-line region with high excitation. This is very similar to Mrk 573, in which the emission regions are interpreted as bow-shocks around radio lobes (Capetti et al. 1996).

A VLA radio map at 3.6cm shows typical triple-lobed structure. If the central lobe is assumed to be coincident with the optical nucleus, the other two lobes fit neatly within the S-shaped arms. This does, how ever, require an unusually large shift ($\sim 2.5''$) of the HST coordinate system.

Ground-based long-slit spectroscopy gives temperature and density measurements across the galaxy. In the central region it is possible, with our highest resolution data, to fit a two component model (broad and narrow lines) to the data.

Imaging Fabry-Perot spectroscopy with the CTIO 4m telescope shows that the inner region is rotating as a solid-body, in a plane markedly different from the outer regions. We fit a two component model (justified by the ground-based long-slit spectroscopy) to these data: a narrow component described by solid body rotation with a fitted FWHM of 190 km/sec; a broad component divided into three regions (NE, centre and SW) each having its own velocity and width. The intensity maps for the narrow and broad components are shown in relation to the HST image (Fig. 1).

3 Discussion

The low temperatures of the optically emitting gas, measured from the long-slit spectroscopy, indicate that the NLR gas must be mainly photoionized, but not whether the the ionizing photons come from a central source or from shock fronts. It is possible that the broader component arises in the turbulent interface between a radio outflow and the surrounding galactic gas.

We find that while the kinematics and general structure appear to be heavily influenced by a radio outflow, the gas ionization is probably dominated by radiation from an anisotropic central source. We expect a paper, covering this work in much greater detail, to be submitted by mid February 1997.

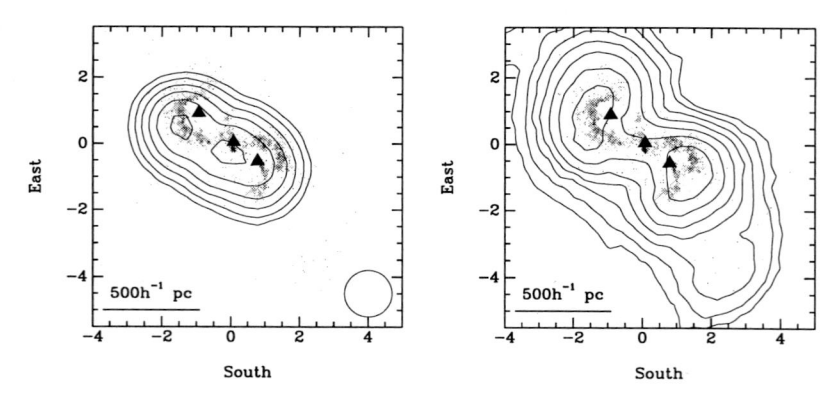

Fig. 1. Broad (left-hand panel) and narrow (right-hand panel) [N II] components from the Fabry Perot data, shown as contours superimposed on the HST [O III] image. The circle shows the FWHM resolution of the Fabry Perot data. Contour levels are 0.088, 0.125, 0.18, 0.25, 0.35, 0.5 and 0.71 of the peak flux in the narrow component. Triangles mark the positions of the three radio components.

References

Capetti, A., Axon, D. J., Macchetto, F., Sparks, W. B., & Boksenberg, A. 1996, ApJ, 469, 554.

AGNs with Composite Spectra

A.C. Gonçalves[1,2], P. Véron[1] and M.-P. Véron-Cetty[1]

[1] Observatoire de Haute-Provence (CNRS), F-04870 St. Michel l'Observatoire, France
[2] Centro de Astrofísica da Universidade do Porto, Rua do Campo Alegre 823, 4150 Porto, Portugal

Abstract. The use of diagnostic diagrams allows the unambiguous classification of the nuclear emission line regions of most galaxies; however, a small fraction of them have a "transition" spectrum, hard to classify. Spectral observations (3.4 Å resolution) of 15 "transition" objects around the Hα and/or Hβ emission lines show that most of these spectra are in fact "composite", due to the simultaneous presence on the slit of a Seyfert/Liner nucleus and a HII region.

1 Introduction

Several combinations of easily-measured emission lines can be used to separate emission-line galaxies into one of three categories according to the principal excitation mechanism: nuclear HII regions, Seyfert 2 galaxies and Liners. The use of such diagnostic diagrams generally yields an immediate classification of the nuclear emission line clouds[1],[2]. "Transition" objects exist however, which cannot be classified unambiguously from their line ratios; when observed with sufficient spectral resolution, such objects show different profiles for the permitted and forbidden lines[3],[4], this being due to the fact that the line ratios and line widths are different in nuclear HII regions, Liners and Seyferts and that two types of nebulosities may be simultaneously present on the slit. If all "transition" spectra are in fact "composite" spectra, rather than resulting from peculiar excitation mechanisms, the regions occupied by Liners and Seyfert 2 clouds in the diagnostic diagrams will have a smaller area, restricting the range of possible physical parameters in these clouds.

2 Observations and Data Analysis

In order to verify this, we have chosen 15 emission line galaxies suspected to have "transition" spectra. Spectroscopic observations were carried out around the Hα and/or Hβ emission lines with CARELEC (66Åmm^{-1} dispersion) at the OHP 1.93m telescope. In each case, the galaxy nucleus was centered on the slit (slit width: 2.1") and usually $\sim$ 5" were extracted. The spectral resolution was 3.4 Å FWHM. The spectra were analysed in terms of Gaussian components[4],[5]. The emission lines were fitted by one or several sets of three Gaussian components; the width and redshift of each component in a set were assumed to be the same and the intensity ratios of the [NII] and [OIII] lines were taken to be equal to 3 and 2.96, respectively; when necessary, we have added an Hα or Hβ absorption component.

3 Results and Conclusions

Out of the 15 observed objects, 10 have a "composite" spectrum with a HII region and a Seyfert-like nucleus projected on the spectrograph aperture - their measured relative line intensities cannot be used in the diagnostic diagrams. Nuclear spectra of 4 other objects are either "normal", the published line ratios being affected by unusually large errors, or not conclusive; only one object (UM 85) is unusual in presenting (at our resolution) lines with the same profile and redshift, although the [NII] lines are too strong for a HII region and too weak for a Seyfert-like nebulosity.

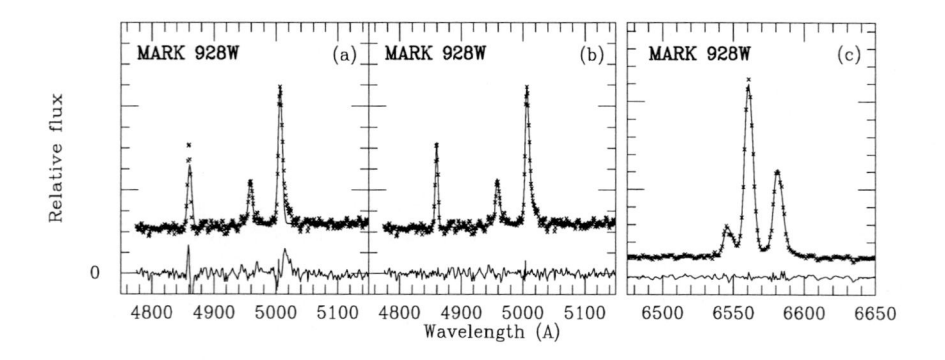

Fig. 1. An example of a composite object: Mark 928W. We present the spectra taken around Hβ (a,b) and Hα (c). The solid line is the best fit with Gaussian components; the lower line shows the residuals. For the Hβ, [OIII] lines we also give, for comparison, the unsatisfactory fit with a single set of Gaussian components (a).

In conclusion, we have shown that most AGN "transition" spectra are in fact "composite", a HII region and a Seyfert or Liner nebulosity being simultaneously present on the slit; our data confirm that Seyfert 2s and Liners occupy well separated regions in the line ratio space. The results of this work will be published in [6]. A.C.G. was supported by the JNICT, Portugal (PRAXIS XXI/BD/5117/95 PhD. grant).

References

[1] Baldwin J.A., Phillips M.M., Terlevich R., 1981, PASP 93, 5
[2] Veilleux S., Osterbrock D.E., 1987, ApJS 63, 295
[3] Heckman T.M., Miley G.K., van Breugel W.J.M., Butcher H.R. 1981, ApJ 247, 403
[4] Véron P., Véron-Cetty M.-P., Bergeron J., Zuiderwijk E.J., 1981, A&A 97, 71
[5] Véron P., Linblad P.O., Zuiderwijk E.J., Véron M.-P., Adam G., 1980, A&A 87, 245
[6] Véron P., Gonçalves A.C., Véron-Cetty M.-P., 1996, A&A *in press*

Nuclear Activity in Very Nearby Galaxies

Paulina Lira[1], Rachel Johnson[1,2,3], Andy Lawrence[1]

[1] Institute for Astronomy, University of Edinburgh, Royal Observatory, Blackford Hill, Edinburgh EH9 3HJ, UK
[2] Institute of Astronomy, University of Cambridge, The Observatories, Madingley Road, Cambridge CB3 0HA, UK
[3] Department of Physics, Queen Mary and Westfield College, Mile End Rd., London E1 4NS, UK

Abstract. We are carrying out a multiwavelength and spectroscopic study of a volume-limited sample of very nearby galaxies. Our main aim is to assess the prevalence and nature of activity in normal galaxies to the lowest possible level. In this poster we present some results from our multiwavelength imaging and some very preliminary spectroscopic results.

1 Introduction

To study the occurrence of nuclear activity in galaxies at the lowest possible levels we have defined a sample of galaxies which includes the nearest examples of various morphological types with a representative range of intrinsic luminosities. The sample is complete within a distance of ~ 7 Mpc and almost certainly includes all galaxies down to $M_B \sim -13$. The observations were restricted to objects with $\delta > -35°$. Because we want to observe objects with well formed nuclei, galaxies classified as Sdm, Sm and Irr were removed from the sample. This gives a basic sample of 46 galaxies (26 spirals, 5 E/S0, 15 dwarfs).

To carry out a comprehensive study of our sample, a multiwavelength search project was started in 1993, collecting radio, optical, IR and X-Ray data. High signal to noise long-slit spectroscopy is being carried out during 1996-97.

The research will also cover other topics related to galaxy nuclei, such as star formation activity and history and the nature of nuclear stellar populations.

2 Example Results

A general lesson from our multiwavelength imaging so far is that nuclear regions (especially for low luminosity spirals) show complex structures and so it is not always obvious where the true nucleus is.

We made an unexpected discovery in the dwarf elliptical **A 0951+68**, a very small galaxy with $M_B \sim -13$. The CCD image showed what appears to be a pair of close spherical galaxies, each with a distinct nucleus. Service spectroscopy of the nuclei showed that one has a stellar spectrum (knot A), while the other (knot B) has high excitation emission lines (O III/H$\beta \sim 6$) as shown in **Fig. 1**

(Johnson et al. 1997). HII region models suggest that this is a very low-metallicity nuclear starbust ionized by a single very luminous O7 star.

In **NGC 4150** a strong X-ray source consistent with the galaxy nucleus had been found in the ROSAT-PSPC survey, and was assumed to be a low luminosity AGN by Boller et al. (1992). However our HRI observations show that the X-ray source is in fact 16″ away from the nucleus and consistent with an unresolved object of magnitude $B = 18.4$. INT service spectroscopy revealed that the object is a background quasar with a redshift of 0.52 as shown in **Fig. 2**.

Our new HRI observations of **NGC 5204** show a strong X-ray source with $L_X \sim 10^{39}$ erg/sec, consistent with the earlier Einstein IPC detection. However our data show that the X-ray source is not consistent with either the center of symmetry or any obvious bright spot. Its true nature remains a puzzle.

Profile fitting to the spectra of **NGC 4395**, the least luminous Seyfert 1 galaxy known today, gives equivalent widths of 215 Å and 60 Å for the broad components of Hα and Hβ respectively, compared with 270 Å and 31 Å found by Filippenko & Sargent (1989). Our preliminary flux calibration also shows evidence of a change in the continuum level by a factor of 2 compared with data acquired in 1988 and 1990 (Shields & Filippenko 1992). The variability in the continuum and the broad line components strongly supports the hypothesis that **NGC 4395** is the low-luminosity version of a classical AGN. The observations are not consistent with a single luminous supernova remnant evolving in a dense medium which would have declined by a large factor over a lapse of 8 years.

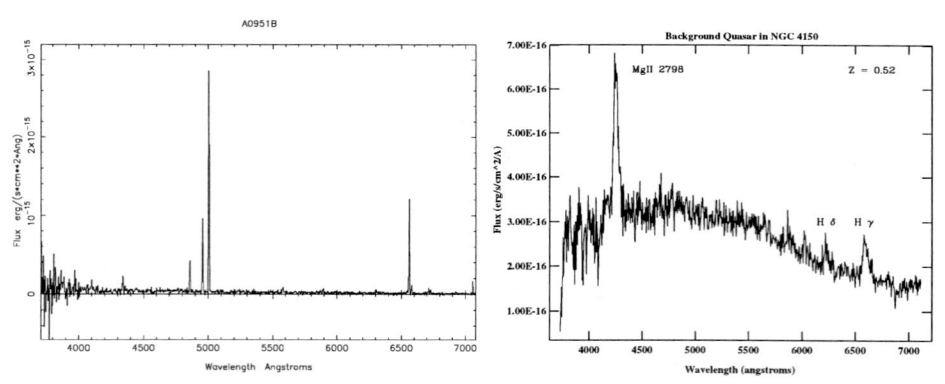

Fig. 1 (left): Emission-line spectrum of knot B in A 0951+68; **Fig. 2** (right): Spectrum of the Background Quasar in NGC 4150.

References

Boller, Meurs, Brinkmann, Fink, Zimmermann & Adorf (1992), A&A **261** 57.
Filippenko & Sargent (1989), Ap J (Letters) **342** L11.
Filippenko, Ho & Sargent (1993), Ap J (Letters) **410** L75.
Johnson, Lawrence, Terlevich & Carter (1997), in press.
Shields & Filippenko (1992), in *Relationship Between AGN and Starbust Galaxies*, ed. Filippenko, p. 267.

The Spectral Variability of QSOs in the Optical Bands

Dario Trèvese[1], Domenico Nanni[2], Richard G. Kron[3], and Alessandro Bunone[4]

[1] Istituto Astronomico, Università di Roma "La Sapienza", via G.M. Lancisi 29, I-00161 Roma, Italy
[2] Osservatorio Astronomico di Roma, via dell'Osservatorio, I-00040 Monteporzio, Italy
[3] Fermi National Accelerator Laboratory, Box 500, Batavia, IL 60510, U.S.A.
[4] Dipartimento di Fisica, Università di Roma "Tor Vergata", via della Ricerca Scientifica, I-00133 Roma, Italy

Abstract. We present preliminary results of the structure function analysis of the variability of the faint QSOs sample in SA 57, in the U and F bands, compared with previous studies in the B_J band. This multicolor analysis indicates a hardening of the spectra in the bright phase and provides an explanation of the correlation of variability with redshift.

1 Introduction

The statistical analysis of optical-UV light curves of QSOs shows that the amplitude of variability is inversely correlated with the intrinsic luminosity and directly correlated with redshift (Giallongo, Trèvese, Vagnetti 1991; Hook et al. 1994; Trèvese et al. 1994 (T94); Cristiani et al. 1996). An explanation of the positive correlation with redshift, based on the spectral variability, has been proposed by Giallongo, Trèvese and Vagnetti (1991): *in a given band, higher z objects are observed at higher rest-frame frequencies, where the variability is stronger.* This explanation was confirmed by Di Clemente et al. (1996) (D96) through a statistical comparison of the variability indexes of different QSO samples with the average increase of variability with redshift measured by Cristiani et al. (1996). A more precise characterization of the spectral variability can be obtained through a direct comparison of the light curves of individual objects in different bands. So far this has been possible only for the sample of Cimatti, Zamorani and Marano (1993), who measured U, B_J, F magnitudes of their sample at three epochs with 1 yr time lags. We present some preliminary results of the analysis of the QSO sample of the Selected Area 57, in the U, B_J, F bands, which show that the variability is indeed stronger at shorter wavelengths. The increase of variability with frequency can entirely account for the correlation with redshift.

2 Data and Analysis

The data are derived from an homogeneous collection of 24 prime focus plates of the SA 57 in the U, B_J, F, bands, obtained with the Mayall 4-m telescope at the

Kitt Peak National Observatory since 1974. The digitization, object detection, classification and photometry is described in Trèvese et al (1989). The methods applied to derive the relative photometry with optimal signal to noise ratio are described in T94 where the B_J plates were analyzed. In the present work the same methods are applied to five U plates taken during 1980, 1984, 1985, and five F plates taken during 1975, 1984, 1985 and 1986. The sample consists of the 35 objects with measured redshift, having stellar images and broad spectral lines, discussed in T94. Although ten of these objects have $M_B < -23$ we call all of them "quasars" ignoring the traditional distinction in absolute magnitude. The first statistical analysis is obtained by computing, for each photometric band, the first-order structure function (Bonoli et al. 1979; Simonetti, Cordes, and Heeschen 1985):

$$S_2(\tau) = \langle [m(t+\tau) - m(t)]^2 \rangle, \qquad (1)$$

where $m(t)$ is the magnitude at the epoch t, τ is the time lag, and the angular brackets indicate the ensemble average over the sample of variable objects. To show the redshift dependence of the variability, the sample is split in the low-redshift and high-redshift halves. The results are shown in Fig. 1. The B_J data are the same of T94 and are reported here for comparison. Although the number of observations in each of the other bands is smaller and the time sampling is less uniform, still the structure functions show the same characteristics of the B_J band. In particular, at each time the variability is stronger for the high-redshift half of the sample. A more robust statistics is obtained by (see D96):

$$S_1(\tau) = \sqrt{\frac{\pi}{2} \langle |m(t+\tau) - m(t)| \rangle^2 - \sigma_n^2}, \qquad (2)$$

where σ_n is the r.m.s. noise associated with the magnitude difference between two plates. The dependence of variability on frequency is analyzed as follows. In each observing band we compute all the magnitude differences in a given interval of time lags, together with the relevant rest-frame frequency. Then we compute S_1 taking the average of all the magnitude differences in this time interval. The relevant frequency is computed as the ensemble average of the rest-frame frequencies. For comparison with Di Clemente et al. (1996) we adopt as reference interval of time lags $0.21\ yr \leq \tau \leq 0.39\ yr$. Moreover, to make a close comparison it has been necessary to restrict all the samples to the same range of absolute magnitudes, because of the dependence of variability on absolute luminosity. Fig. 2 shows that the results of the present analysis are fully consistent with those of other samples, as already found for the B_J band. In particular, within the present QSO sample there is an increase of variability with the rest-frame frequency, which is also consistent with the previous analysis of D96. Spectral changes of individual quasars have been found by Cutri et al. (1985) who directly compared $UBVRIJHK$ observations of a sample of 7 quasars, finding some evidence of a hardening of the spectrum during the bright phase. Such a behavior has also been found by Kinney et al. 1991 from IUE observation of a subsample of PG quasars. It should be noted that the analysis of the correlation between the spectral slope α, defined by assuming a power-law spectrum $f_\nu \propto \nu^\alpha$, and the brightness requires some care to avoid spurious correlations discussed by Massaro and Trèvese (1996), who find that the data of Cutri et al. (1996) do in fact imply a statistically significant positive correlation of the spectral slope with brightness. As discussed in D96, the slope of a straight line representing the data in Fig. 2 is about $\Delta S_1(\Delta t)/\Delta \log \nu \simeq 0.28$. If we assume that, during the brightness changes, the power-law shape of the individual QSO spectra $f_\nu \propto [\nu_{obs} \times (1 + z)]^\alpha$ is preserved, this corresponds to the same slope $\Delta S_1(\Delta t)/\Delta \log(1+z) \simeq 0.28$ in the variability-redshift relation. A rough estimate of the latter slope $\Delta S_1(\Delta t)/\Delta \log(1+z) \simeq 0.25 - 0.30$ can be obtained from Fig. 5 of Cristiani et al. (1996) who analyze the dependence of the QSO structure function on redshift and luminosity. As discussed in D96, this means that the spectral changes can entirely account for the redshift dependence of variability.

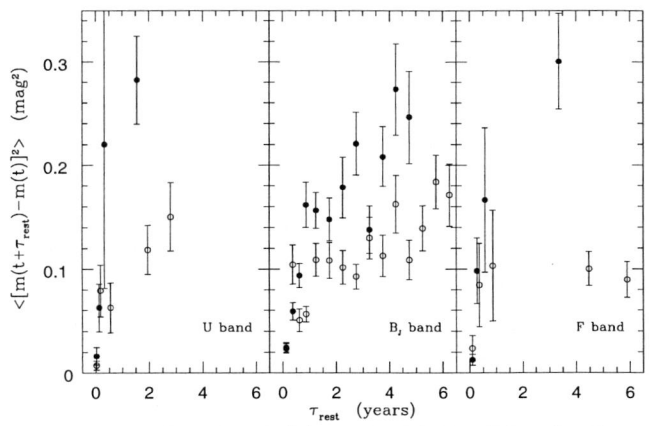

Fig. 1. The structure function $S_2(\tau)$ for the U, B_J, F bands. Open circles refer to the low-redshift half of the sample and filled circles to the high-redshift half. In each band the variability is larger for the high-redshift half of the sample.

3 Conclusions

- The structure function analysis of the variability of our QSO sample, performed in the U, B_J, F bands confirms the result, previously obtained by T94 for the B_J band, that higher redshift QSOs appear more variable.
- The variability increases with the rest frame spectral frequency, consistently with a hardening of the spectrum in the brighter phases.
- The present results, obtained by multi-band observations of the faint QSOs sample of SA 57, are consistent with previous findings of Di Clemente et al. (1996).
- Assuming that the power law shape of the spectrum is preserved during the brightness variations, the spectral changes can entirely account for the increase of variability with redshift.

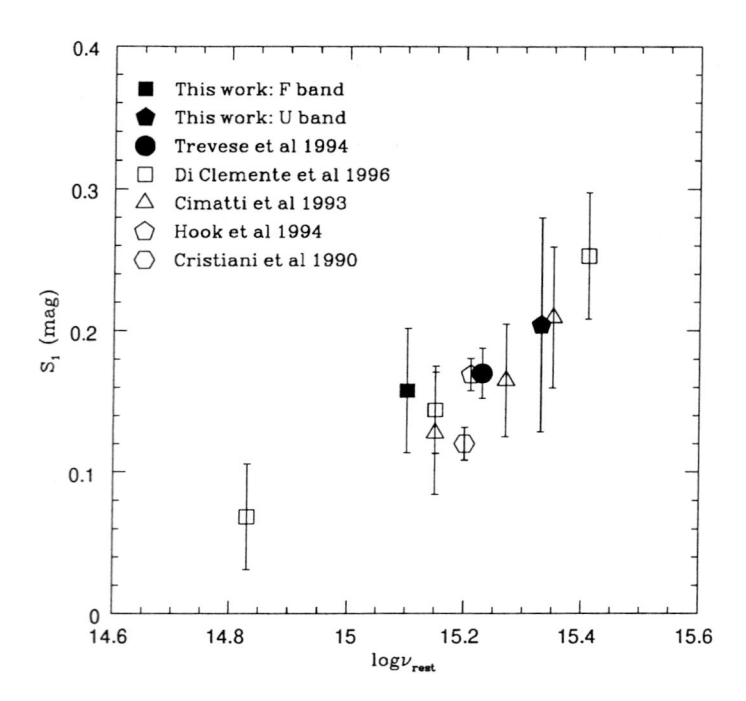

Fig. 2. The function $S_1(\tau)$ (eq. 2) as a function of the rest-frame frequency. Filled symbols refer to the data of SA 57. The variability increases with the rest-frame frequency.

References

Cimatti, A., Zamorani, G., Marano, B. (1993): MNRAS **263**, 236
Cristiani, S., Trentini, S., La Franca, F., Aretxaga, I., Andreani, P., Vio, R., Gemmo, A. (1996): A&A **306** , 395
Cutri, R. M., Winsniewski, W. Z. , Rieke, G. H., Lobofsky, H. J. (1985): ApJ **296**, 423
Di Clemente, A., Natali, G., Giallongo, E., Trévese, D., Vagnetti, F. (1996): ApJ **463**, 466 (**D96**)
Giallongo, E., Trévese, D., Vagnetti, F. (1991), ApJ **377**, 345
Hook, I. M., McMahon, R. G., Boyle, B. J., Irwin, M. J. (1994): MNRAS **268**, 305
Kinney, A. L., Bohlin, R. C., Blades, J. C., York, D. G. (1991): ApJS **75**, 645
Massaro, E., Trévese, D. (1996): A&A **312**, 810
Simonetti, J. H., Cordes, J. M., Heeschen, D. S. (1985): ApJ **296**, 46
Trévese, D., Pittella, G., Kron, R. G., Koo, D. C., Bershady, M. (1989): AJ **98**, 108
Trévese, D., Kron, R. G., Majewski, S. R., Bershady, M. A., Koo, D. C. (1994): ApJ **433**, 494 (**T94**)

Part 6

RADIO GALAXIES AT HIGH REDSHIFT

Radio Galaxies at High Redshift: Unification and Host Galaxies

Patrick McCarthy[1]

Carnegie Observatory, 813 Santa Barbara St, Pasadena, CA 91101

Abstract. I present new results on the unification of radio galaxies and quasars and the hosts of distant radio galaxies. Using the now completely identified MRC/1Jy sample, I compare the linear sizes of radio galaxies and quasars and find no significant differences. Emission-line images of 3CR radio galaxies obtained with HST reveal evidence for ionization cones with a wide range of apparent opening angles. Finally, I show our most recent K-band Hubble diagram and argue that radio galaxy, and radio quasar hosts, can be identified with the progenitors of present day gE galaxies.

1 Introduction

In this paper I will present recent results relating to the unification of radio galaxies and quasars and the degree to which our understanding of the host of radio loud AGN can be applied to the population of early type galaxies. In recent years there has been an increased awareness of the importance of orientation in the observed phenomenology of AGN. The Antonucci & Miller (1985) work on NGC1068 altered our view of the Seyfert I & II dichotomy and a similar paradigm is now in place for the broad-lined (quasars) and narrow-lined (galaxies) populations of radio-loud AGN. The Barthel (1989) orientation-based "unified" model of radio sources combines the Antonucci & Miller anisotropic UV source with the relativistic effects observed at centimeter wavelengths to produce a model linking radio galaxies and quasars. One of the success of the Barthel model is that it reproduces the linear size distributions of radio sources in the 3CRR sample, albeit in a restricted range of redshifts. This simple geometric demonstration of the orientation differences between the galaxies and quasars is quite appealing, but it is based on a single sample and there have been suggestions that it does not hold well when tested in other redshift ranges, or in other samples (Kapahi 1990; Singal 1993a,b; Nesser et al. 1995). Complete samples of radio sources selected at frequencies low enough to avoid significant corruption from biases arising from relativistic boosting are few in number.

2 The MRC/1 Jy Sample

Over the past several years my collaborators and I (Kapahi, van Breugel, Persson, Athreya and Subrahmanya) have carried out a program to produce a large and complete sample of sources selected near the peak of the low frequency source counts, at S(408MHz) = 1Jy. This sample, the MRC/1Jy survey, contains 543 sources selected from the Molonglo Reference Catalog (Large et al. 1981) with the restriction S(408)> 0.95Jy, and sky location. More than 98% of the sources have now been identified. Of these, 20% are quasars, 1% are BL Lac objects and the remainder are radio galaxies. Obtaining a high completeness fraction required observations in the near IR, as roughly 2% of the sources have identifications fainter than R= 25. These sources have been identified with galaxies having K $\sim$ 18 − 19. Kapahi et al. (1997) have obtained redshifts for nearly all of the quasars, while McCarthy et al. have obtained spectroscopic redshifts for $\sim$ 70% of the radio galaxies and redshift estimates from K-band photometry for the remaining sources. In Figure 1 I show the run of 408 MHz power with redshift for the 3CRR (Laing et al. 1981) and MRC/1Jy samples. The tight correlations are the result of the Malmquist bias and the gap between the 3CRR and MRC/1Jy samples reflects the very different sky areas/volumes sampled by the two surveys.

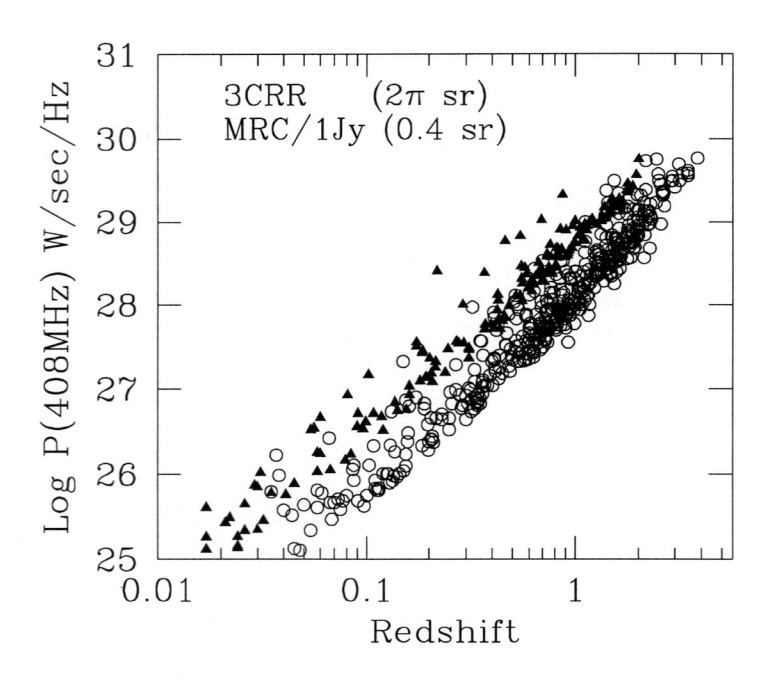

Fig. 1. The run of 408MHz power with redshift for the 3CRR (filled symbols) and MRC/1Jy (open symbols) surveys.

We have used this sample to carry out the projected linear size test of the Barthel unification scheme. In Figure 2 I show histograms of the projected linear sizes for the radio galaxies and radio quasars in the range $0.5 < z < 1.0$, analogous to Figure 3 in Barthel (1989). As one can see, we find no significant difference between the projected sizes of radio galaxies and quasars in this redshift range in the MRC/1Jy. Similar results hold for the redshift ranges $0.25 - 0.5, 1.0 - 2.0$, and > 2.0. In contrast to this null result for the unification model, we find that the fraction of sources that are quasars (20%) is similar to the 3CRR and is constant for all $z > 0.25$ and that the redshift distributions of the radio galaxies and quasars are similar for $z > 0.3$, where we are dominated by FR II sources.

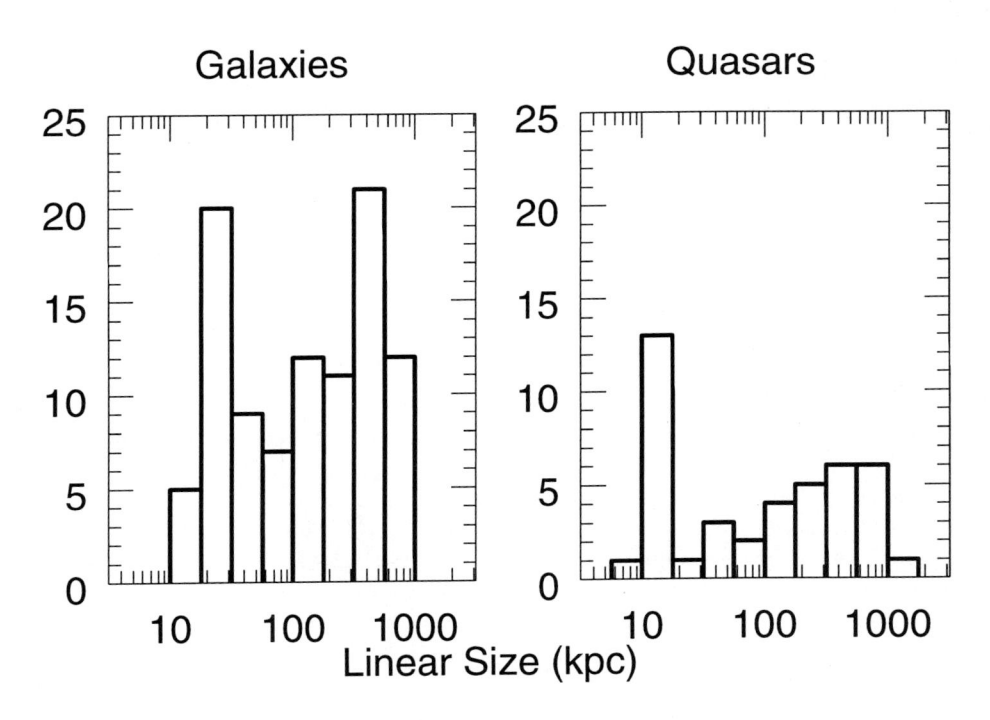

Fig. 2. The projected linear sizes of radio galaxies and radio quasars in the MRC/1-Jy sample in the redshift range $0.5 - 1.0$. The quasars are in the right panel, the radio galaxies in the left panel. In this redshift range all of the radio galaxy redshifts are spectroscopic.

Taken at face value our result, and those of Singal (1993a, b) would suggest that the orientation based unification model is ruled out at a high confidence level. Alternatively, the null result could be telling us that the projected source size is a poor orientation indicator. The preponderance of evidence in support of the unification model to at least zero-th order argues for the later interpretation of our result. As discussed by Gopal-Krishna & Wiita (1996) temporal evolution in the sizes and luminosities of radio sources can mask the signal expected from modest orientation differences.

3 HST Emission-Line Imaging of 3CR Radio Sources

S. Baum & I have imaged three 3CR radio galaxies in the light of [OIII]5007 using the linear ramp filters on WFPCII for exposure times of roughly 3000 seconds. Snap-shot exposures (600 seconds) of a large number of 3CR galaxies and quasars have been obtained by Biretta et al. (1997). These reveal substantial structure in the inner few arcseconds of the extended emission-line nebulae. In Figure 3 I show [OIII]5007,4959 images of two objects, 3C 330 ($z = 0.55$) and 3C 79 ($z = 0.23$). The image of 3C 330 shows fairly clear evidence for an edge-brightened cone of emission with a narrow opening angle. The symmetry axis of the cone is well aligned with the radio source axis. The image of 3C 79 is more complex in its morphology and does not suggest anisotropy at first glance. For both images I include a pair of solid lines to demark the extent of the emission line regions. These show that the [OIII] emission in 3C 79 is confined to a symmetric "cone" with an opening angle of roughly $120°$. The larger sample of 3CR galaxies and quasars imaged in the snap-shot survey show similar structures, some showing obvious cones of emission, while others are less structured but are confined to a reasonably well defined range of azimuthal angles with respect to the nucleus. Matching up the HST emission line regions ($0.3'' - 5''$ scales) to the larger scale emission line regions as imaged from the ground often reveals the anisotropy more clearly.

We suggest that these apparent emission-line cones are a strong piece of evidence in favor of one basic tenet of the quasar/radio galaxy unification model, namely that the radio galaxies have powerful UV continuum sources that are not visible to the observer. The apparent increase in projected opening angle as one moves from classic narrow lined radio galaxies (like 3C 330) to BLRGS (like 3C 79) and lobe dominated quasars is tantalizing. Perhaps with a sufficiently large sample the cone opening angle may prove to be an orientation indicator.

4 The Host of Distant Radio Galaxies

There is a growing body of evidence supporting the identification of high redshift radio galaxies with the population of progenitors of present day gE galaxies. Some of the evidence is circumstantial, some is more direct. Among the former are the propensity of high luminosity radio galaxies for the centers of rich clusters (e.g. Hill & Lilly 1991; Dickinson & Spinrad 1997) and the similarities in luminosity and color between radio galaxies and first ranked cluster ellipticals (Eisenhardt & Lebofsky 1987; Lilly 1988; Dunlop et al 1989). The color and luminosity evolution of radio-quiet ellipticals has recently come into focus. Im et al. (1996) find evidence for $\sim 0.5 - 1$ magnitude of luminosity evolution for the reddest field galaxies at $z \sim 1$, while Lilly et al. (1995) find a slightly smaller value at $z \sim 0.6 - 0.8$ and only modest color evolution. These results are similar to that seen in near-IR studies of radio galaxy evolution (e.g. Lebofsky & Eisenhardt 1986; Lilly 1988).

More direct evidence that distant radio galaxies reside in elliptical hosts, or their progenitors come from surface photometry both from HST and from

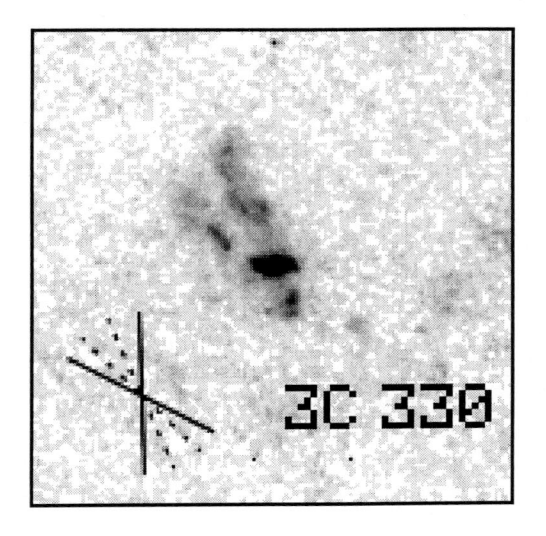

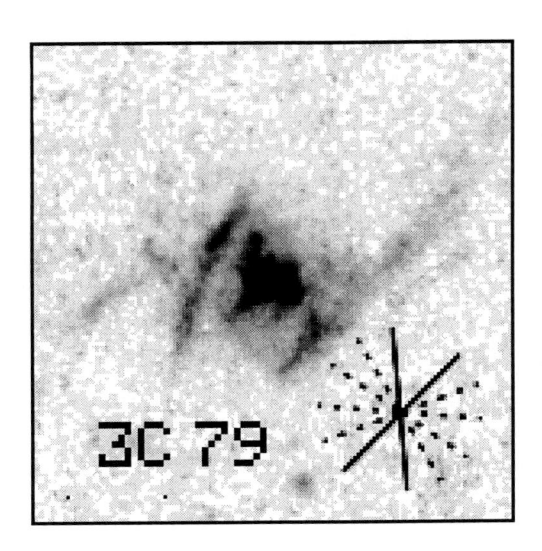

Fig. 3. [OIII]5007,4959 images of 3C 330 and 3C 79 obtained with the WFPC2/LRF and HST. The markings on the lower right corner of each figure are intended to show the size and orientation of the emission-line cones. Each image is 15″ on a side

the ground. Windhorst et al. (1995) show clear evidence for an $r^{1/4}$ profile in 53W002, a weak radio galaxy at $z = 2.39$. In the near-IR Rigler & Lilly (1994) show convincing evidence for a $r^{1/4}$ profile in 3C 65 ($z = 1.2$). In figure 4 I show a surface brightness profile obtained from a Keck/NIRC K-band image of 3C 265 ($z = 0.811$) obtained by van Breugel & Graham. This too shows a clear de Vaucouleurs profile. Thus radio galaxies and first ranked cluster ellipticals are from the same structural family, inhabit similar environments at intermediate and high redshifts, and have similar color (for $\lambda > 4000$Å) and luminosity evolution.

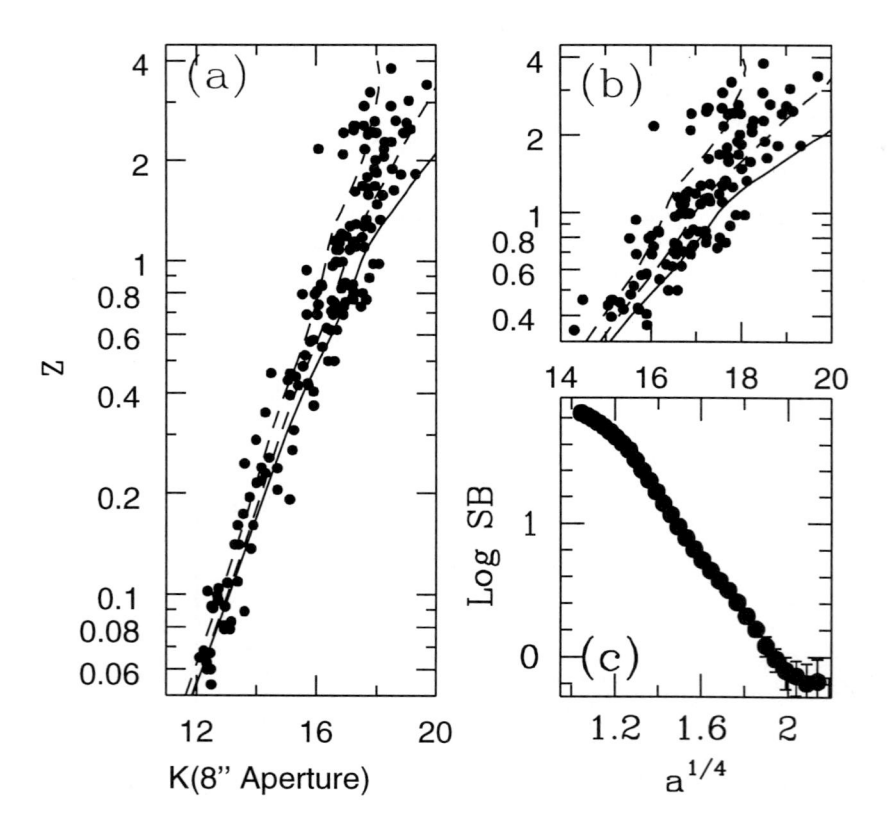

Fig. 4. (a) The K-band Hubble diagram for radio galaxies from the 3CRR, MRC/1-Jy surveys and a number of $z > 3$ radio galaxies taken from the literature. The solid line is a no-evolution model with $H_0 = 50$ and $q_0 = 0.1$; dotted lines are passive evolution models with $H_0 = 50$ $q_0 = 0.1 - 0.5$, and $z_f = 20$; (b): same as (a), but enlarged about the high-redshift o objects (c): The surface brightness profile obtained from a Keck K-band image of 3C 265. The profile is well fit by an $r^{1/4}$-law with $r_e = 1.8''$

One of the most powerful, and least understood, tools for examining the evolution of radio loud AGN hosts is the K-band Hubble diagram. The radio galaxy $K-z$ relation, as emphasized by Lilly (1988), shows little scatter at redshifts less than about 1.5 ($\sigma \sim 0.3$ magnitudes). This is all the more remarkable as the galaxies are selected solely on the basis of their flux densities at meter wavelengths. One explanation for the small scatter and high continuity in the $K-z$ relation is based on a rough scaling between nuclear and host galaxy luminosity that saturates at the top of the galaxy luminosity function. The recent work on Seyfert and Quasar host galaxies by McLeod (1997) and McLeod and Rieke (1996) offers support for this view. The apparent increase in the elliptical fraction among quasar hosts shown by Bachall et al, McLeod, Disney and others at this meeting provide additional support for the view that the 3CRR and 1Jy class radio galaxy hosts are the extreme end of the host luminosity and elliptical fraction correlations with AGN luminosity.

If this view of radio galaxy hosts is correct then we are on firmer ground in using the radio galaxies as probes of early galaxy evolution, provided we steer clear of the rest-frame ultraviolet continuum. The recent success of Steidel et al. (1996) and others in identifying a large population of star forming galaxies at high redshift does not necessarily reduce the importance of the AGN host galaxies at high redshifts. On the contrary, the AGN hosts continue to offer our only glimpses of apparently evolved massive stellar systems at high redshifts. The colors and luminosities of radio loud AGN host at high redshifts, and their apparent consistency with passive evolution suggest that a fraction of the progenitors of massive galaxies formed their stars at very early epochs. The question of how widely this inference can be applied rests on our very poor knowledge of the duty cycle of the AGN phenomenon and the fraction of massive galaxies that undergo episodes of nuclear activity.

References

Antonucci, R. R., & Miller, J. S., 1985, ApJ, 297, 621
Barthel, P. D., 1989, ApJ, 336, 606
Biretta, J., Baum, S., et al. 1997, ApJS, submitted
Dickinson, M., & Spinrad, H. 1997, in prep.
Eisenhardt, P. R. M. & Lebofsky, M. J. 1987, ApJ, 316, 70
Gopal-Krishna, & Wiita, P. J., 1996, ApJ, 467, 191
Hill, G. J., & Lilly, S. G. 1991 ApJ 367, 1
Im, M., Griffiths, R. E., Ratantunga, K. & Sarajedini, V. 1996, ApJ, 461, L97
Kapahi, V. K. 1990, in Parsec-Scale Radio Jets, ed. J. A. Zensus & T. J. Pearson (Cambridge: Cambridge Univ. Press), 304
Kapahi, V. K., et al. 1997, ApJS, in press.
Large, M. I., Mills. B. Y., Little, A., Crawford, D., & Sutton, J. 1981, MNRAS, 194, 693
Lilly, S. J. 1988 ApJ 340, 77
Lilly, S. J., Tresse, L., Hammer, F., Crampton, D., & Le Fevre, O. 1995, ApJ, 455, 108
McCarthy, P., Kapahi, V. K., van Breugel, W., Persson, S. E., Athreya, R. M., & Subrahmanya, C. R. 1996, ApJS, 107, 19

McLeod, K. K., & Rieke, G. H. 1995, 454, 77
McLeod, K. K. 1997, this volume
Nesser, M. J., Eales, S. A., Law-Green, J. D., Leahy, J. P., & Rawlings, S. 1995, ApJ, 451, 76
Rigler, M. A., & Lilly S. J. 1994, ApJ, 427, 79
Singal, A. K., 1993a, MNRAS, 262, L27
Singal, A. K., 1993b, MNRAS, 263, 139
Steidel, C. C., Giavalisco, M., Pettini, M., Dickinson, M., & Adelberger, K. 1996, ApJ, 462, L17
Windhorst, R. A., Mathis, D., & Keel, W. C. 1992, ApJ, 400, L1

The Nature of the UV Excess

C.N. Tadhunter[1], R. Dickson[1], R. Morganti[2,3] and M. Villar-Martin[1]

[1] Department of Physics, University of Sheffield, Sheffield S3 7RH
[2] Instituto di Radioastronomia, CNR, via Gobetti 101, I-40129 Bologna, Italy
[3] Australia Telescope National Facility, CSIRO, PO Box 76, Epping, NSW 2121, Australia

Abstract. The nature of the UV excess is one of the key problems for our understanding of the extended envelopes around radio galaxies and quasars. We present here preliminary results from the first combined spectroscopic/polarimetric survey of an *optically unbiased* sample of powerful radio galaxies. We show that the UV continuum is multi-component in nature, comprising to varying degrees scattered AGN light, direct AGN light, the light from young stellar populations and nebular continuum. The incidence of large UV polarization — a signature of scattered AGN light — is lower than reported in previous studies, and at least some sources show clear evidence for a starburst component. The results serve to emphasize that high quality spectroscopic and polarimetric data are *essential* for unravelling the various components which contribute to extended envelopes around quasars.

1 Introduction

Shortly after the discovery of $z > 1$ radio galaxies using CCD detectors in the early 1980's, measurements of their optical/IR colours revealed substantial UV excesses compared with passively evolving elliptical galaxies (Lilly & Longair 1984). The nature of this UV excess is one of the key outstanding issues for extragalactic radio sources, relevant to our understanding of the stellar populations of the host galaxies, the origin of the activity and the relationships between radio galaxies and quasars.

Given that the high-z radio sources are among the most distant known galaxies, they are potentially important probes of galaxy evolution in the early universe. However, the presence of close alignments between the UV structures and the large-scale radio structures (e.g. McCarthy et al. 1987) indicates that a substantial fraction of the UV excess is related to the activity, rather than being a consequence of the normal evolution of the stellar populations in the host galaxies.

What is the origin of the UV excess? Two main models have been proposed: **starbursts** — jet-induced, merger-induced, or associated with the evolution of the host galaxies (e.g. Lilly & Longair 1984, Rees 1989, Smith & Heckman 1989); and **scattered AGN light** (e.g. Tadhunter et al. 1988, Fabian 1989). The great attraction of the scattering idea is that it ties in with the unified schemes for radio sources which propose that all powerful radio galaxies have luminous quasars hidden in their cores (e.g. Barthel 1989). Recent attempts to search for the polarized features characteristic of the scattered AGN light have met with considerable success: not only have imaging polarimetry observations revealed UV polarization at the 5 – 20% level in many $z > 0.5$ radio galaxies (e.g. Tadhunter et al. 1992, Cimatti et al. 1993), but detailed spectropolarimetric observations also show evidence for scattered broad quasar features in several of the polarized sources (e.g. Antonucci 1984, Cimatti et al. 1996). However, because polarization measurements of such faint sources are difficult, the early polarization studies of the high-z radio galaxies concentrated on the brightest, most spectacular objects in a given redshift range. Consequently, there remain uncertainties about the true occurrence rate of the scattered AGN component and its proportional contribution to the UV excess in the *general population* of powerful radio galaxies.

In order to remove these uncertainties we have recently completed a major spectroscopic/polarimetric survey of an optically unbiased sample of radio galaxies. We report below some preliminary results from this survey, along with the implications for understanding the nature of the extended UV envelopes around powerful radio galaxies and quasars.

1.1 The Sample and Observations

Our sample consists of a complete sample of radio galaxies from Wall & Peacock (1985) with $S_{2.7} > 2$Jy, $0.15 < z < 0.7$, $13^{hr} < RA < 05^{hr}$ and $\delta < +10°$. Optical, and radio data for all the sample members are presented in Tadhunter et al. (1993), Morganti et al. (1993), the completeness of the sample is discussed in Tadhunter et al. (1993), and some preliminary results are presented in Morganti et al. (1995).

The data were taken using the ESO Faint Object Spectrograph and Camera (EFOSC) in runs on the ESO3.6m telescope in 1993, 1994 and 1995. Deep optical spectra (covering 3500 – 5500Å in the rest-frame) were obtained for all of the sample objects, and B-band imaging polarimetry observations were obtained for all but three of the objects. The advantage of the B-band is that allows us to sample the rest-frame UV continuum for all the sample members and thus reduce the problems with dilution by the (unpolarized) old stellar populations of the host galaxies which have affected most previous studies (e.g. Tadhunter et al. 1992, Cimatti et al. 1993). 1648+05 and 2211-17 were not observed polarimetrically because they have unusually weak emission lines and their optical spectra are dominated by the absorption line features of the old stellar populations of the host galaxies (dubbed ALO — absorption line objects — in Table 1), while the CSS source 0252-71 was left out at random because of lack of time.

The polarimetric data consist of cycles of four exposures taken with EFOSC in polarimetric mode. For data taken in 1993 and 1994 the polarized signal was modulated by rotating the telescope rotator successively by 0,45,90,135 degrees, while for the run in 1995 a half-wave plate was available, and the modulation was affected by rotating the half-wave plate through 0, 22.5, 45, 67.5 degrees. The data reduction and analysis procedures and some preliminary results are presented in Tadhunter et al. (1994) and Shaw et al. (1995). Errors on the polarization measurements were calculated by popagating the statistical uncertainties in the standard way, and all the polarization measurements have been corrected for the Ricean bias. Note that it all cases in which we detected significant polarization (i.e. $P/\sigma > 3$) we have checked that the 'o' and 'e' ray intensities follow pattern expected for linearly polarized light in each cycle. For all the but one of the polarized sources (1934-63) we have at least two independent cycles of data. Where we fail to detect significant polarization we give conservative 3σ upper limits.

The resulting polarization measurements are listed in Table 1, and plotted against redshift in Figure 1. For reference, we also give estimates in Table 1 of the fraction of light below 4000Å which is not contributed by the old stellar populations (F_{ns}). These latter estimates are based on measurements of the 4000Å break from our deep spectra.

2 Results

The results of our survey are disappointing for advocates of the scattered light hypothesis. Despite the fact that most of the objects in our sample show substantial UV excesses (i.e. large F_{ns} in Table 1), significant polarization is detected in only 7 (36%) of the measured sources; only 3 objects are polarized at the >5% level, and none are polarized at the >10% level. Note that these measurements refer to the continuum, since our spectra show that the emission line contamination in the filter bandpasses is small in all cases (<20%).

In contrast, for a dominant scattered AGN component we would expect levels of polarization in the range 10 — 20%, even allowing for the geometrical dilution of the polarization which results from integrating the polarization E-vectors over the broad radiation cones of the illuminating quasars. Thus, we can conclude that *scattered AGN light is not the main cause of the UV excess in most of the powerful radio galaxies in our sample.*

Optical/UV spectra can give futher clues to the nature of the UV excess, and we now go on to consider the relationship between the polarization and the spectral characteristics of the sources in our sample.

Broad Line Radio Galaxies We find that that a significant fraction (36%) of the sample show weak broad lines in our deep optical spectra (i.e. they are broad line radio galaxies: BLRG). We have considered the possibility that we are detecting scattered broad lines in these objects (e.g. 3C234: Antonucci 1984). However, none of the BLRG in our sample is highly polarized in the UV. Thus, the UV excesses in these objects must represent direct- rather than scattered AGN light.

This result has implications for our understanding of the nature of BLRG. Although it has been proposed that BLRG are the low-z, low radio power counterparts of high-z, high radio power quasars (e.g. Urry & Padovani 1995), we find evidence for BLRG out to the redshift limit of our sample, and overlapping with the redshift range of the radio-loud quasars. One explanation for this overlap is that there exists a considerable range in the *instrinsic* luminosities of the central

Object	Optical	Radio	z	F_{ns}	P_B (%)
0023-26	NLRG	CSS	0.322	0.63	<2.0
0035-02 (3C17)	BLRG	(FRII)	0.220	0.80	2.8±0.3
0038+09 (3C18)	BLRG	FRII	0.188	1.0	<1.8
0039-44	NLRG	FRII	0.346	0.87	4.8±1.2
0105-16 (3C32)	NLRG	FRII	0.400	0.28	<4.4
0117-15 (3C38)	NLRG	FRII	0.565	0.50	8.2±0.9
0235-19	(BLRG)	FRII	0.620	0.95	<4.3
0252-71	NLRG	CSS	0.566	–	–
0347+05	BLRG	FRII	0.339	0.87	<4.3
0409-75	NLRG	FRII	0.693	1.0	<6.3
1306-09	NLRG	CSS	0.464	0.60	6.3±1.3
1547-79	BLRG	FRII	0.483	1.0	2.8±1.0
1549-79	NLRG	FSD	0.150	0.59	<1.8
1602+01 (3C327.1)	BLRG	FRII	0.462	0.84	<4.5
1648+05 (Her A)	ALO	(FRI)	0.154	0.26	–
1932-46	NLRG	FRII	0.231	0.67	<1.6
1934-63	NLRG	GPS	0.183	0.62	3.5±0.5
1938-15	BLRG	FRII	0.452	0.82	<5.5
2135-20	BLRG	CSS	0.635	0.90	<2.7
2211-17 (3C444)	ALO	FRII	0.153	0.60	–
2250-41	NLRG	FRII	0.310	0.66	4.9±0.7
2314+03 (3C459)	NLRG	(CSS)	0.220	0.66	<1.23

Table 1. B-band polarization measurements and basic information for the complete sample of radio galaxies. Column 3: optical spectral classification; column 4: radio morphology classification; column 5: redshift; column 6: fraction of the total light below 4000Å not contributed by old (15Gyr) stellar populations; column 6: B-band polarization measurements or 3σ upper limits (the polarization measurements were made in 3 – 5 arcsec. diameter circular apertures centred on the nuclei of the host galaxies).

AGN at a given radio power/redshift. Alternatively, the BLRG may represent partially obscured quasars, which — on the basis of the unified schemes — we are observing at intermediate angles to the line of sight.

Polarized Narrow Line Radio Galaxies Only 5 (~25%) of the sources are highly polarized in the UV ($P_B > 3\%$), and all of these polarized sources are classified as narrow line radio galaxies (NLRG), with no evidence for broad lines in their optical/UV spectra. Note that, even for these polarized sources the relatively low level of UV polarization ($P_B < 10\%$) suggests dilution of the polarized emission by an unpolarized UV continuum source.

We have have been able to measure an accurate polarization angle for four of the polarized NLRG. We find that for 0039-44, 1934-63 and 2250-41 the polarization E-vectors line up close to the perpendicular to the radio axes — as expected in the case of scattered light — while for 1306-09 the E-vector is within 40° of the parallel to the radio axis.

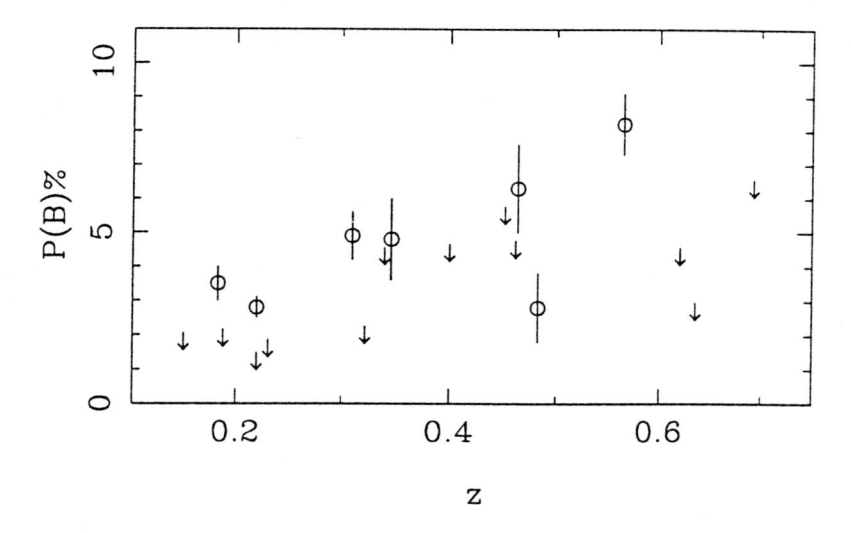

Fig. 1. B-band percentage polarization plotted against redshift for the objects in the complete sample of southern radio galaxies. Note the relatively low levels of polarization measured in most of the objects

Interestingly, two of the compact steep spectrum radio sources in our sample show large UV polarization (1306-09 and 1934-63). The detection of significant optical polarization in these objects supports the application of the unified schemes to the compact radio sources, as well as to the extended radio sources for which they were originally proposed (see also Tadhunter et al. 1994).

Narrow Line Radio Galaxies with Low UV Polarization For the majority of the NLRG we have failed to detect significant UV polarization (~30% of total sample). Although we cannot rule out the possibility that some of these objects are unrecognised BLRG, our spectra provide direct evidence for the following two unpolarized components which contribute to the UV excess.

- *Nebular Continuum (recombination+2-photon+free-free).* Using direct measurements of the Balmer lines we calculate that this component makes up 5 – 40% of the UV continuum in the nuclear regions, and more off-nucleus (see Dickson et al. 1995). Note that this component is present in all sources with strong emission lines, and its presence may help to explain the relatively low levels of polarization observed in the polarized NLRG.
- *Starbursts* Spectra of at least two of the unpolarized sources — 3C459 and 1549-79 — show clear evidence for young (<2Gyr) stellar populations in the

form of a Balmer break and/or the higher Balmer lines in absorption. In the case of 3C459 the starburst component dominates the optical/UV continuum (see also Miller 1981). This component may also contribute in some of the polarized sources.

3 The Multi-Component UV Continuum

It is clear that no single component dominates the UV continuum in the powerful radio galaxies. The main components which contribute to the UV excess are as follows.

- **Scattered AGN light:** this component is significant in ~25% of the objects in our complete sample.
- **Direct AGN light:** a large proportion (~36%) of our sample are BLRG in which the UV continuum is dominated by direct AGN light.
- **Nebular continuum:** the nebular continuum is significant in all objects with strong emission lines.
- **Starbursts:** clear evidence for a starburst component is found in two objects. Since the spectral features of a young stellar population can be subtle, detailed modelling will be required to determine whether this component is also important in the other objects.

By way of illustrating the multi-component nature of the UV continuum, Figure 2 shows the results of a multi-component fit to the continuum SED of the northern radio galaxy 3C321 (see Tadhunter et al. 1996). Although there is clear evidence for a scattered quasar in this source from the detection of scattered broad Hα (Young et al. 1996) and a steep rise in the polarization to the UV (Figure 3), attempts to model the SED in terms of a combination of scattered quasar/power-law and an old stellar population proved unsuccessful, because a positive residual always remained in the 3700 — 4200Å region (the position of the Balmer break for a young stellar population). The best-fitting model consists of a combination of 15Gyr old stellar population, nebular continuum, 1Gyr-old starburst and scattered quasar/power-law components. We emphasise that the young stellar population in 3C321 only came to light through careful modelling of high-quality spectropolarimetric data.

A particularly interesting point about 3C321 is that the young stellar population cannot be less than 10^8 yr old (else the Balmer break would not be so strong). This is considerably older than the age of the radio source ($\sim 10^{6-7}$ yr). Thus, if both the radio and starburst activity have been triggered by a merger (see Heckman et al. 1985, Smith & Heckman 1989)), there must be a substantial delay ($>10^8$ yr) between the start of the merger and onset of the activity triggered by the merger. The difference in age between the starburst and the radio source also rules out the jet-induced star formation model for this object.

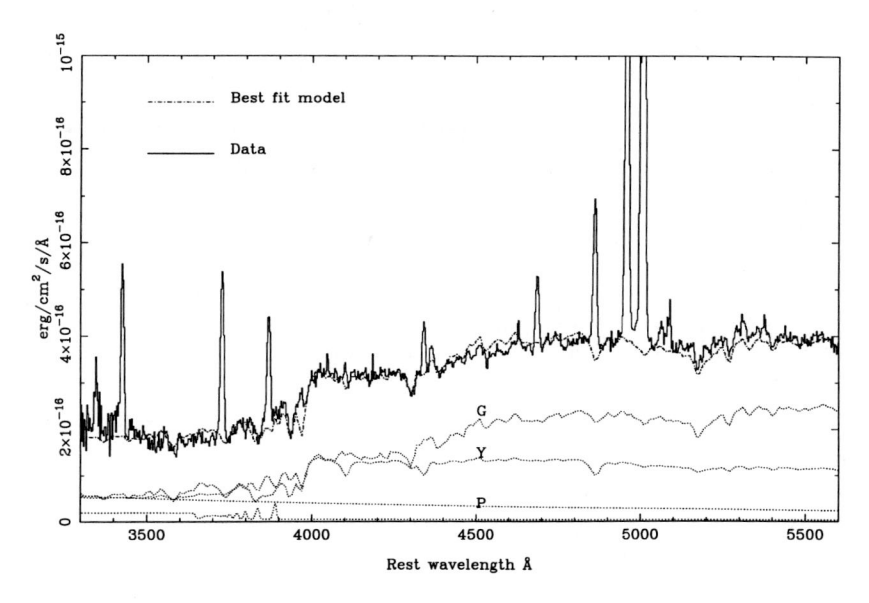

Fig. 2. Multi-component fit to the optical/UV continuum of the powerful northern radio galaxy 3C321 ($z = 0.096$): G — 15 Gyr old stellar population; Y — 1 Gyr young stellar population; P — scattered power-law or quasar component.

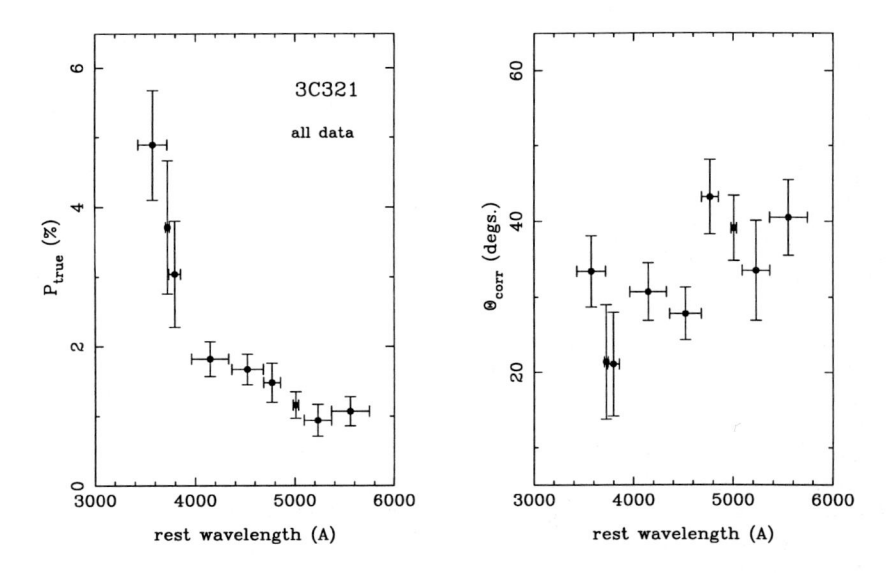

Fig. 3. The wavelength dependence of the percentage polarization (left) and polarization angle (right) for 3C321.

4 Implications for Quasars

On the basis of the unified schemes we expect to the extended envelopes of quasars and radio galaxies to be similar. The question then arises as to whether the extended features observed in high resolution images of quasars are truly the host galaxies or simply the manifestations of activity already observed in the radio galaxies. If the main scattering medium is Mie dust grains, then the scattered AGN component will appear stronger in the quasars because of the forward scattering properties of the grains. The nebular continuum will also be strong in the quasar envelopes wherever the extended emission lines are strong.

The main lesson from the radio galaxies is that high-quality spectroscopic and polarimetric data are essential for distinguishing the host galaxies from the activity-related components in the quasar nebulosities. Broad-band colours alone are inadequate.

References

Antonucci, R.R.J., 1984, ApJ, 297, 621
Barthel, P.D., 1989, ApJ, 336, 606
Cimatti A., di Serego Alighieri S., Fosbury R., Salvati M., Taylor D., 1993, MNRAS, 264, 421
Cimatti, A., Dey, A., van Breugel, W., Antonucci, R., Spinrad, H., 1996, ApJ, 465, 145
Dickson, R., Tadhunter, C., Shaw, M., Clark, N. & Morganti, R., 1995, MNRAS, 273, L29
Fabian, A.C., 1989, MNRAS, 238, 41p
Heckman T.M., Smith E.P., Baum S.A., van Breugel W.J.M., Miley G.K., 1986, 311, 526
Lilly, S.J., Longair, M.S., 1984, MNRAS, 211, 833
McCarthy, P.J., van Breugel, W., Spinrad, H., Djorgovski, S. 1987, ApJ, 321, L29
Miller, J.S., 1981, PASP, 93, 681
Morganti, R., Killean, N., Tadhunter, C.N., 1993, MNRAS, 263, 1023
Morganti, R. et al., 1995, Publ.Astron.Soc.Aust., 12, 3
Rees, M.J., 1989, MNRAS, 239, 1p
Shaw M., Tadhunter C., Dickson R., Morganti R., 1995, MNRAS, 275, 703
Siebert, J., Brinkmann, W., Morganti, R., Tadhunter, C.N., Danziger, I.J., Fosbury, R.A.E., di Serego Alighieri, S., 1996, MNRAS, 279, 1331
Smith, E.P., Heckman, T.M., 1989, ApJ, 341, 658
Tadhunter C., Fosbury, R.A.E., di Serego Alighieri S., 1988, in Maraschi L., Maccacaro T., Ulrich M.-H., eds, Proc. Como Conf. 1988, BL Lac Objects. Springer-Verlag, Berlin, p.79
Tadhunter, C., Scarrott S., Draper P., Rolph C., 1992, MNRAS, 256, 53p
Tadhunter C., Morganti R., di Serego Alighieri S., Fosbury R., Danziger I., 1993, MNRAS, 263, 999
Tadhunter, C., Shaw, M.A., Morganti, R., 1994, MNRAS, 271, 807
Tadhunter, C.N., Dickson, R.C., Shaw, M.A., 1996, MNRAS, 281, 591
Urry, C.M., Padovani, P., 1995, PASP, 107, 803
Young, S. et al., 1996, MNRAS, 279, L72
Wall, J.V., Peacock, J.A., 1985, MNRAS, 216, 173

Clive Tadhunter

The Role of Shocks in the Extended Emission Line Regions of Powerful Radio Galaxies

N.E. Clark[1,2,3], C.N. Tadhunter[1], D.J. Axon[2,3,4] & A. Robinson[4]

[1] Department of Physics, University of Sheffield, Hicks Building, Hounsfield Road, Sheffield S7 8RH, UK
[2] Space Telescope Science Institute, 3700 San Martin Drive, Baltimore, MD21218, USA
[3] Affiliated to: Astrophysics Division, Space Science Department of ESA, ESTEC, NL-2200 AG Nordwijk, The Netherlands
[4] On leave from: The Nuffield Radio Astronomy Laboratory, University of Manchester, Jodrell Bank, Macclesfield, Cheshire, SK11 9DL, UK
[5] Dept. of Physics & Astronomy, University of Hertfordshire, College Lane, Hatfield, Herts, AL10 9AB, UK

Abstract. A detailed optical spectroscopic study of the extended line-emitting gas in a sample of powerful radio galaxies has revealed clear evidence that fast shocks have a major impact on the state of the warm gas along the radio axes of the host galaxies. The effects of the shocks on the morphology, kinematics, physical conditions and ionization of the extended gas are outlined below. We also consider the likely implications for radio galaxies in general.

1 Introduction

It has been proposed that the extended emission line regions (EELR) around powerful radio galaxies represent ambient gas in the haloes of the host galaxies which has been illuminated by a cone of ionizing radiation from a central obscured AGN (e.g. Fosbury 1989). This hypothesis generally appears to be consistent with the observed properties (morphologies, kinematics and emission line ratios) of the majority of *low-redshift* radio galaxies (Robinson et al. 1987; Tadhunter et al. 1989; Binette et al. 1996; Simpson & Ward 1996).

The properties of the EELR associated with *high redshift* radio galaxies are markedly different. Firstly, the precise alignments with the axes of the extended radio emission (Chambers et al. 1987; McCarthy et al. 1987), and the detailed radio-optical associations and highly collimated jet-like EELR seen in a number of sources (Miley et al. 1992; Longair et al. 1995), are inconsistent with the broad ionization cones predicted by the unified shemes for radio galaxies. Secondly, extreme emission line kinematics are observed along the radio axes in the majority of high redshift radio galaxies (van Ojik 1995; McCarthy et al. 1996). In these sources, the shocks induced by violent interactions between the advancing radio jet and the ambient ISM provide a compelling alternative to AGN-illumination (see e.g. Dopita & Sutherland 1996). The expansion of the radio source through the ISM drives strong shock waves into the warm clouds of gas embedded in the hot halo of the host galaxy. The shocks rapidly heat, compress and accelerate the gas clouds and the enveloping halo gas. The shocked gas is likely to be a strong source of line emission because: i) the shock-heated clouds cool by emitting line radiation; and ii) the post-shock clouds are subjected to an intense local ionizing radiation field from the hot, shocked gas. Also, the covering factor of the gas will be increased as the clouds of gas are swept-up, flattened and then shredded by the passage of the shock (as seen in numerical simulations of cloud shocks: Klein et al. 1994).

Although the extreme properties of the EELR in high redshift radio galaxies appear to be consistent with the jet-cloud interaction model outlined above, these distant sources are difficult to study because: i) they have small angular sizes and the emission line fluxes are low; and ii) the rest-frame optical spectra (including important diagnostic emission lines) are redshifted to the infra-red. These problems disappear when we move to lower redshifts. Therefore, in an attempt to learn more about the physical processes occurring in the EELR of powerful radio galaxies, we have made a detailed spectroscopic study of a sample of relatively nearby ($z < 0.5$) radio galaxies. These sources (PKS 2250−41, 3C 171, 4C 29.30, and Coma A) share many of the properties of high redshift radio galaxies, but are close enough to study in detail. The long-slit spectra obtained—supplemented by further optical, radio and X-ray data—allow us to study the physics of jet-cloud interactions, and the effects of these interactions upon the morphology, kinematics, physical conditions and ionization of the extended line-emitting gas. The main results of this study are summarized in the following section (see also: Clark 1996, Clark et al. 1997a,b).

2 Results

2.1 Morphology

The extended emission line regions in all of the sources studied are closely associated with the radio structures, and the spatial extents of the EELR (over 100 kpc in some cases) are defined by the boundaries of the radio emission. The EELR appear to have two distinct morphologies: those coincident with the hot spots at the head of the advancing radio jets have an arc-like morphology (e.g. PKS 2250−41: Fig. 1a), whereas the EELR along the radio jets have elongated jet-like structures (e.g. 3C 171: Fig. 1b). The different structures are interpreted as being due to line-emission from shocked gas associated with the main bow-shock at the head of the radio jet, and from shocks along the length of the radio jet, respectively. A similar dichotomy in emission-line morphologies has been observed on a much smaller scale in HST narrow-band images of Seyfert NLR (Capetti et al. 1996).

2.2 Kinematics

The emission line-kinematics are disturbed in the EELR, with line-splitting, large linewidths (FWHM $\sim$ 1000 km s^{-1}) and a clear anti-correlation between

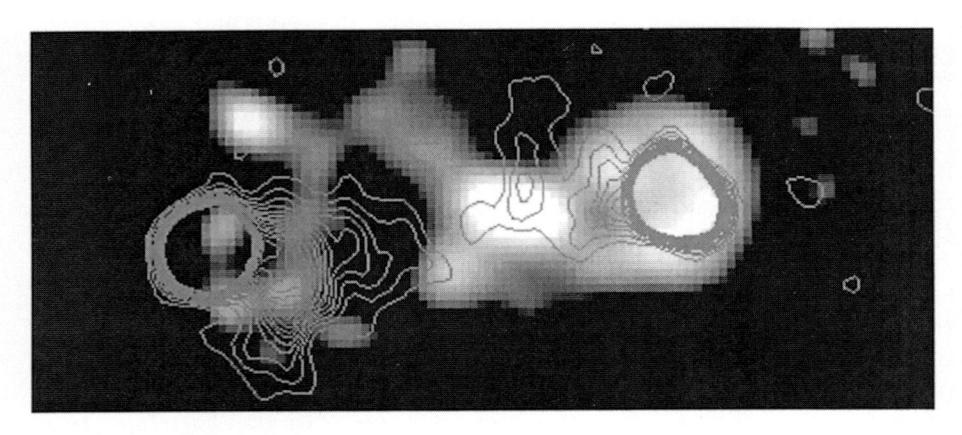

Fig. 1. Grey-scale representations of: (a) the ESO 3.6m narrow-band [OIII] image of PKS 2250−41, with contours of the 8 GHz ATCA radio map superimposed (top), showing the bright line-emitting arc which circumscribes the western radio lobe; and (b) the broad-band WFPC2 HST image of 3C 171 (bottom), with filaments of line emission extending along the radio jets. North is up, west is to the right, and the image sizes are 33 × 14 and 11 × 5 arcseconds respectively.

linewidth and ionization state consistent with a shock structure (as observed in shock-excited supernova remnants: Shull 1983). For further discussion, see Clark et. al. 1997a,b.

2.3 Physical Conditions

The electron temperatures and densities were measured in the EELR of three of the sources studied using optical diagnostic line ratios. The pressures in the EELR can be deduced from these measurements, allowing an investigation of the

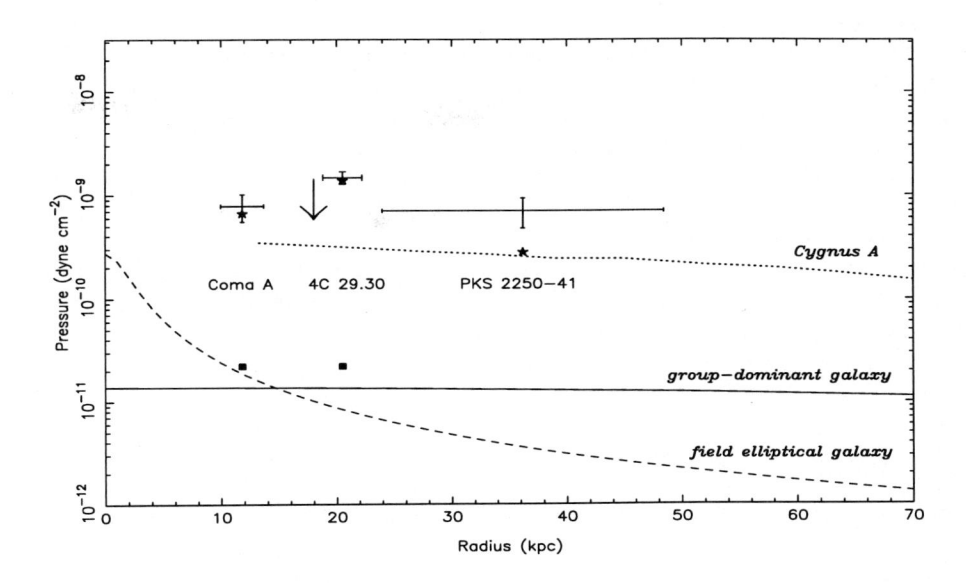

Fig. 2. Calculated pressures against radius for the EELR in 4C 29.30, PKS 2250−41 and Coma A, derived from the measured [OIII] temperatures and [SII] densities (the arrow indicates the effect of using the [NII] temperature for the pressure calculation in the case of 4C 29.30). The minimum-energy pressures for the associated radio hot spots/knots (stars) and the radio lobes (squares) are plotted at the same radii as the EELR. Also plotted are the radial variations in the hot halo pressures for a typical X-ray bright field elliptical galaxy (Forman et al. 1985: dashed line), a group-dominant galaxy (Ponman et al. 1994: solid line) and the X-ray luminous central-cluster galaxy Cygnus A (Reynolds & Fabian 1996: dotted line).

confinement mechanism for the warm line-emitting gas. In Fig. 2, the pressures are plotted for the EELR in 4C 29.30, PKS 2250−41 and Coma A. For comparison, the variation in the pressure of the hot phase of the ISM is plotted for: a) a typical field elliptical galaxy with $L_x \sim 10^{41}$ erg s^{-1} (Forman et al. 1985); b) a group-dominant galaxy with $L_x = 4.5 \times 10^{43}$ erg s^{-1} (Ponman et al. 1994); and c) Cygnus A, a central cluster galaxy surrounded by a dense halo of hot gas with $L_x \sim 10^{45}$ erg s^{-1} (Reynolds & Fabian 1996).

It is clear that the pressures of the three line-emitting regions are too high for the EELR to be confined by the hot ISM in either a field elliptical galaxy or a galaxy at the centre of a small group. In fact, the EELR pressures are as high as those measured in the dense halo of the Cygnus A cluster at the same radius. However, the low X-ray luminosities and apparent isolation of host galaxies are inconsistent with them being situated in rich clusters, and it is therefore highly unlikely that confinement by the hot X-ray haloes can explain the high pressures in the EELR.

The minimum-energy pressures of the associated radio knots/hot spots and the diffuse radio lobe material are also plotted in Fig. 2. It can be seen that the estimated pressures of the radio knots/hot spots are similar to those of the associated line-emitting nebulae. This is consistent with the theory that the emission line gas has been compressed by shocks associated with the radio jet (as ram-pressure confinement of the advancing radio jet leads to approximate pressure equilibrium between the post-shock gas and the radio hot spot at the head of the radio jet). In contrast, the minimum-energy pressures estimated for the diffuse lobe material are much lower and are consistent with those measured in the haloes of isolated galaxies or groups of galaxies.

Overall, *the relatively high pressures measured in the EELR provide convincing evidence for the compression effect of shocks.*

2.4 Ionization

Two strands of evidence point to the ionizing effect of shocks in the EELR. Firstly, the ionizing photon luminosities of the central AGN required to maintain the ionization levels measured in the EELR are orders of magnitude higher than the luminosities expected from the extended radio power. Secondly, although most of the line ratios are consistent with either shock-ionization (Dopita & Sutherland 1996) or photoionization by an $\alpha \approx -1.5$ power-law continuum (Robinson et al. 1987), the low HeII(4686)/Hβ ratios measured in the extended gas are more consistent with shock-ionization than AGN-photoionization. Although a model has recently been proposed which can explain a spread in HeII(4686)/Hβ ratios in terms of AGN-photoionization (Binette et al. 1996), the fact that the HeII(4686)/Hβ ratio is systematically lower in the EELR which show the other signs of shocks outlined above supports the shock-ionization theory. The issue of shock-ionization is discussed further in Clark (1996).

3 Comparison with High Redshift Radio Galaxies

From the above discussion it is apparent that that shocks can have a major impact on the energetics of the EELR in powerful radio galaxies. The powerful radio galaxies observed at high redshift exhibit many of the properties observed in the relatively nearby sources discussed above (see Section 1). Since the effects of jet-cloud interactions dominate over central-source photoionization in our sample sources, the number of common features shared by more distant powerful radio galaxies strongly suggests that jet-cloud interactions are also important at higher redshifts.

Consider the example of the powerful high redshift radio galaxy 3C 368 ($z = 1.132$). There is a striking similarity between the optical strucures observed in the broad-band HST images of 3C 368 (Longair et al. 1995) and in 3C 171 ($z = 0.2381$), and the relative positions of the radio hot spots and the optical line/continuum emission are also similar in the two sources. Recent Keck spectra of 3C 368 (Stockton et al. 1996) again reveal a striking resemblance with our spectra of 3C 171. In both sources the level of ionization falls substantially off-nucleus, and the ionization states in extended line-emitting gas are also similar ([OII](3727)/H$\gamma \sim$ 8.3). Finally, the detailed kinematics of the gas in 3C 368 also resemble those in 3C 171: in both sources, a central narrow component is observed extending the length of the nebulosities with a smooth and relatively small change in velocity, and superimposed on this component we observed symettrical line-splitting (most clearly resolved in 3C 171) and broad diffuse emission.

Given the evidence discussed in Section 2 for the influence of shocks in 3C 171, and the similarity of 3C 368 and 3C 171, it is natural to suppose that shocks are also important in 3C 368. Indeed, a close examination of the Keck long-slit spectrum appears to confirm this hypothesis. A faint, kinematically-unresolved high ionization state line-emitting component is observed extending $\sim$ 3-6 arcseconds north of the nucleus. This kinematically undisturbed component clearly represents line emission from gas beyond the shocks driven by the expanding radio jet, and the contrasting properties of the shocked and unshocked gas are immediately apparent:

Kinematics. The central velocity of the precursor gas ahead of the shock front changes by $<$ 100 km s^{-1}, and no line-splitting or broad components are observed. The full-width-zero-maximum of the [OII] profile for the precursor gas, the diffuse shocked gas and the northern knot are $<$ 100, 1000–2000 and 4000 km s^{-1} respectively.

Ionization state. The line emission from the precursor gas has a much higher ionization state than that from the shocked gas ([NeIII]/[NeV] = 1.3 for the precursor, compared to 6.5 and 6.1 for the diffuse extended emission and northern knot respectively). In contrast, assuming that both the intensity of the AGN radiation and the ambient gas density fall approximately as r^{-2}, AGN-photoionization would predict little change in ionization state with radius. The low ionization state in the shocked gas must be largely due to shock-compression.

Surface brightness. The emission-line surface brightness of the precursor component is much lower than that of the shocked gas—the ratio of surface brightnesses deduced from the measured line fluxes and the relevant apertures is 1:14:125 for the precursor, diffuse shocked gas and northern knot respectively.

4 Conclusions

Our detailed study of a sample of low-redshift radio galaxies has revealed the major role shocks can play in the host galaxies of powerful radio sources. While the effects of shocks are obvious only in a minority of nearby sources, shocks appear to become increasingly influential as the radio power/redshift increases, as demonstrated by recent observations of the high redshift radio galaxy 3C 368.

Acknowledgments

NEC is grateful to the University of Sheffield and the Astrophysics Division, Space Science Department of ESA for financial support.

References

Binette, L. et al. (1996): A&A, **312**, 365
Capetti, A. et al. (1996): ApJ, 466, 169
Chambers, K.C. et al. (1987): Nat, **329**, 604
Clark, N.E. (1996): PhD thesis, University of Sheffield.
Clark, N.E. et al. (1997a): accepted for publication in MNRAS
Clark, N.E. et al. (1997b): in preparation.
Dopita, M.A. & Sutherland, R.S. (1996): ApJS, **102**, 161
Forman, W. et al. (1985): ApJ, **293**, 102
Fosbury, R.A.E. (1989): *ESO Workshop on Extranuclear Activity in Galaxies* (eds. Meurs E.J.A. & Fosbury R.A.E.), ESO scientific publications, p169
Longair, M.S. et al. (1995): MNRAS, **275**, L47
Klein, R. et al. (1994): ApJ, **420**, 213
McCarthy, P.J. et al. (1987): ApJ, **321**, L29
McCarthy, P.J. et al. (1996): ApJS, **106**, 281
Miley, G.K. et al. (1992): ApJ, **401**, L69
Ponman, T.J. et al. (1994): Nat, **369**, 46
Reynolds, C.S. & Fabian, A.C. (1996): MNRAS, **278**, 479
Robinson, A. et al. (1987): MNRAS, **227**, 97
Shull, P. (1983): ApJ, **275**, 611
Simpson, C. & Ward, M. (1996): MNRAS, **282**, 797
Stockton, A. et al. (1996): AJ, **112**, 902
Tadhunter, C.N., Fosbury R.A.E. & Quinn P. (1989a): MNRAS, **240**, 225
van Ojik, R. (1995): PhD thesis

The Hosts of High-z Powerful Radio Sources: Keck Observations[*]

Andrea Cimatti

Osservatorio Astrofisico di Arcetri, Largo E. Fermi 5, I-50125 Firenze, Italy

Abstract. We present recent results of Keck observations of powerful 3C radio sources at $z \sim 1$, and we show how deep spectroscopy and spectropolarimetry can be successfully used to study their host galaxies. We find that the UV–optical continuum is the result of a combination of different components (stellar light, nebular continuum and scattered radiation), and we quantify their relative contributions to the observed continuum. The whole picture is complicated by the presence of significant reddening that affects the overall spectral shape, and demonstrates that the host galaxies have a dusty interstellar medium. Finally, we discuss our results in the framework of the unified model of radio-loud AGN.

1 Introduction

The study of the host galaxies of distant AGN is important in order to understand the relationship between the nuclear activity, the stellar populations and the interstellar medium in active galaxies. In practice, this is very arduous because the radiation emitted from the quasar makes the observation of the host galaxies extremely difficult from the ground, and feasible from space mostly in low to medium z objects.

High redshift radio galaxies (HzRGs) are observable to cosmological distances competitive with the most distant quasars (see McCarthy 1993 for a review). Their continuum is spatially resolved and observable from the ground, and as not dominated by the direct nuclear radiation as in quasars. Therefore HzRGs can be used to study the host galaxies more easily than in quasars. Moreover, if the unification of radio-loud quasars and powerful radiogalaxies is valid (Antonucci 1993; Urry & Padovani 1995), HzRGs may be used to investigate the hosts of *hidden* quasars, taking advantage of the obscuration of the nucleus.

However, the use of HzRGs is heavily complicated by their multi-component nature, where both stellar and non-stellar radiation are present. This problem is critical in the most powerful HzRGs, like the ones selected from the 3C sample, where the strong polarization of the UV continuum suggests the presence of a significant contribution of non-stellar radiation (Cimatti 1996). The problem of the contamination by non-stellar light may be less important in HzRGs with lower radio power, where the host galaxy radiation seems to be dominated by stars (Dunlop et al. 1996).

---

[*] The Keck program is in collaboration with Wil van Breugel, Ski Antonucci, Hy Spinrad, Arjun Dey, and Todd Hurt.

Therefore it is critical to investigate the nature of the HzRGs with a tool capable to separate the different components. Spectropolarimetry is one of the most powerful tools because it provides information on the polarized and total fluxes separately for the the continuum and the emission lines. Spectropolarimetry is also the best technique to search for hidden quasar nuclei visible through scattering, and to test the validity of the unification model of radio-loud quasars and powerful radiogalaxies (see Antonucci 1993; Urry & Padovani 1995).

2 The Need of Keck Observations

The typical magnitudes of powerful HzRGs at $z \geq 1$ are $V \geq 21$, and the 4m class telescopes cannot perform optical spectropolarimetry of such faint objects. The advent of the Keck 10m telescope has rendered spectropolarimetry of faint galaxies feasible.

We observed 6 HzRGs selected from the 3CR sample with $0.7 < z < 1.8$. The observations were made at the Cassegrain focus of the 10-m W.M. Keck Telescope using the Low Resolution Imaging Spectrometer (LRIS) in March and July 1995. The spectropolarimeter is a dual beam instrument which uses a calcite analyser and a rotatable waveplate. We used a 300 line/mm grating and a 1 or 1.5 arcsec wide slit. The LRIS detector is a Tek 2048^2 CCD with $24\mu m$ pixels which correspond to a scale of 0.214 arcsec pix^{-1}. The dispersion is ≈ 2.5 Å pix^{-1} and the spectral region covered is $\lambda_{obs} \sim 3970$–9010 Å, corresponding to the UV rest-fram region ~ 1500–4000 Å. The observations were made under good seeing conditions (≈ 0.5-0.7 arcsec). A complete observation includes 4 integrations made at four half-wave plate positions (0°, 45°, 22.°5, 67.°5). The integration times per position angle ranged from 30 to 60 minutes. The observations and data analysis are described in details in our first papers (Cimatti et al. 1996, 1997; Dey et al. 1996).

3 Hidden Quasars and Unification

The high S/N ratio reached, and the excellent seeing conditions allowed us to perform *spatially resolved* spectropolarimetry. This is crucial if we want to understand not only what are the spectral and polarization properties, but also from what region of the galaxy they originate. High linear polarization of the UV continuum is detected in all the observed galaxies (with the possible exception of 3C 368; Dey et al. in preparation). The polarized flux is spatially extended along the UV continuum axis. The perpendicularity of the electric vector to the UV continuum axis, and its constancy with λ suggest that scattering is the dominant polarization mechanism. The detection of the broad MgIIλ2800 emission line in polarized flux suggests that the incident radiation comes from an obscured quasar nucleus and is emitted anisotropically along the radio axis. In particular, the broad and polarized MgIIλ2800 in 3C 324 and 3C 356a has velocity and equivalent widths consistent with those observed in radio-loud quasars (Cimatti et al. 1996, 1997). On the other hand, the always lower or null polarization of the forbidden narrow lines implies that they are emitted isotropically outside the obscuring region. Figure 1 shows the case of 3C 356a, a highly polarized galaxy at $z=1.079$. These results are in agreement with the requirement of the unified model of powerful radio-loud AGN, where the differences between Type 1 and Type 2 AGN are mainly due to orientation effects and not to intrinsic diversities. Our results also imply that scattered light is a necessary ingredient to explain the alignment effect (i.e., the alignment between the radio and UV continuum axis).

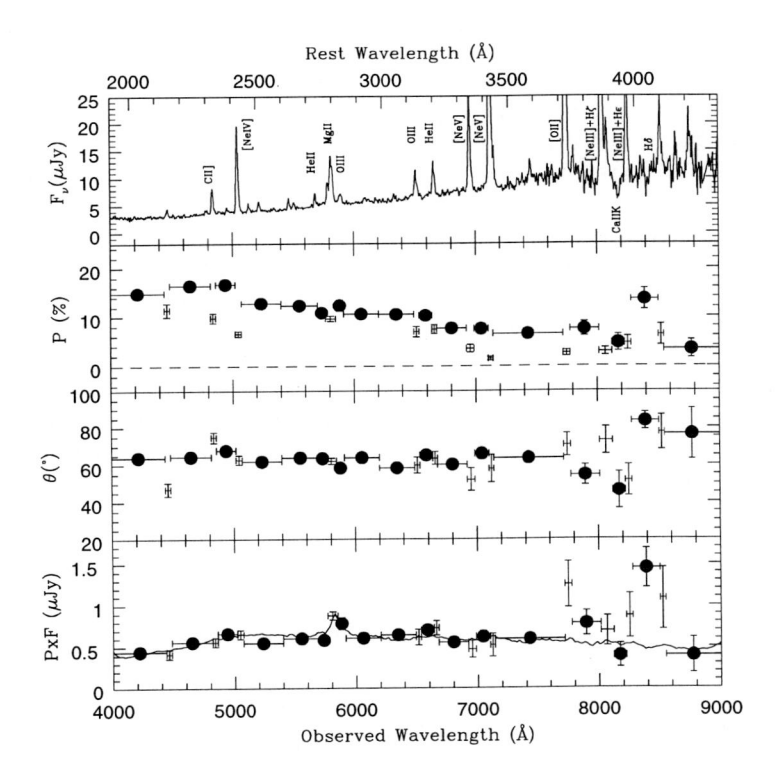

Fig. 1. The spectral and polarization properties of 3C 356a. From top to bottom: the total flux spectrum, the percentage polarization, the position angle of the electric vector and the polarized flux spectrum. Filled circles and crosses indicate respectively continuum, and emission lines with their underlying continuum. The continuum line in the $P \times F$ plot is a radio–loud quasar average spectrum (Cristiani & Vio 1990) scattered by Galactic type dust ($E_{B-V}=0.05$) (see Manzini & di Serego Alighieri 1996 for the dust scattering model).

4 The Components of the UV–Optical Continuum

Our observations allow us for the first time to quantify the components contributing to the UV continuum of HzRGs. The ratio between the unpolarized radiation and the total flux at $\lambda \sim 2800$ Å can be estimated by comparing the equivalent widths of the broad MgIIλ2800 in total and polarized flux (see Cimatti et al. 1996, 1997 for more details). In the cases analysed so far, 3C 324 and 3C 356a, we find that typically 50% of the total UV continuum at $\lambda \sim 2800$ Å is due to unpolarized radiation and the other 50% is due to scattered light. The nebular continuum is an important ingredient of the UV continuum. For instance, we find that a significant fraction of the unpolarized light of 3C 356a at $\lambda \sim 2800$ Å (about 30%) is due to nebular continuum emission (the Balmer limit at 3646 Å is clearly visible in the total flux spectrum of 3C 356a, Fig. 1). Finally, we derive that an old stellar population (~ 1.5–2 Gyr) can account for the rest of the light of 3C 356a at 2800 Å, in agreement with the detection of the stellar CaII K absorption line. To summarize, at 2800 Å the continuum of 3C 356a is made by 50% scattered light, 30% nebular continuum and 20% stellar light from an evolved population of stars. Similar results have been found for 3C 324.

These results show how spectropolarimetry can be used to disentangle the different radiative components present in HzRGs. However, they also demonstrate that the non-stellar radiation can be dominant, and render the study of the host galaxies and their stellar populations more difficult than it was previously thought.

Another approach to derive the ingredients of the UV-optical continuum is to fit the Spectral Energy Distributions (SEDs) of HzRGs combining the deep Keck spectrophotometry of the continuum (rest-frame UV) with broad-band photometry obtained in the near-IR (rest-frame optical). Figure 3 shows an example of fitting of the SED of 3C 356a (Cimatti & di Serego Alighieri, in preparation). We find that the SED can be reproduced reasonably well with three components : a dust scattered quasar spectrum, the nebular continuum, and a stellar population with an age of 2 Gyr. The observed SED is corrected for reddening with $E_{B-V}=0.11$, as estimated from the HeII line ratios. For the stellar populations we used the synthetic spectra of Bruzual & Charlot (1993), with solar metallicity, Salpeter IMF and instantaneous burst. The nebular continuum is not a free parameter, but is computed from the Hβ flux estimated using the observed Hδ line flux. The dust scattering function is taken from Manzini & di Serego Alighieri (1996) assuming Galactic dust, and the quasar spectrum is the average radio-loud quasar spectrum of Cristiani & Vio (1990). The discrepancy of the fit in the region around 2200 Å is likely due to the poorly known scattering properties of the dust across the 2200 Å feature (see Manzini & di Serego Alighieri 1996 for a detailed discussion of this issue).

It is interesting to note that from the fit we derive that at 2800 Å the scattered light contributes to about 41% of the total flux, while the nebular continuum and the stellar light contribute to about 38% and 21% respectively. These results are, within the uncertainties, in good agreement with those derived independently with spectropolarimetry.

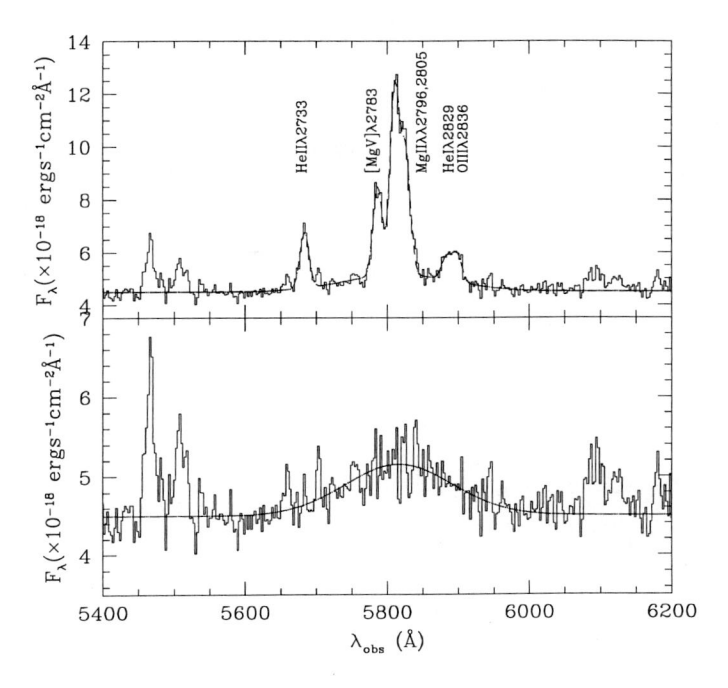

Fig. 2. The broad component of the MgIIλ2800 line in total flux as derived by fitting and subtracting the narrow lines. The ratio of the equivalent width of the broad component of MgIIλ2800 to that observed in the polarized flux spectrum, allows to measure the ratio of the unpolarized light to the total flux at 2800 Å (see Cimatti et al. 1997 for more details).

We notice that a significant amount of reddening is often required to fit the SEDs of HzRGs. The typical values of E_{B-V} are in the range $\sim$0.1–0.3. This indicates that dust is an important ingredient of the ISM of HzRGs at $z \sim 1$. Consistent amounts of reddening are also derived using the ratios of the HeII lines thanks to the high S/N ratio of the Keck spectra. Finally, dust particles are often required to explain the properties of the scattered radiation (see Cimatti 1996 for a review). Other evidences of dust in HzRGs come from Lyα/Hα ratios (McCarthy 1993) and from sub-mm observations of the dust thermal emission (Hughes 1996).

References

Antonucci, R. 1993, ARAA, 31, 473.
Bruzual, G., Charlot, S. 1993, ApJ, 405, 538
Cimatti, A. 1996, in "New Extragalactic Perspectives in the New South Africa – Changing Perceptions of the Morphology, Dust Content and Dust-Gas Ratios in Galaxies", ed. D. Block et al., Kluwer Academic Publishers, p. 493

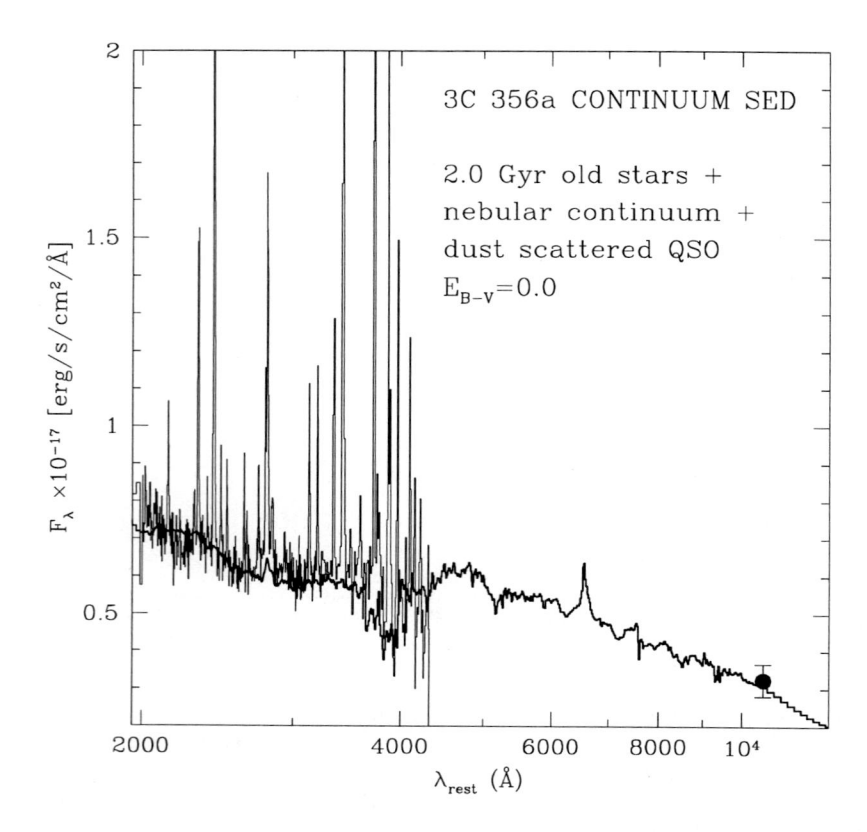

Fig. 3. The observed SED of 3C 356a in the rest-frame region ~2000-11000 Å. The K-band flux is taken from Eales & Rawlings (1990). The continuous line is the fit obtained with old stars, nebular continuum, and quasar radiation scattered by dust (see text).

Cimatti, A., Dey, A., van Breugel, W., Antonucci, R. & Spinrad, H. 1996, ApJ, 465, 145
Cimatti, A., Dey, A., van Breugel, W., Hurt T., Antonucci, R. 1993, ApJ, in press
Cristiani S., Vio R. 1990, A&A, 227, 385
Dey A., Cimatti A., van Breugel W., Antonucci R., Spinrad H. 1996, ApJ, 465, 157
Dunlop J.S. et al. 1996, Nature, 381, 581
Eales S.A., Rawlings S. 1990, MNRAS, 243, 1p
Hughes D.H. 1996, in "Cold Gas at High Redshift", ed. M. Bremer, P. van der Werf, H. Röttgering and C. Carilli, Kluwer, p.311
Manzini A., di Serego Alighieri S. 1996, A&A, 311, 79
McCarthy P.J. 1993, ARAA, 31, 693
Urry C.M., Padovani P. 1995, PASP, 107, 803

The volcano caldera

Author Index

Acosta-Pulido, J. 266, **288**
Akiyama, M. 110
Aretxaga, I. 84
Arimoto, N. 122
Arribas, S. 171
Asif, M.W. **168**
Axon, D.J. 320
Bahcall, J.N. **37**
Baldwin, J. 291
Barthel, P. **63**
Barvainis, R. **101**
Benn, C. 21, 260
Bershady, M.A. 278
Blades, J.C. 76
Boksenberg, A. 76
Boroson, T. 177
Boyce, P.J. **76**
Boyle, B.J. 84
Bremer, M.N. **70**
Bunone, A. 297
Cabrera-Guerra, F. **266**
Carballo, R. 21, 260
Cimatti, A. **327**
Clark, N.E. **320**
Cooke, A. **291**
Crane, P. 76
Deharveng, J.M. 76
del Burgo, C. **171**
Devillers, A. 92
Dickson, R. 311
di Serego Alighieri, S. **5**
Disney, M.J. 76
Dunlop, J.S. 177, 185
Eckart, A. 116, **282**
Evans, A.S. 122
Falomo, R. **194**, **212**
Ferland, G. 291
Foltz, C.B. 206
Fosbury, R.A.E. **3**
Fricke, K.J. 90
Fried, J.W. **215**

García-Lorenzo, B. 171
Garrington, S.T. 13
Geffert, M. 92
Genzel, R. 128, 282
Gerritsen, J. 63
Giavalisco, M. 194
Goldschmidt, P. **162**
Gonçalves, A.C. **293**
González-Serrano, J.I. 21, **260**
Gutiérrez, C.M. 173
Heidt, J. 90, **200**, 215, 223, 225
Heinämäki, P. 223
Holmes, G.F. 13
Hooper, E.J. **206**
Hughes, D.H. 177, 185
Hutchings, J.B. **51**, **58**
Impey, C.D. 206
Jäger, K. **90**
Jeyakumar, S. 19
Johnson, R. 295
Köhler, T. **254**
Kawabe, R. 110
Kawara, K. 122
Kirhakos, S. 37
Kohno, K. 110
Krabbe, A. 128
Kroker, H. 128
Kron, R.G. 278, 297
Kukula, M.J. **177**, **185**
Lacy, M. 188
Lawrence, A. 295
Lira, P. **295**
Macchetto, F.D. 76
Mackay, C.D. 76
Maiolino, R. 128
McCarthy, P. **303**
McLeod, K.K. **45**
Mediavilla, E. 171
Miley, G. 70
Miller, L. 162
Monnet, G. 70

Morganti, R. 311
Mundell, C.G. **156**, 168
Murayama, T. 122
Nakai, N. 116
Nakanishi, K. 110
Nanni, D. 297
Nass, P. **272**
Netzer, H. 291
Nilsson, K. 223, 225
Ohta, K. 110
Okumura, S.K. 116
Örndahl, E. **217**
Pérez Garcia, A.M. 144
Pérez-Fournon, I. 266, 288
Pedlar, A. 168
Pesce, J. 194
Prada, F. 173
Pursimo, T. 223, 225
Rönnback, J. 217
Rawlings, S. 185, 188
Rigopoulou, D. **242**
Robinson, A. 168, 320
Rodriguez Espinosa, J.M. 144
Sánchez, S.F. **21**, 260
Saikia, D.J. **13**, **19**
Sanders, D.B. 122, **229**, 236
Scarpa, R. 194
Schinnerer, E. **116**
Schneider, D.P. 37
Serjeant, S. **188**
Sillanpää, A. **223**, 225
Sinachopoulos, D. **92**

Sparks, W.B. 76
Srianand, R. **150**
Surace, J.A. 229, **236**
Tacconi, L.J. 116
Tadhunter, C.N. **311**, 320
Takalo, L.O. 223, **225**
Taniguchi, Y. **122**
Taylor, G. 177
Terlevich, R.J. 84
Thatte, N. 116, **128**
Trèvese, D. **278**, **297**
Treves, A. 194
Tsvetanov, Z.I. 266
Ulrich, M.-H. 212
Unger, S.W. 168
Urry, C.M. 194
Véron, P. 293
Véron-Cetty, M.-P. **27**, 293
Vagnetti, F. **23**
van Groningen, E. 217
Vigotti, M. 21, 260
Vila-Vilaró, B. 288
Villar-Martin, M. 311
Walton, N.A. 168
Wieringa, M. 70
Wilkes, B.J. **136**
Wills, B. 291
Wilson, A. 266, 291
Wisotzki, L. **96**, 254
Woltjer, L. 27
Yamada, T. **110**, **248**

ESO ASTROPHYSICS SYMPOSIA
European Southern Observatory

Series Editor: Philippe Crane

P. Crane (Ed.), **The Light Element Abundances**
Proceedings, 1994. XVI, 432 pages. 1995.

J. R. Walsh, I. J. Danziger (Eds.), **Science with the VLT**
Proceedings, 1994. XXV, 477 pages. 1995.

C. G. Tinney (Ed.), **The Bottom of the Main Sequence – And Beyond**
Proceedings, 1994. XVII, 309 pages. 1995.

G. Meylan (Ed.), **QSO Absorption Lines**
Proceedings, 1994. XXIII, 471 pages. 1995.

D. Minniti, H.-W. Rix (Eds.), **Spiral Galaxies in the Near-IR**
Proceedings, 1995. X, 350 pages. 1996.

H. U. Käufl, R. Siebenmorgen (Eds.), **The Role of Dust in the Formation of Stars**
Proceedings, 1995. XXII, 461 pages. 1996.

P. A. Shaver (Ed.), **Science with Large Millimetre Arrays**
Proceedings, 1995. XVII, 408 pages. 1996.

J. Bergeron (Ed.), **The Early Universe with the VLT**
Proceedings, 1996. XXII, 438 pages. 1997.

F. Paresce (Ed.), **Science with the VLT Interferometer**
Proceedings, 1996. XXII, 406 pages. 1997.

D. L. Clements, I. Pérez-Fournon (Eds.), **Quasar Hosts**
Proceedings, 1996. XVII, 336 pages. 1997.

L. N. da Costa, A. Renzini (Eds.), **Galaxy Scaling Relations: Origins, Evolution and Applications**
Proceedings, 1996. XX, 404 pages. 1997.

Druck: Druckhaus Beltz, Hemsbach
Verarbeitung: Buchbinderei Schäffer, Grünstadt